MANAGEMENT AND ENGINEERING OF CRITICAL INFRASTRUCTURES

MANAGEMENT AND ENGINEERING OF CRITICAL INFRASTRUCTURES

Edited by

BEDIR TEKINERDOGAN

Information Technology Group, Wageningen University & Research, Wageningen, The Netherlands

MEHMET AKŞIT

Department of Computer Science, University of Twente, Enschede, The Netherlands

CAGATAY CATAL

Department of Computer Science and Engineering, Qatar University, Doha, Qatar

WILLIAM HURST

Information Technology Group, Wageningen University & Research, Wageningen, The Netherlands

TAREK ALSKAIF

Information Technology Group, Wageningen University & Research, Wageningen, The Netherlands

ACADEMIC PRESS

An imprint of Elsevier

Academic Press is an imprint of Elsevier
125 London Wall, London EC2Y 5AS, United Kingdom
525 B Street, Suite 1650, San Diego, CA 92101, United States
50 Hampshire Street, 5th Floor, Cambridge, MA 02139, United States
The Boulevard, Langford Lane, Kidlington, Oxford OX5 1GB, United Kingdom

Notices
Knowledge and best practice in this field are constantly changing. As new research and experience broaden our understanding, changes in research methods, professional practices, or medical treatment may become necessary.

Practitioners and researchers must always rely on their own experience and knowledge in evaluating and using any information, methods, compounds, or experiments described herein. In using such information or methods they should be mindful of their own safety and the safety of others, including parties for whom they have a professional responsibility.

To the fullest extent of the law, neither the Publisher nor the authors, contributors, or editors, assume any liability for any injury and/or damage to persons or property as a matter of products liability, negligence or otherwise, or from any use or operation of any methods, products, instructions, or ideas contained in the material herein.

ISBN 978-0-323-99330-2

For information on all Academic Press publications
visit our website at https://www.elsevier.com/books-and-journals

Publisher: Mara E. Conner
Editorial Project Manager: Aera F. Gariguez
Production Project Manager: Jayadivya Saiprasad
Cover Designer: Christian J. Bilbow

Typeset by STRAIVE, India

Working together
to grow libraries in
developing countries

www.elsevier.com • www.bookaid.org

Contents

PART II Management of critical infrastructures

PART III   Engineering of critical infrastructures

PART IV   Application domains

14. Deep learning for agricultural risk management: Achievements and challenges   **307**

Saman Ghaffarian, Yann de Mey, João Valente, Mariska van der Voort, and Bedir Tekinerdogan

Contributors

Mehmet Akşit
Department of Computer Science, University of Twente, Enschede, The Netherlands

Tarek Alskaif
Information Technology Group, Wageningen University & Research, Wageningen, The Netherlands

Maria Carmela Annosi
Business Management & Organisation Group, Wageningen University & Research, Wageningen, The Netherlands

Rabia Arslan
Department of Engineering, TOBB University of Economics and Technology, Ankara, Turkey

Cigdem Avci
Information Technology Group, Wageningen University & Research, Wageningen, The Netherlands

Salih Bıçakcı
Department of International Relations, Kadir Has University, Istanbul, Türkiye

Cagatay Catal
Department of Computer Science and Engineering, Qatar University, Doha, Qatar

Abide Coskun-Setirek
Department of Social Sciences, Wageningen University & Research, Wageningen, The Netherlands

Yann de Mey
Business Economics Group, Wageningen University & Research, Wageningen, The Netherlands

Wilfred Dolfsma
Business Management & Organisation Group, Wageningen University & Research, Wageningen, The Netherlands

Mehmet Arda Eren
Department of Engineering, TOBB University of Economics and Technology, Ankara, Turkey

Ayhan Gücüyener Evren
Department of International Relations, Kadir Has University, Istanbul, Türkiye

Saman Ghaffarian
Business Economics Group; Information Technology Group, Wageningen University & Research, Wageningen, The Netherlands; Institute for Risk and Disaster Reduction, University College London, London, United Kingdom

Wim Hardyns
Faculty of Law and Criminology, Institute for International Research on Criminal Policy (IRCP), Ghent University, Ghent; Faculty of Social Sciences, Master of Safety Sciences, University of Antwerp, Antwerp, Belgium

William Hurst
Information Technology Group, Wageningen University & Research, Wageningen, The Netherlands

Miguel Ángel Pardo Picazo
Department of Civil Engineering, University of Alicante, Alicante, Spain

Genserik Reniers
Safety and Security Science Group, TUDelft, Delft, The Netherlands

Mathias Reveraert
Advisor Security Management at Infrabel, Brussels, Belgium

Marlies Sas
Attaché Noodplanning Nationaal Crisiscentrum, Brussels, Belgium

Tom Sauer
Department of Political Science, University of Antwerp, Antwerp, Belgium

Hanne Say
Department of Engineering, TOBB University of Economics and Technology, Ankara, Turkey

Nathan Shone
School of Computing and Mathematics, Liverpool John Moores University, Liverpool, United Kingdom

Bedir Tekinerdogan
Information Technology Group, Wageningen University & Research, Wageningen, The Netherlands

João Valente
Information Technology Group, Wageningen University & Research, Wageningen, The Netherlands

Mariska van der Voort
Business Economics Group, Wageningen University & Research, Wageningen, The Netherlands

İskender Yakın
ASELSAN A.Ş., Ankara, Turkey

Umur Togay Yazar
Department of Engineering, TOBB University of Economics and Technology, Ankara, Turkey

PART I

Key concepts

CHAPTER 1

Introduction

Bedir Tekinerdogan[a], Mehmet Akşit[b], Cagatay Catal[c], Tarek Alskaif[a], and William Hurst[a]
[a]Information Technology Group, Wageningen University & Research, Wageningen, The Netherlands
[b]Department of Computer Science, University of Twente, Enschede, The Netherlands
[c]Department of Computer Science and Engineering, Qatar University, Doha, Qatar

1 Objectives

In today's interconnected and technology-driven world, critical infrastructure systems are the backbone of modern society. These systems include transportation, energy, water, telecommunications, and other essential facilities and services that enable our daily lives. The reliable and secure operation of critical infrastructure systems is essential for national security, public safety, economic growth, and the well-being of individuals and communities.

The management and engineering of critical infrastructures are a complex and interdisciplinary field that requires a deep understanding of the technological, social, economic, and political factors that influence their design, operation, and maintenance. It involves the application of engineering principles, management practices, and policy frameworks to ensure the resilience, sustainability, and efficiency of critical infrastructure systems.

This book provides a comprehensive overview of the state-of-the-art in the management and engineering of critical infrastructures. It covers a wide range of topics, including but not limited to the following:

System design and optimization: The book covers the design and optimization of critical infrastructure systems, including transportation networks, energy grids, and water distribution systems. The authors discuss various approaches and techniques for designing and optimizing these systems to ensure their reliability, efficiency, and sustainability.

Risk assessment: The book covers the importance of risk assessment in identifying and evaluating the vulnerabilities of critical infrastructure systems. It discusses various methods and tools for conducting risk assessments and highlights the challenges and opportunities in this area.

Performance evaluation: The book covers the importance of performance evaluation in assessing the effectiveness of critical infrastructure systems. It discusses various performance evaluation methods and tools and highlights the challenges and opportunities in this area.

Cybersecurity: The book covers the importance of cybersecurity in protecting critical infrastructure systems from cyber-attacks. It discusses various cybersecurity methods and tools and highlights the challenges and opportunities in this area.

Emergency management: The book covers the importance of emergency management in responding to critical infrastructure disruptions, such as natural disasters or cyber-attacks. It discusses various emergency management approaches and techniques and highlights the challenges and opportunities in this area.

Sustainability: The book covers the importance of sustainability in ensuring the long-term viability of critical infrastructure systems. It discusses various sustainability principles and frameworks and highlights the challenges and opportunities in this area.

Resilience: The book covers the importance of resilience in ensuring that critical infrastructure systems can continue to function in the face of disruptions or disasters. It discusses various approaches and techniques for building resilience in these systems and highlights the challenges and opportunities in this area.

Policy and regulation: The book covers the importance of policy and regulation in managing and engineering critical infrastructure systems. It discusses various policy and regulatory frameworks and their impact on critical infrastructure management and engineering.

Interdependencies: The book covers the interdependencies among different critical infrastructure systems and the need to consider these interdependencies in the management and engineering of these systems. It discusses various methods and tools for analyzing and managing these interdependencies.

Innovation and emerging technologies: The book covers the importance of innovation and emerging technologies in the management and engineering of critical infrastructure systems. It discusses various technologies, such as artificial intelligence, blockchain, and the Internet of Things, and their potential impact on critical infrastructure management and engineering.

The book brings together contributions from leading experts in the field, including engineers, managers, policymakers, and academics, to provide a multidisciplinary perspective on critical infrastructure management and engineering.

The book is intended for a wide audience, including professionals in engineering, management, and policy fields, as well as graduate students and researchers interested in critical infrastructure systems. It provides a comprehensive reference for understanding the challenges and opportunities in managing and engineering critical infrastructure systems, as well as the latest techniques, methods, and tools for addressing these challenges.

We have divided the book into four key parts, grouping chapters by their link to key themes. Part I Key Concepts of Critical Infrastructures covers key concepts related to critical infrastructure systems. Part II encompasses chapters that focus on approaches for the management of critical infrastructures. Part III includes the chapters related to the engineering of critical infrastructures. Finally, Part IV includes the chapters related to specific application domains of critical infrastructures.

2 Outline

2.1 Part I—Key concepts

This section provides a comprehensive introduction to the fundamental concepts and principles of critical infrastructure systems. The chapters cover topics such as the definition and classification of critical infrastructure, the interdependencies among different infrastructure systems, and the role of risk assessment and management in ensuring the resilience and reliability of these systems. This part consists of the following chapters:

2.1.1 Chapter 1—Introduction

Bedir Tekinerdogan, Mehmet Akşit, Cagatay Catal, Tarek Alskaif, William Hurst
    This chapter provides an overview of the other chapters in the book.

2.1.2 Chapter 2—Critical infrastructures: Key concepts and challenges

Bedir Tekinerdogan, Mehmet Akşit, Cagatay Catal, Tarek Alskaif, William Hurst
    This chapter presents a comprehensive overview of the key concepts and challenges associated with critical infrastructures. The chapter uses a domain analysis approach to identify and analyze the main features, components, and challenges related to critical infrastructures. To enhance the understanding of critical infrastructures, a feature model is provided that describes the common and variant features of critical infrastructures. The chapter also examines the various stakeholders involved in critical infrastructures, such as government agencies, regulators, industry associations, and private companies. Key terms related to critical infrastructures, including resilience, sustainability, risk, and vulnerability, are defined in the terminology section. Additionally, a survey of important critical infrastructures is presented, such as energy systems, transportation networks, water and wastewater systems, telecommunications, and financial systems.

    Finally, the chapter delves into the primary challenges associated with critical infrastructures, such as cybersecurity, complexity, resilience and continuity, regulation and policy, aging infrastructure, digitalization, public awareness and perception, funding and resources, and supply chain security. Addressing these challenges is essential for promoting the resilience and sustainability of critical infrastructures, and safeguarding societal well-being and national security.

2.1.3 Chapter 3—Insider threats to critical infrastructure: A typology

Mathias Reveraert, Marlies Sas, Genserik Reniers, Wim Hardyns, Tom Sauer
    The chapter examines the vulnerability of critical infrastructure to insider threats. The chapter provides an innovative typology of the various types of intentional misconduct that employees can commit, based on an interplay between insider threat literature and publicly available examples of insider threat incidents found in (inter)national media. The typology is validated and reassessed using a spin-off version of the who, what, where,

when, why, and how (5W1H) methodology. The seven-part insider threat typology answers elementary questions, such as what the insider wants to achieve by committing intentional misconduct, who suffers or benefits from the insider threat, why the insider wants to commit intentional misconduct, when the insider becomes untrustworthy, how the insider commits intentional misconduct, how serious the impact of the insider threat is, and how many insiders are involved with the insider threat. The authors argue that this typology is a valuable starting point in the development of tailor-made approaches to insider threat mitigation. Organizations can use the typology to prioritize their risk management budget, paving the way for a risk-based approach to insider threat mitigation.

2.2 Part II—Management of critical infrastructures

This section presents various approaches and strategies for managing critical infrastructure systems. The chapters cover topics such as performance evaluation, maintenance and repair, emergency management, and cybersecurity. The authors discuss best practices and case studies from different sectors. This part consists of the following chapters:

2.2.1 Chapter 4—Health management of critical digital business ecosystems: A system dynamics approach

Abide Coskun-Setirek, William Hurst, Maria Carmela Annosi, Bedir Tekinerdogan, Wilfred Dolfsma

The chapter focuses on the health management and resilience of digital business ecosystems (DBEs) in the commercial and governmental facility sectors. As business ecosystems increasingly rely on digital platforms for efficiency, scalability, and security, ensuring the resilience of critical DBEs against adversarial attacks and vulnerabilities has become crucial. However, the factors that impact the health and resilience of critical DBEs are often not explicitly defined, making their health management challenging. In this chapter, the authors adopt a systems dynamic approach to support the health management and resilience of critical DBEs. They identify the variables related to the health management of DBEs and model the cause–and–effect relationships among them using causal loop diagrams. The study provides a holistic systems view of critical DBEs, including the variables, relations, and dynamics that define health and resilience. This approach enhances practitioners' visualization of decision-making on DBE administration, especially in critical infrastructure sectors. The chapter highlights the importance of adopting systems thinking approach to address the challenges associated with the health management and resilience of critical DBEs. It provides valuable insights for practitioners involved in the design, operation, and maintenance of DBEs in the commercial and governmental facility sectors, enabling them to make informed decisions to enhance the resilience of these systems.

2.2.2 Chapter 5—Key performance indicators of emergency management systems
Mehmet Akşit, Mehmet Arda Eren, Hanne Say, Umur Togay Yazar

In this chapter, the authors address the importance of efficient emergency management systems for critical infrastructures. The negative impact of damages to these infrastructures on the people relying on them can be significant. Key performance indicators (KPIs) have been introduced in the literature to measure and improve the efficiency of business processes, but there is a lack of rigorously defined KPIs for emergency management processes. The absence of a mathematical basis makes it difficult to measure and analyze these processes objectively and computationally. Therefore, this chapter proposes a novel set of KPIs specifically designed for evaluating emergency management processes. To demonstrate the utility of their approach, the authors also introduce a simulation framework for modeling and assessing emergency monitoring and management processes along with their associated KPIs. They evaluate the effectiveness of their approach using various disaster scenarios. The development of these KPIs and simulation frameworks provides a valuable contribution to the field of emergency management, especially in the context of critical infrastructures. The ability to objectively measure and analyze emergency management processes can significantly improve their efficiency and effectiveness in minimizing the negative impact of disasters on critical infrastructures and the people relying on them.

2.2.3 Chapter 6—Responding cyber-attacks and managing cyber security crises in critical infrastructures: A sociotechnical perspective
Salih Bıçakcı, Ayhan Gücüyener Evren

This chapter explores the vulnerability of Critical Infrastructures (CIs) to cyber threats, which can result in disruptions and require the implementation of effective cybersecurity crisis management strategies. The chapter argues that crisis decision-making, sense-making, and termination efforts in cyberspace present new challenges that cannot be addressed solely by technical solutions. Instead, the chapter emphasizes the need for human-centric approaches, such as re-considering crisis decision-making strategies, leadership structures, and organizational cultures. The chapter proposes that continuous crisis simulations and training can increase the preparedness level of CIs for cybersecurity crises. As CIs are complex Sociotechnical Systems (STS), effective cybersecurity crisis management requires the simultaneous orchestration of both human-centric and technical solutions.

2.3 Part III—Engineering of critical infrastructures

This section focuses on the engineering aspects of critical infrastructure systems. The chapters cover topics such as system design and optimization, modeling and simulation, control and automation, and sustainability. The authors discuss the latest techniques and tools for designing and operating critical infrastructure systems in a reliable, efficient, and sustainable manner. This part consists of the following chapters:

2.3.1 Chapter 7—Event-based digital-twin model for emergency management
Rabia Arslan, Mehmet Akşit

The chapter explores the use of distributed digital-twin–based control architectural styles for emergency management systems in critical infrastructures. Emergency management systems are essential for monitoring and rescuing critical infrastructures in the event of natural or man-made disasters. The distributed nature of these systems makes them inherently linked with Geographic Information Systems (GIS). However, current GIS systems lack the expressivity required to support emergency management activities and digital-twin–based control. This chapter addresses this challenge by extending the GIS system CityGML with a set of new abstractions. Additionally, a novel architecture is proposed to address the problem of data inconsistency caused by the distributed and heterogeneous nature of digital-twin–based architectures and the lack of standards. The architecture includes an atomic publisher-subscriber protocol and an event-based layer on top of a GIS database. To verify the proposed architecture, the authors simulate and test it using a model checker. The chapter highlights the importance of extending GIS systems to support emergency management activities in critical infrastructures and proposes a novel architecture to address the challenges associated with the distributed and heterogeneous nature of digital-twin–based control architectures. It highlights the importance of considering the expressivity and consistency of GIS systems in supporting digital-twin–based control and proposes a novel architecture to address these challenges.

2.3.2 Chapter 8—Analyzing systems product line engineering process alternatives for physical protection systems
Bedir Tekinerdogan, İskender Yakın

Physical protection systems (PPS) play an important role in safeguarding critical infrastructures against malevolent attacks. A PPS is a comprehensive solution that integrates people, procedures, and tools to provide effective protection. Developing an effective PPS is a complex process that requires consideration of various factors. Although PPSs may differ, they all share a similar structure and set of features that can be developed through a systematic approach to large-scale reuse. One such approach is product line engineering (PLE), which has been successfully used in various application domains to shorten the time to market, reduce costs, and enhance overall system quality. This paper provides an overview of the PPS domain, with a specific focus on its importance for critical infrastructures. We describe the three different types of PLE methods, including single, multiple, and feature-based PLE, and present a method for selecting the appropriate PLE method. To illustrate the practical application of the method, we apply it to a real industrial case study for a large-scale systems company.

2.3.3 Chapter 9—Reference architecture design for machine learning supported cybersecurity systems

Cigdem Avci, Bedir Tekinerdogan, Cagatay Catal

This chapter discusses the importance of cybersecurity in software-intensive systems and the potential benefits of integrating machine learning into cybersecurity systems. It addresses the issue of potential gaps in overall security by presenting a reference architecture design for machine learning-supported cybersecurity systems. The proposed architecture integrates engineering and machine learning perspectives to enhance the security of software-intensive systems. The effectiveness of the reference architecture is demonstrated through a domain analysis and architecture design stage, which synthesize architecture design approaches and machine learning models used in the cybersecurity domain. The resulting solution approach is holistic and effective for machine learning-supported cyber-secure systems.

2.3.4 Chapter 10—An architectural framework for the allocation resources in emergency management systems

Umur Togay Yazar, Mehmet Akşit

Emergency management systems are critical infrastructures that play a crucial role in protecting public safety and responding to disasters. These systems need to operate with high effectiveness and efficiency to accurately detect emergency conditions and allocate available resources optimally. However, in the case of large-scale disasters and insufficient resources, manual prioritization and compromising of resources can be extremely difficult. To address these challenges, this chapter proposes a novel architectural framework that automatically converts emergency reports to a set of rescue tasks that can be scheduled. If tasks cannot be scheduled as desired due to conflicts in allocating resources, control algorithms are applied. This architecture allows for the introduction of new disaster types and control strategies as needed. The proposed framework is designed and implemented in a simulated environment, and its utility is tested with various emergency scenarios. The chapter presents the architecture of the framework, the methods for emergency report detection and task scheduling, and the control algorithms used to resolve conflicts in resource allocation.

2.4 Part IV—Application domains

2.4.1 Chapter 11—Urban water distribution networks: Challenges and solution directions

Miguel Ángel Pardo Picazo, Bedir Tekinerdogan

This chapter focuses on the challenges faced by drinking water utilities in ensuring a reliable and safe supply of water while meeting increasing demands and dealing with limited resources. Aging infrastructure and water leakages pose significant challenges, particularly in urban areas. Moreover, recent threats such as cyber-physical attacks and terrorist attacks on water distribution networks have emerged, posing a risk to their operation. The effective management of water infrastructure is crucial to meet the demands of consumers and ensure sustainable development. The chapter outlines the methodologies and tools available to address these challenges and highlights the importance of managing water distribution networks as critical infrastructure.

2.4.2 Chapter 12—Critical infrastructure security: Cyber-threats, legacy systems, and weakening segmentation

William Hurst, Nathan Shone

This chapter addresses the pressing issue of cyber threats to critical infrastructures in the digital age. As critical infrastructures increasingly rely on digital technologies, they become more vulnerable to sophisticated cyber-attacks that can have a significant impact on society's health, safety, security, and economic well-being. The interconnected nature of these infrastructures means that attacks can spread and cause a cascading effect, posing a significant challenge to their effective management. Using case studies, this chapter highlights different attack types and their potential impact. It also explores the barriers that hinder the adoption of more advanced security solutions, such as Artificial Intelligence, and proposes ways to overcome these obstacles to enhance the protection of critical infrastructures.

2.4.3 Chapter 13—Energy systems as a critical infrastructure: Threats, solutions, and future outlook

Tarek Alskaif, Miguel Ángel Pardo Picazo, Bedir Tekinerdogan

The energy system has undergone significant changes with the digitalization trend and the integration of small-scale distributed energy resources. However, this transformation has also made the energy system more vulnerable to cyber-physical attacks and natural disasters caused by climate change. This chapter provides an overview of the different sources of criticality in energy systems, including man-made and naturally caused threats, and discusses potential technological and digital design solutions to enhance energy system resilience. The chapter concludes by outlining the challenges associated with these solutions and providing an outlook for energy system resilience. Its goal is to improve understanding of the energy system as a critical infrastructure and the need for effective management and protection.

2.4.4 Chapter 14—Deep learning for agricultural risk management: Achievements and challenges

Saman Ghaffarian, Yann de Mey, João Valente, Mariska van der Voort, Bedir Tekinerdogan

This chapter focuses on the application of deep learning (DL) in agricultural risk management (ARM). Agriculture is a sector that is vulnerable to various risks such as natural uncertainties, financial constraints, volatile prices, and changes in policies and regulations.

Farmers need to make informed decisions to mitigate these risks and maintain essential functions. DL is a subcategory of machine learning, which has been increasingly used in ARM to extract useful information from data to support decision-making. This review presents a systematic literature review of DL-ARM, highlighting the current trends, challenges, and achievements in the field. The chapter provides insights into the different types of risks addressed, the DL algorithms used, and the data types employed. The authors discuss the potential for DL to further contribute to ARM, beyond assessing production risk, and recommend areas for future research in this field.

Acknowledgments

We would like to thank all the authors and contributors of the chapters.

- Bedir Tekinerdogan, Information Technology Group, Wageningen University & Research, Wageningen, The Netherlands
- Mehmet Akşit, Department of Computer Science, University of Twente, Enschede, The Netherlands
- Cagatay Catal, Department of Computer Science and Engineering, Qatar University, Doha, Qatar
- Tarek Alskaif, Information Technology Group, Wageningen University & Research, Wageningen, The Netherlands
- William Hurst, Information Technology Group, Wageningen University & Research, Wageningen, The Netherlands

CHAPTER 2

Critical infrastructures: Key concepts and challenges

Bedir Tekinerdogan[a], Mehmet Akşit[b], Cagatay Catal[c], Tarek Alskaif[a], and William Hurst[a]

[a]Information Technology Group, Wageningen University & Research, Wageningen, The Netherlands
[b]Department of Computer Science, University of Twente, Enschede, The Netherlands
[c]Department of Computer Science and Engineering, Qatar University, Doha, Qatar

1 Introduction

Critical infrastructures are the backbone of modern society, providing essential services and systems that are crucial to the functioning of our economies, societies, and governments [1–4]. These infrastructures include energy systems, transportation networks, water and sanitation systems, telecommunications, financial systems, and many others. However, these systems are also vulnerable to a wide range of threats, including natural disasters, cyber-attacks, terrorism, and human error. To ensure the resilience and sustainability of critical infrastructures, it is important to understand the key concepts and challenges that are involved in their design, operation, and maintenance.

The key objective of this chapter is to provide an overview of the key concepts and challenges that are involved in critical infrastructures. The chapter will focus on identifying and defining key terms and concepts related to critical infrastructures, as well as exploring the main challenges that are involved in ensuring their resilience and sustainability. By providing this overview, the chapter aims to provide a foundation for further research and discussion on critical infrastructures and to inform policy and decision-making related to their design, operation, and maintenance. To achieve this objective, the concrete research questions are the following:

1. What are the key concepts and definitions related to critical infrastructures, and how are they related to each other?
2. What are the existing critical infrastructures?
3. What are the key challenges in the development of critical infrastructure, including planning, design, construction, and operation?

To answer these research questions, a domain analysis approach will be adopted. Domain analysis is a systematic approach to understanding a particular field or domain of knowledge and involves identifying and analyzing the key concepts, relationships, and dependencies within the domain. By applying domain analysis to the field of critical

Management and Engineering of Critical Infrastructures
https://doi.org/10.1016/B978-0-323-99330-2.00011-8

infrastructure, we can identify the key components of the domain, such as the sectors, industries, and stakeholders involved, their relationships, the terminology and definitions used, and the challenges and risks associated with critical infrastructure. We will analyze the key components of the domain and identify patterns and themes related to the research questions. The insights and findings generated through the domain analysis process have been used to provide a comprehensive overview of the key concepts and challenges of critical infrastructure.

The chapter is organized as follows: In Section 2, we elaborate on and describe the adopted domain analysis process. Section 3 describes the domain of critical infrastructures by considering the currently adopted definitions. Section 4 describes the key stakeholder categories and stakeholders of critical infrastructures. Section 5 describes the key terms based on the domain analysis process. Section 6 provides a feature diagram that defines the common and variant features of critical infrastructures. Section 7 provides an overview of various critical infrastructure types. Section 8 lists the identified key challenges, and finally, Section 9 concludes the chapter.

2 Domain analysis

Domain analysis is an important activity for organizations and individuals who want to develop a deeper understanding of a particular domain and identify potential opportunities for innovation, improvement, or disruption [5]. By identifying the key concepts, terms, and relationships within a domain, as well as any opportunities and risks, domain analysis can help to inform decision-making, identify areas for innovation, and ensure that organizations and individuals are well-positioned to succeed within their chosen domain.

The key objective of domain analysis is to identify and describe the key elements and relationships that make up a particular domain, as well as to understand how these elements and relationships interact with each other. This includes identifying the key concepts, terms, and relationships that define the domain, as well as identifying any gaps or inconsistencies in current understanding or practices.

There are several key activities involved in domain analysis, which can vary depending on the domain and the objectives of the analysis (Fig. 1). We focus on the following activities:

- *Identification of the domain*: The first step in domain analysis is to identify the domain that will be analyzed. This could be a particular business or industry, a scientific field, or any other area of knowledge or practice.
- *Identification of key stakeholders*: Once the domain has been identified, it is important to identify the key stakeholders who are involved in or affected by the domain. This could include customers, suppliers, regulators, industry associations, and other relevant parties.

- *Identification of key concepts and terms*: The next step is to identify the key concepts and terms that define the domain. This includes both technical terms and more general concepts and may involve developing a glossary or ontology to capture the relationships between these concepts and terms.
- *Identification of relationships and dependencies*: Once the key concepts and terms have been identified, the next step is to analyze the relationships and dependencies between these elements. This may involve developing models or diagrams to illustrate these relationships and to identify any gaps or inconsistencies in understanding.
- *Identification of the challenges*: Finally, domain analysis can be used to identify the challenges within the domain.

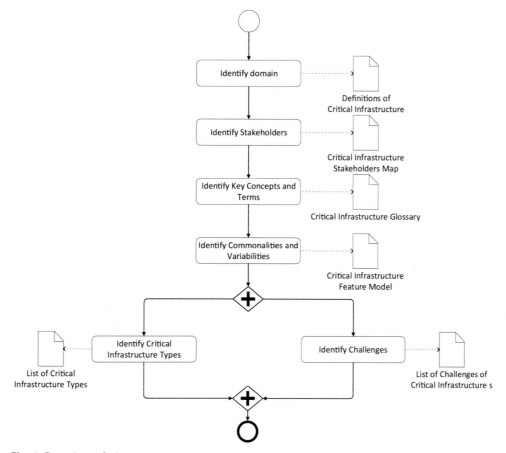

Fig. 1 Domain analysis steps.

3 Domain description and definitions

The first step in the domain analysis process is to define the notion of critical infrastructures, which is essential for understanding the scope and complexity of the domain. As shown in Table 1, there are various definitions of critical infrastructure in the literature, reflecting different perspectives and priorities of the organizations and institutions that have developed them. However, all definitions share a common understanding that critical infrastructure is crucial for the security, safety, and economic well-being of a nation and its people.

The definitions of critical infrastructure in Table 1 highlight the criticality, complexity, and interdependence of the various physical and cyber systems and assets that make up

Table 1 Selected definitions of critical infrastructure.

Definition critical infrastructure	References
Assets, systems, and networks are critical to a country's economy, security, and social well-being	US Department of Homeland Security [1,6]
Physical and virtual systems and assets are vital for the functioning of modern society	European Commission [7]
Infrastructure whose incapacity or destruction would have a debilitating impact on national security, governance, public health, safety, or economic security	National Infrastructure Protection Plan [4]
Systems and assets, whether physical or virtual, are so vital to the nation that the incapacity or destruction of such systems and assets would have a debilitating impact on security, national economic security, national public health or safety, or any combination of those matters	Presidential Policy Directive 21 [8]
Essential services and systems that underpin modern society	International Atomic Energy Agency [9]
Physical and cyber systems and assets are essential for the functioning of the economy and the delivery of critical services to the public	Australian Government [10]
Physical and virtual systems are essential for the functioning of the state and society, the disruption or destruction of which would have a significant impact on the security and well-being of the population	Ministry of Internal Affairs and Communications (Japan) [11]
The essential facilities, systems, and networks, whether physical or virtual, are so vital that their incapacity or destruction would have a debilitating impact on national security, economic security, public health, or safety	National Institute of Standards and Technology [12,13]
Physical and cyber systems and assets are critical to the reliable functioning of the essential services of a society	Canadian Cyber Incident Response Center [14]
The infrastructure is critical to the functioning of society and the economy, the disruption or destruction of which would have significant impacts on public safety, national security, and economic prosperity	World Economic Forum [15]

critical infrastructure. These systems and assets are essential for providing essential services and maintaining the functioning of modern society. The definitions also acknowledge that critical infrastructure is vulnerable to various threats, including cyber threats, physical threats, and natural disasters, which can have severe consequences for society and the economy. The definitions of critical infrastructure provided in the table highlight the fact that it is essential to the security, safety, and economic well-being of a nation and its people. These definitions emphasize the interconnectivity and interdependence of critical infrastructure, meaning that the failure of one system or asset can have a domino effect on others, leading to cascading failures. This highlights the criticality, complexity, and vulnerability of these systems and assets, which are essential for providing essential services and ensuring the security, safety, and economic well-being of a nation and its people. The definitions also recognize the various threats that critical infrastructure faces, such as cyber threats, physical threats, and natural disasters, which can have significant consequences for society and the economy.

The definitions vary slightly in their wording and also reflect the different perspectives and priorities of the organizations and institutions that have developed them. For instance, some definitions, such as those from the World Economic Forum and the National Institute of Standards and Technology, focus on the impact that the disruption or destruction of critical infrastructure would have on society and the economy. Other definitions, such as those from the European Commission and the International Atomic Energy Agency, focus on the essential services and systems that underpin modern society. Despite these differences, all the definitions emphasize the criticality, complexity, and interdependence of the various physical and cyber systems and assets that make up critical infrastructure. This understanding is crucial for identifying and protecting critical infrastructure from various threats, including cyber threats, physical threats, and natural disasters.

4 Stakeholders of critical infrastructures

A stakeholder is an individual, group, or organization that have a vested interest in a particular project, organization, or system. They are individuals or groups who may be impacted by the decisions, actions, or outcomes related to the project or system, or who may have the ability to impact those decisions, actions, or outcomes.

In the context of critical infrastructure, stakeholders are important because they play a critical role in ensuring the resilience and sustainability of these systems. As critical infrastructures are complex systems that provide essential services to communities, it is essential to understand the needs, interests, and perspectives of all stakeholders involved in their development, operation, and maintenance. By involving stakeholders in decision-making processes, critical infrastructure owners and operators can better understand the potential impacts of their decisions and actions, and make more informed choices that are more likely to meet the needs and expectations of all stakeholders involved.

Stakeholder engagement is a critical component of building and maintaining resilient critical infrastructures. Effective stakeholder engagement ensures that critical infrastructure systems are designed, built, and operated with the input and feedback of all relevant stakeholders, including those who may be affected by the system. By engaging stakeholders early and often in the process, critical infrastructure owners and operators can identify potential issues or concerns early on, and work collaboratively to address these issues before they become significant problems.

Furthermore, stakeholders can also play a critical role in responding to and recovering from disruptions or disasters affecting critical infrastructures. By involving stakeholders in emergency planning and response activities, critical infrastructure owners and operators can better understand the needs and concerns of the communities they serve and work collaboratively to develop effective response and recovery plans that meet the needs of all stakeholders involved.

Various categories of stakeholders can be identified for critical infrastructures, including government agencies, private sector companies, local communities, nongovernmental organizations (NGOs), emergency responders, and others. However, the specific categories of stakeholders and their relative importance may vary depending on the type of critical infrastructure being considered. For example, stakeholders for a transportation system may include government agencies responsible for transportation policy and regulation, transportation companies, trade unions representing transportation workers, local communities impacted by transportation routes, emergency responders, and others. On the other hand, stakeholders for a water treatment plant may include government agencies responsible for water policy and regulation, the company operating the plant, local communities that rely on the plant for clean water, NGOs concerned with water quality and conservation, and others.

Based on the domain analysis process, we can identify the following categories of stakeholders related to critical infrastructure:

- *Government*: Governments play a critical role in the development, regulation, and oversight of critical infrastructure. They are responsible for setting policies and standards, allocating funding, and ensuring compliance with regulations.
- *Infrastructure owners and operators*: These are the entities that own and operate critical infrastructure, including utilities, transportation networks, telecommunications companies, and other organizations. They are responsible for ensuring the reliable and safe operation of their infrastructure.
- *Users and customers*: These are the individuals and businesses that rely on critical infrastructure to carry out their daily activities. They may include residents, tourists, transportation users, and other consumers.
- *Industry associations*: Trade and professional associations that represent the interests of critical infrastructure sectors, promoting industry standards, best practices, and collaboration among members.

- *Suppliers and contractors*: These are the companies and individuals that provide goods and services to critical infrastructure owners and operators, including construction firms, technology vendors, and maintenance contractors.
- *Emergency responders*: Emergency responders, including police, fire, and medical personnel, are critical stakeholders in the event of an emergency or disaster that affects critical infrastructure.
- *Regulators and standards organizations*: These entities are responsible for setting and enforcing standards and regulations related to critical infrastructure. They may include government agencies, industry associations, and international organizations.
- *Local communities*: Local communities play a critical role in the development and operation of critical infrastructure, as they may be impacted by the location, design, and operation of infrastructure within their communities.
- *Academia and research organizations*: Researchers and academics play a key role in developing and advancing knowledge related to critical infrastructure, including best practices for design, construction, and operation.

Each stakeholder group has unique perspectives, priorities, and needs and must be considered in the development and operation of critical infrastructure.

After the identification of the stakeholders, stakeholder mapping can be used to identify and analyze stakeholders based on their level of interest and power over a particular project or system. The power-interest matrix is a commonly used tool for this analysis. This matrix plots stakeholders on a graph based on their level of power (or influence) and their level of interest in the project or system. The matrix is divided into four quadrants.

High Power, High Interest (Key Players): Stakeholders with significant influence and interest in the project or decision. These stakeholders should be closely engaged and managed to ensure their support and cooperation.

High Power, Low Interest (Keep Satisfied): Stakeholders with considerable influence but limited interest in the project or decision. These stakeholders should be kept informed and satisfied to prevent them from becoming potential obstacles.

Low Power, High Interest (Keep Informed): Stakeholders with high interest but limited influence over the project or decision. These stakeholders should be kept informed and consulted to ensure their concerns are addressed and to gain their support.

Low Power, Low Interest (Monitor): Stakeholders with minimal influence and interest in the project or decision. These stakeholders should be monitored, but extensive engagement may not be necessary.

To perform a power-interest matrix analysis for critical infrastructures, the following steps can be taken:
- *Identify stakeholders*: Begin by identifying all stakeholders who have an interest or involvement in the critical infrastructure, including those who are directly and indirectly affected.

- *Assess power*: Consider the level of power each stakeholder holds over the critical infrastructure. Power can be assessed by looking at factors such as their ability to influence decision-making, control resources, or impact the system's operations.
- *Assess interest*: Evaluate the level of interest each stakeholder has in the critical infrastructure. Interest can be assessed by looking at factors such as how closely their work is tied to the infrastructure, the degree to which they depend on the infrastructure, or their potential to be impacted by its operation.
- *Plot stakeholders on the matrix*: Once power and interest have been assessed, stakeholders can be plotted on the power-interest matrix. The matrix typically consists of four quadrants: High power-high interest, high power-low interest, low power-high interest, and low power-low interest.
- *Analyze the results*: Once stakeholders have been plotted on the matrix, they can be analyzed based on their location in the quadrants. This analysis can be used to identify key stakeholders, prioritize stakeholder engagement efforts, and inform decision-making processes.

Table 2 shows an example power-interest matrix for critical infrastructure stakeholders. In this example, stakeholders with high power and high interest, such as government agencies, infrastructure owners and operators, and emergency management agencies, are considered key players. They should be closely engaged and managed to ensure their support and cooperation. Stakeholders with high power but low interest, such as regulatory bodies and industry associations, should be kept satisfied and informed to prevent them from becoming potential obstacles. Stakeholders with low power but high interest, such as research and development institutions, law enforcement, and security agencies, should be kept informed and consulted to ensure their concerns are addressed and to gain their support. Finally, stakeholders with low power and low interest, such as private sector companies, international organizations, and NGOs, should be monitored, but extensive engagement may not be necessary.

Table 2 Example power-interest matrix for critical infrastructure stakeholders.

Power	Interest	High	Low
High		<ul><li>Government agencies</li><li>Infrastructure owners and operators</li><li>Emergency management agencies</li></ul>	<ul><li>Regulatory bodies</li><li>Industry associations</li></ul>
Low		<ul><li>Research and development institutions</li><li>Law enforcement</li><li>Security agencies</li></ul>	<ul><li>Private sector companies international organizations</li><li>NGOs</li></ul>

5 Key terms

A feature model of critical infrastructures can help to identify the different components and dependencies that are critical to the functioning of these complex systems. We have identified and described the key concepts as shown in Appendix.

6 Feature model of critical infrastructures

A feature model is a tool used in software engineering and product management to describe the different features and options that a product or software system can have [16]. A feature model describes the different options and constraints for a particular product or system and can help to guide the development process by ensuring that all requirements and constraints are taken into account.

A feature model typically consists of a tree-like structure, where the root node represents the overall product or system, and each branch represents different features or options. Each node in the tree represents a different feature or option and may have child nodes that represent subfeatures or options.

Features may be mandatory or optional and may have different levels of complexity or specificity. Constraints may also be included in the feature model, which describes dependencies or relationships between different features or options.

A feature model is often used to help to guide the development process by ensuring that all requirements and constraints are considered, and by providing a clear framework for understanding the different features and options that a product or system can have. By modeling the different features and options in a structured way, a feature model can help to identify potential conflicts or inconsistencies and can help to ensure that the final product or system meets all requirements and constraints.

The feature diagram that we have derived for critical infrastructures is shown in Fig. 2.

From a systems engineering perspective, critical infrastructure can be viewed as a complex system of interdependent components that work together to provide essential services to society [5,17]. The elements of a critical infrastructure can be broadly categorized into physical, cyber, human components, and environmental components as outlined below

- *Physical components*: These are the tangible and visible parts of a critical infrastructure system, including buildings, facilities, equipment, vehicles, and other physical assets. Physical components also include the various networks and systems that enable the transmission, distribution, and delivery of essential services, such as power grids, water distribution systems, and transportation networks.
- *Cyber components*: These are the intangible and often invisible parts of a critical infrastructure system, including the various information and communication technologies (ICTs) that underpin the functioning of the system. Cyber components include

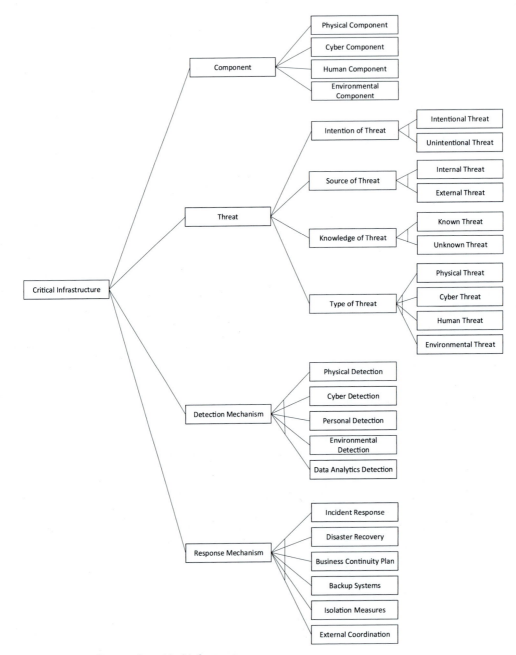

Fig. 2 Feature diagram for critical infrastructures.

computer networks, software systems, data centers, sensors, and other ICT infrastructures that enable the monitoring, control, and coordination of physical components.

- *Human components*: These are the people who operate, manage, and maintain the critical infrastructure system. Human components include engineers, technicians, operators, managers, and other personnel who are responsible for ensuring the smooth and safe operation of the system. Human components also include the various stakeholders who are affected by the system, such as customers, suppliers, regulators, and policymakers.
- *Environmental components*: These are the natural or environmental factors that can impact critical infrastructures, such as weather events or geological hazards. These components may be further subdivided into different types or categories, depending on the specific environmental factors and their impact on critical infrastructures.

In the context of critical infrastructures, a threat refers to any potential danger, hazard, or risk that could cause harm or disrupt the normal functioning of the infrastructure. Threats can come from a wide variety of sources, including natural disasters, accidents, human error, and deliberate acts of sabotage or terrorism. The impact of a threat on critical infrastructure can be severe and far-reaching, potentially affecting not only the infrastructure itself but also the people and communities that rely on it. For example, a cyber-attack on a power grid could result in widespread power outages and disrupt the functioning of essential services such as hospitals, transportation, and communication systems. The feature diagram in Fig. 2 includes the following features:

- *Physical threats*: These are threats that are physically present and can cause harm to the organization or system, such as natural disasters, vandalism, or theft.
- *Cyber threats*: These are threats that come from the digital realm and can harm the organization or system, such as malware, phishing attacks, or hacking attempts.
- *Human threats*: These are threats that come from individuals within the organization or from individuals outside the organization, such as disgruntled employees, social engineering attacks, or terrorism.
- *Environmental threats*: These are threats that come from the environment and can impact the organization or system, such as pollution, climate change, or resource depletion.
- *Intentional vs unintentional threats*: This feature represents the distinction between threats that are intentionally caused by humans or organizations, versus those that are caused unintentionally or accidentally.
- *External vs internal threats*: This feature represents the distinction between threats that originate from outside the organization or system versus those that come from within.
- *Known vs unknown threats*: This feature represents the distinction between threats that are known and can be anticipated versus those that are unknown and unexpected.

Detection mechanisms are crucial for critical infrastructure protection because they help to identify and respond to threats promptly, potentially mitigating their impact. In the context of critical infrastructure, threats can come from a variety of sources, such as cyber-attacks, physical attacks, natural disasters, and human error. A detection mechanism is a system or process that is designed to identify the presence of a threat or anomaly and alert relevant stakeholders so that they can take appropriate action to prevent or mitigate the potential impact.

For example, in the case of cyber threats, detection mechanisms can include intrusion detection systems (IDS), which monitor network traffic for suspicious activity, and antivirus software, which scans for and identifies malicious code. In the case of physical threats, detection mechanisms can include surveillance cameras and sensors that detect motion, temperature changes, or other anomalies. In the case of natural disasters, detection mechanisms can include early warning systems, such as seismic sensors, that can alert relevant authorities to the possibility of an earthquake or other natural event.

Detection mechanisms can also help to prevent false alarms or unnecessary disruptions. By accurately detecting and identifying threats, these mechanisms can reduce the likelihood of false positives, which can result in unnecessary disruptions or costly responses. Additionally, detection mechanisms can provide valuable data for analyzing and understanding threats, which can help to inform future prevention and response strategies.

We have identified the following features related to detection mechanisms in critical infrastructures:

- *Intrusion detection*: Systems and technologies designed to detect unauthorized access or attempted access to critical infrastructure systems or networks.
- *Physical security*: Physical barriers, access control systems, and surveillance technologies that protect critical infrastructure facilities from physical threats such as theft, vandalism, or sabotage.
- *Network security*: Technologies such as firewalls, intrusion prevention systems, and secure communications protocols that protect critical infrastructure networks from cyber threats.
- *Monitoring and analytics*: Technologies that enable the continuous monitoring of critical infrastructure systems and networks, as well as advanced analytics tools that can detect anomalous behavior or potential threats.
- *Risk management*: Technologies and processes for identifying and assessing risks to critical infrastructure systems and networks.

Response mechanisms are necessary to minimize the impact of a successful attack or disruption to critical infrastructure. While detection mechanisms identify when an attack or disruption is occurring, response mechanisms are responsible for implementing a coordinated and effective response to minimize the impact and restore functionality as quickly as possible.

Response mechanisms include measures such as incident response plans, disaster recovery plans, and business continuity plans. These plans are put in place to ensure that in the event of an attack or disruption, a well-coordinated response can be executed. They typically include predefined steps and procedures to follow, such as activating backup systems, isolating affected systems, and coordinating with external agencies and stakeholders.

The feature diagram for response mechanisms includes the following key features:

- *Incident response plan*: A predefined plan of action to be executed in the event of an incident.
- *Disaster recovery plan*: A predefined plan of action to be executed in the event of a disaster that causes damage to the infrastructure or facilities.
- *Business continuity plan*: A predefined plan of action to be executed in the event of a disruption to business operations that ensures essential functions are maintained.
- *Backup systems*: Systems and processes that provide redundancy and backup capabilities to restore critical services and systems in the event of an outage or disruption.
- *Isolation measures*: Steps taken to isolate affected systems or areas to prevent the spread of the disruption or attack.
- *External coordination*: Processes for coordinating with external stakeholders, such as emergency services, government agencies, and critical infrastructure partners, to coordinate response efforts and ensure a consistent approach.

Note that while some of these mechanisms may overlap or interact with each other (for example, the incident response may involve access control), they are primarily focused on either detecting or responding to security threats in critical infrastructure.

By modeling the different components and dependencies of critical infrastructures in a feature model, it is possible to identify potential risks and vulnerabilities and to develop strategies for mitigating these risks. Additionally, a feature model can help to ensure that all key components and dependencies are taken into account in the design and operation of critical infrastructures, thereby improving their resilience and reliability. The presented feature model can be expanded by a more detailed study, which we leave for future work.

7 Critical infrastructure types

Several critical infrastructure examples can be found in the literature. We list these in the following subsections:

7.1 Energy infrastructure

Energy infrastructure is a vital component of modern society, encompassing a wide range of facilities and systems that are essential for the production, transmission, distribution, and consumption of energy. The energy sector plays a critical role in powering all aspects of human life, from powering homes, businesses, and hospitals to fueling transportation, communication systems, and manufacturing processes [18–22]. Therefore, ensuring the security, reliability, and resilience of energy infrastructure is of utmost importance to maintain the uninterrupted supply of energy and prevent severe consequences to public health, safety, and economic well-being.

The energy infrastructure includes various types of facilities and systems, each with unique characteristics and challenges. One of the primary components of energy infrastructure is power plants. Power plants generate electricity from various sources, such as coal, natural gas, nuclear, hydroelectric, wind, and solar power plants. Power plants can be large centralized facilities or smaller distributed systems, depending on the type and scale of the power generation technology used.

Oil and gas pipelines are another critical component of energy infrastructure. These pipelines transport oil and natural gas from production sites to refineries, storage facilities, and distribution networks. Oil and gas pipelines are often located underground and can stretch for thousands of miles, making them vulnerable to various threats, such as natural disasters, cyber-attacks, and physical attacks.

Electricity transmission and distribution systems are critical infrastructure components that are responsible for the delivery of electricity from power generation sources to end-users, including homes, businesses, and essential services. The electricity transmission system is responsible for transmitting high-voltage electricity from power generation facilities to substations, where the voltage is then stepped down and distributed to end-users via the electricity distribution system. The transmission system is made up of high-voltage power lines and towers, as well as transformers and other equipment that regulate the flow of electricity. The electricity distribution system, on the other hand, is responsible for delivering low-voltage electricity to end-users. This system includes transformers, power lines, poles, and other equipment that distribute electricity throughout the community.

Renewable energy sources, such as solar and wind farms, are becoming increasingly important as the world seeks to reduce its reliance on fossil fuels. These sources generate electricity from renewable sources of energy and are critical for achieving sustainable and carbon-free energy systems. However, renewable energy sources present unique challenges to energy infrastructure due to their intermittent and variable nature. Energy storage facilities, such as batteries, pumped hydro storage, and other technologies, are essential for storing energy for use during times of peak demand or when renewable sources are not available.

Natural gas storage facilities are also critical infrastructure components that store natural gas for use during times of peak demand or when supply is disrupted. These facilities can include underground storage facilities and liquefied natural gas (LNG) terminals.

Finally, energy trading and financial systems facilitate the buying and selling of energy commodities, such as oil, gas, and electricity, and are critical for ensuring the efficient functioning of energy markets. Energy markets are complex and dynamic, with various actors, such as energy producers, consumers, regulators, and traders, interacting to balance supply and demand and ensure price stability.

The security and resilience of energy infrastructure are of utmost importance, as any disruption or failure of these systems can have severe consequences for public health, safety, and economic well-being. Therefore, it is essential to invest in the development of robust and resilient energy infrastructure that can withstand various types of threats and hazards, including cyber-attacks, natural disasters, and physical attacks.

7.2 Transportation infrastructure

Transportation infrastructure is a critical infrastructure that includes various modes of transportation, such as roads, highways, bridges, tunnels, railways, airports, seaports, and public transportation systems [23–25]. It is essential for the movement of people and goods and plays a vital role in the functioning of many other critical infrastructures, such as healthcare, emergency services, and the energy sector.

Transportation infrastructure includes the following components:

1. Roads and highways: These are networks of paved or unpaved roads and highways that allow for the movement of vehicles, bicycles, and pedestrians. They provide access to communities, businesses, and other critical infrastructure.
2. Bridges and tunnels: These are structures that allow roads and highways to cross over or under natural barriers such as rivers, mountains, or bodies of water. They are essential for maintaining the connectivity of transportation systems.
3. Railways: These are networks of tracks and trains that transport goods and people over long distances. They are used for the transportation of bulk commodities, such as coal and grain, as well as for passenger transportation.
4. Airports: These are facilities that allow for the takeoff, landing, and maintenance of airplanes. They are essential for the transportation of people and goods over long distances and are often critical for emergency response and disaster relief.
5. Seaports: These are facilities that allow for the docking and maintenance of ships. They are essential for the transportation of goods, particularly those that are heavy or bulky and cannot be transported by air or land.
6. Public transportation systems: These include buses, subways, light rail, and other systems that allow for the transportation of people within and between urban areas. They are critical for providing access to employment, education, healthcare, and other services.

The security and resilience of transportation infrastructure are essential for public safety, economic well-being, and national security. Therefore, it is essential to invest in the development of robust and secure transportation infrastructure that can withstand various types of threats and hazards, such as natural disasters, cyber-attacks, and physical attacks. Additionally, transportation infrastructure must be designed to accommodate the changing needs of society, such as the increasing demand for sustainable and environmentally friendly transportation options.

7.3 Telecommunications infrastructure

Telecommunications infrastructure refers to the network of technologies, equipment, and services that enable communication and the exchange of information through electronic means [26–28]. This infrastructure includes a range of communication technologies, such as wired and wireless networks, fiber-optic cables, satellites, and other digital communication technologies [29]. It is essential for enabling businesses, individuals, and governments to communicate with one another and to access and share information across the globe.

Telecommunications infrastructure consists of three main components:

1. Access network: This refers to the network of infrastructure that provides users with access to communication services, such as phone lines, wireless networks, and broadband services. Access networks can be wired or wireless and can support a range of different communication technologies.
2. Transport network: This is the backbone of the telecommunications infrastructure and provides high-speed connectivity between different parts of the network. Transport networks use a combination of fiber-optic cables, microwave links, and other technologies to transmit large volumes of data over long distances.
3. Core network: This is the central network that manages and directs the flow of information across the telecommunications infrastructure. It is responsible for routing data between different networks and services, and for ensuring that information is transmitted efficiently and securely.

Telecommunications infrastructure plays a vital role in supporting economic development, enabling remote work and education, and facilitating the delivery of critical services such as healthcare and emergency response. However, like other critical infrastructure, telecommunications infrastructure is vulnerable to a range of threats, including cyber-attacks, natural disasters, and physical attacks. As a result, it is essential to invest in the development of robust and secure telecommunications infrastructure that can withstand these threats and maintain its functionality under adverse conditions. Additionally, the growing importance of telecommunications infrastructure in modern society has led to increasing concerns about issues such as data privacy, security, and equitable access to communication services.

7.4 Water and wastewater infrastructure

Water and wastewater infrastructure refer to the network of facilities, equipment, and processes that are involved in the supply, treatment, and distribution of clean water and the collection, treatment, and disposal of wastewater [30–33]. This infrastructure is critical to public health, environmental protection, and economic development.

Water infrastructure consists of three main components
1. Source: This refers to the location where water is collected for treatment and distribution. Water sources can include surface water (lakes, rivers, and reservoirs), groundwater (wells and aquifers), and rainwater harvesting systems.
2. Treatment: This involves the process of treating water to remove impurities, such as bacteria, viruses, chemicals, and sediment before it is distributed to consumers. The treatment process may include physical, chemical, and biological treatment methods, depending on the water source and the quality of the water.
3. Distribution: This involves the network of pipes, pumps, and storage tanks that transport treated water from the treatment plant to consumers.

Wastewater infrastructure also consists of three main components
1. Collection: This involves the network of pipes and pumps that collect wastewater from homes, businesses, and other facilities and transport it to the treatment plant.
2. Treatment: This involves the process of treating wastewater to remove contaminants and pollutants before it is released back into the environment. The treatment process may include physical, chemical, and biological treatment methods, depending on the type of pollutants present in the wastewater.
3. Disposal: This involves the safe and environmentally responsible disposal of treated wastewater. Treated wastewater may be discharged into surface water bodies or used for irrigation purposes, depending on the local regulations and environmental conditions.

Water and wastewater infrastructure are vulnerable to a range of threats, including natural disasters, aging infrastructure, and contamination. To ensure the security and resilience of these systems, investments are made to improve their efficiency, reliability, and security. Additionally, new technologies such as smart water grids and water reuse systems are being developed to improve the sustainability and resilience of the water and wastewater infrastructure. A reliable and resilient water and wastewater infrastructure are essential to ensure public health, support economic activity, and maintain environmental quality.

7.5 Financial infrastructure

Financial infrastructure refers to the network of institutions, systems, and technologies that enable financial transactions and activities to take place. It encompasses a wide range of services and processes, including payment systems, clearing and settlement systems, credit bureaus, stock exchanges, and regulatory bodies. Financial infrastructure is critical to the functioning of the global economy, as it facilitates the movement of funds and the allocation of resources across borders and industries [34–36].

Payment systems are a key component of financial infrastructure, enabling individuals and businesses to transfer funds securely and efficiently. These systems include credit cards, debit cards, electronic fund transfers, and mobile payment systems. Clearing and settlement systems are also essential to financial infrastructure, providing a framework for the settlement of financial transactions and the management of counterparty risk.

Credit bureaus play an important role in financial infrastructure by providing information on the creditworthiness of individuals and businesses. This information helps lenders to make informed decisions about the risks associated with lending and enables borrowers to access credit on favorable terms.

Stock exchanges are another key component of financial infrastructure, providing a platform for the trading of securities and other financial instruments. These exchanges help to facilitate capital formation and allocation, allowing investors to invest in businesses and governments to raise funds.

Regulatory bodies are responsible for overseeing and regulating the financial infrastructure, ensuring that it operates safely and efficiently. These bodies set standards for the operation of payment systems, credit bureaus, and stock exchanges, and enforce rules and regulations designed to protect consumers and maintain financial stability.

Financial infrastructure is vulnerable to a range of threats, including cyber-attacks, fraud, and systemic risks. To mitigate these risks, significant investments are made in the development of robust and secure financial systems, as well as in the education and training of financial professionals. Additionally, innovations such as blockchain and distributed ledger technology are being developed to improve the security and efficiency of financial infrastructure. A reliable and resilient financial infrastructure are essential to support economic growth and development and to ensure the stability and integrity of financial systems around the world.

7.6 Healthcare infrastructure

Healthcare infrastructure refers to the network of facilities, institutions, and personnel involved in the delivery of healthcare services. It encompasses a range of services and processes, including hospitals, clinics, medical equipment, pharmaceuticals, public health systems, and regulatory bodies. Healthcare infrastructure is critical to the health and well-being of individuals and communities and plays a vital role in promoting public health, preventing and managing disease, and improving health outcomes [37–40].

Hospitals and clinics are key components of healthcare infrastructure, providing a range of services from emergency care to long-term treatment and rehabilitation [41]. These facilities employ a wide range of healthcare professionals, including doctors, nurses, and allied health professionals, and are equipped with medical equipment and technology to support diagnosis, treatment, and monitoring.

Medical equipment and pharmaceuticals are also critical components of healthcare infrastructure, supporting the diagnosis and treatment of illness and disease. Medical equipment includes diagnostic tools, such as X-ray and MRI machines, and treatment devices, such as ventilators and dialysis machines. Pharmaceuticals include drugs and other medical products used to treat illness and disease.

Public health systems are another key component of healthcare infrastructure, encompassing a range of activities related to the prevention and control of disease. These systems include vaccination programs, disease surveillance and response, and health promotion activities designed to promote healthy lifestyles and behaviors.

Regulatory bodies are responsible for overseeing and regulating the healthcare infrastructure, ensuring that it operates safely and efficiently. These bodies set standards for healthcare facilities, medical equipment, and pharmaceuticals, and enforce rules and regulations designed to protect consumers and maintain the integrity of healthcare systems.

Healthcare infrastructure is vulnerable to a range of threats, including natural disasters, disease outbreaks, and inadequate funding. To ensure the security and resilience of healthcare infrastructure, investments are made to improve the quality and accessibility of healthcare services, as well as to develop and deploy new medical technologies and treatments. Additionally, innovations such as telemedicine and electronic health records are being developed to improve the efficiency and effectiveness of healthcare delivery, yet this has further increased the attack vector making the infrastructure more digitally vulnerable. A reliable and resilient healthcare infrastructure are essential to promote public health and well-being and to ensure that individuals and communities have access to the care and support they need to live healthy and productive lives.

7.7 Emergency services infrastructure

Emergency services infrastructure refers to the network of institutions, systems, and personnel involved in providing emergency response and support during crises and disasters. It encompasses a range of services and processes, including emergency medical services, firefighting, law enforcement, search and rescue, and disaster management. Emergency services infrastructure is critical to public safety and plays a vital role in protecting lives and property during emergencies and disasters [42–44].

Emergency medical services (EMSs) are a key component of emergency services infrastructure, providing emergency medical care and transportation to patients in need. EMS personnel include emergency medical technicians (EMTs) and paramedics, who are trained to provide advanced medical care in the field. EMS systems are typically coordinated through a central dispatch system, which helps to ensure that patients receive prompt and appropriate care.

Firefighting services are another critical component of emergency services infrastructure, providing fire suppression and prevention services to communities. Firefighters are trained to respond to a range of emergencies, including building fires, wildfires, and hazardous materials incidents. Firefighting services may also include rescue services, such as extricating people from wrecked vehicles or collapsed buildings.

Law enforcement agencies are responsible for maintaining public safety and order and play an important role in emergency services infrastructure. Law enforcement personnel include police officers, sheriffs, and state troopers, who are responsible for responding to emergencies and enforcing the law during disasters and other crises.

Search and rescue teams are also a key component of emergency services infrastructure, assisting people who are lost or in danger during emergencies and disasters. These teams may include specialized personnel, such as mountain rescue or swift water rescue teams, who are trained to respond to specific types of emergencies.

Disaster management agencies are responsible for coordinating emergency response and recovery efforts during disasters and other crises. These agencies may include federal, state, and local government entities, as well as nongovernmental organizations (NGOs) and volunteer groups. Disaster management agencies are responsible for planning and preparedness efforts, as well as response and recovery efforts during and after disasters.

Emergency services infrastructure is vulnerable to a range of threats, including natural disasters, terrorism, and cyber-attacks. To mitigate these risks, significant investments are made in the development of robust and resilient emergency response systems, as well as in the education and training of emergency services personnel. Additionally, innovations such as unmanned aerial vehicles (UAVs) and advanced communication technologies are being developed to improve the effectiveness and efficiency of emergency services infrastructure. A reliable and resilient emergency services infrastructure are essential to protect public safety and security and to ensure that communities can respond effectively to emergencies and disasters.

7.8 Food and agriculture infrastructure

Food and agriculture infrastructure refer to the network of facilities, systems, and processes involved in producing, distributing, and processing food and agricultural products. It encompasses a range of activities, including farming, ranching, fishing, food processing, transportation, and distribution. Food and agriculture infrastructure are critical to ensuring food security and supporting the food and beverage industry, which are a major contributor to the global economy [34,45–47].

Farming and ranching are key components of food and agriculture infrastructure, producing crops and livestock for food and other products. These activities involve a range of processes, including planting, harvesting, and animal husbandry, and often require specialized equipment and technology to improve efficiency and productivity.

Fishing is another important component of food and agriculture infrastructure, providing a source of protein and other essential nutrients. Fishing activities include commercial and subsistence fishing and may involve harvesting fish from the ocean, freshwater sources, or aquaculture systems.

Food processing is a critical component of food and agriculture infrastructure, transforming raw agricultural products into food and other products. Food processing facilities may include meat processing plants, dairy processing facilities, and grain mills and may involve a range of activities, such as slaughtering, processing, packaging, and distribution.

Transportation and distribution systems are also key components of food and agriculture infrastructure, facilitating the movement of food and agricultural products from farms and processing facilities to consumers. These systems may include trucks, trains, ships, and planes and require specialized infrastructure, such as ports and transportation hubs, to support efficient and reliable transportation.

Food and agriculture infrastructure are vulnerable to a range of threats, including natural disasters, disease outbreaks, and cyber-attacks. To mitigate these risks, significant investments are made in the development of robust and resilient food and agriculture systems, as well as in research and development efforts to improve agricultural productivity and sustainability. Additionally, innovations such as precision agriculture and vertical farming are being developed to improve efficiency and sustainability in food production. A reliable and resilient food and agriculture infrastructure are essential to ensuring food security and supporting economic development and growth.

7.9 Chemical and hazardous materials infrastructure

Chemical and hazardous materials infrastructure refer to the network of facilities, systems, and processes involved in the production, storage, transportation, and disposal of chemicals and other hazardous materials. This infrastructure is essential to many industrial and manufacturing processes but also poses significant risks to public safety and the environment [2,32,40,48].

Chemical production facilities are a key component of chemical and hazardous materials infrastructure, producing a range of chemicals and chemical products for use in manufacturing and other industries. These facilities may involve the use of hazardous materials and may pose risks of fire, explosion, and chemical releases.

Storage and transportation systems are also critical components of chemical and hazardous materials infrastructure, providing safe and secure storage and transportation of chemicals and hazardous materials. These systems may include tanks, pipelines, and rail and truck transportation systems and require careful planning and management to minimize risks of spills and accidents.

Disposal and remediation systems are another important component of chemical and hazardous materials infrastructure, providing safe and effective disposal of hazardous waste and remediation of contaminated sites. These systems may include hazardous waste landfills, incinerators, and treatment facilities, as well as soil and groundwater remediation systems.

Chemical and hazardous materials infrastructure are vulnerable to a range of threats, including natural disasters, accidents, and intentional attacks. To mitigate these risks, significant investments are made in the development of robust and resilient chemical and hazardous materials systems, as well as in the education and training of workers and emergency responders. Additionally, innovations such as advanced sensors and remote monitoring technologies are being developed to improve safety and security in chemical and hazardous materials infrastructure. A reliable and resilient chemical and hazardous materials infrastructure are essential to protecting public safety and the environment, while also supporting economic development and growth.

7.10 Government infrastructure

Government infrastructure refers to the network of facilities, systems, and processes that support the functioning of government and public services. It includes a range of physical and digital assets, such as government buildings, roads, bridges, public transit systems, communication networks, and information technology systems [4].

Government buildings and facilities serve as the physical infrastructure for government operations and public services, such as courthouses, town halls, police stations, and public libraries. These buildings provide space for government employees to work, as well as for the public to access government services and resources.

Roads, bridges, and public transit systems are essential components of government infrastructure, supporting the movement of people and goods within and between communities. These systems may include highways, bridges, railways, buses, subways, and other forms of public transportation.

Communication networks, such as telephone and internet systems, are critical to supporting the flow of information among government agencies, public institutions, and individuals. These systems enable government employees to communicate and collaborate with each other, as well as with the public, and support a range of functions, such as emergency response, public safety, and healthcare.

Information technology systems are also key components of government infrastructure, providing the digital infrastructure for government operations and public services. These systems may include databases, software applications, and other digital platforms that support government functions, such as tax collection, voting systems, and public health management.

Government infrastructure is essential to the functioning of society and the provision of public services and is typically funded by tax revenues and public funds. To ensure the reliability and resilience of government infrastructure, significant investments are made in infrastructure planning, maintenance, and upgrades. Additionally, innovation and technological advancements are continually being developed to improve the efficiency and effectiveness of government infrastructure, such as the development of smart city technologies and digital government platforms.

7.11 Information technology infrastructure

Information technology (IT) critical infrastructure refers to the technological systems, networks, and assets that are essential to the functioning of various sectors of society, including government, finance, healthcare, transportation, and communication. These systems provide the necessary support for critical services, such as power grids, emergency services, and supply chains.

The importance of IT critical infrastructure can be seen in the impact of disruptions or failures in these systems. For example, a cyber-attack on a power grid could result in a widespread blackout that could last for days or even weeks, causing significant economic and social disruption. Similarly, a data breach in a hospital's network could lead to compromised patient data and a loss of trust in the healthcare system.

One key challenge in protecting IT critical infrastructure is the constantly evolving nature of cyber threats. Cybercriminals are constantly developing new techniques and methods to exploit vulnerabilities in systems and networks, making it necessary for organizations to stay up-to-date on the latest security measures and technologies. Another challenge is the complexity of IT critical infrastructure systems, which often involve numerous interconnected networks and components. This complexity can make it difficult to identify and address vulnerabilities and risks, particularly when multiple organizations are involved in the operation of the infrastructure.

Despite these challenges, there are a variety of strategies that organizations can use to protect their IT critical infrastructure. One important approach is to prioritize risk management, identifying the most critical systems and assets and implementing targeted security measures to protect them. This can involve conducting risk assessments and developing threat models to identify potential vulnerabilities and risks.

Another approach is to establish partnerships and collaboration among organizations involved in the operation of IT critical infrastructure. This can involve sharing threat intelligence and best practices, coordinating response plans, and establishing joint cybersecurity exercises to test the resilience of the infrastructure.

Overall, IT critical infrastructure plays a critical role in modern society, providing the technological backbone for essential services and systems. Protecting this infrastructure requires a proactive and collaborative approach, with organizations working together to identify and address risks and vulnerabilities and ensure the continued functioning and security of these essential systems.

7.12 Defense industry infrastructure

The defense industry's critical infrastructure refers to the technological systems, networks, and assets that are necessary to support and maintain the defense capabilities of a country. This infrastructure includes a wide range of systems and facilities, including research and development centers, production facilities, testing sites, and communication networks [2,49–51]. The defense industry critical infrastructure is responsible for providing the military with the necessary tools and technologies to operate effectively and protect the country's security interests.

The importance of the defense industry's critical infrastructure cannot be overstated, as it plays a critical role in maintaining the readiness and effectiveness of a country's military forces. The infrastructure supports a range of critical defense capabilities, including missile defense systems, cyber warfare capabilities, intelligence gathering and

analysis, and advanced weaponry. Given the strategic importance of the defense industry critical infrastructure, it is also a prime target for hostile actors seeking to disrupt or
compromise a country's military capabilities. These actors may include foreign governments, terrorist organizations, or criminal groups, all of whom have demonstrated the
ability to launch sophisticated cyberattacks or engage in other types of sabotage to
achieve their objectives.

To ensure the continued effectiveness of the defense industry critical infrastructure,
governments and defense contractors must take a range of steps to protect their systems
and networks from these threats. This includes implementing strong cybersecurity protocols, such as firewalls, intrusion detection systems, and encryption, as well as conducting regular vulnerability assessments and audits. In addition, defense contractors must
comply with strict security standards and protocols to protect their sensitive data and
technologies from unauthorized access or theft. These standards may include implementing physical security measures, such as access controls and surveillance cameras, as well as
strict data handling policies to ensure that sensitive information is not compromised.

One of the unique challenges of protecting the defense industry's critical infrastructure is the constantly evolving nature of the threats facing it. Hostile actors are constantly
developing new techniques and methods to exploit vulnerabilities in systems and networks, making it necessary for governments and defense contractors to stay
up-to-date on the latest security measures and technologies. Another challenge is the
high degree of complexity involved in the defense industry's critical infrastructure, which
often involves numerous interconnected networks and components. This complexity
can make it difficult to identify and address vulnerabilities and risks, particularly when
multiple organizations are involved in the operation of the infrastructure.

Despite these challenges, there are a range of strategies that governments and defense
contractors can use to protect their defense industry's critical infrastructure. One key
approach is to prioritize risk management, identifying the most critical systems and assets
and implementing targeted security measures to protect them. This can involve conducting
risk assessments and developing threat models to identify potential vulnerabilities and risks.

Another approach is to establish partnerships and collaboration between government
agencies and defense contractors involved in the operation of the defense industry's
critical infrastructure. This can involve sharing threat intelligence and best practices,
coordinating response plans, and establishing joint cybersecurity exercises to test the
resilience of the infrastructure.

7.13 Summary of critical infrastructures

In the previous section, we have described several examples of critical infrastructures.
Each infrastructure has its own goals and consists of specific components. In Table 3,
we have provided an example list of major critical infrastructure sectors. For each sector,

Table 3 Selected examples of critical infrastructures.

Critical infrastructure	Specific components	Goals
Energy infrastructure	Power plants, transmission lines, pipelines	Provide reliable and secure energy supply to homes, businesses, and essential services
Transportation infrastructure	Roads, bridges, airports, seaports, public transit systems	Facilitate the movement of people and goods within and between communities
Telecommunications infrastructure	Communication networks, satellites, internet systems	Support the flow of information among individuals, government agencies, and public institutions
Water and wastewater infrastructure	Water treatment plants, distribution systems, sewage systems	Provide access to safe and clean drinking water and ensure the proper treatment and disposal of wastewater
Financial infrastructure	Banks, stock exchanges, payment systems	Facilitate financial transactions and support economic growth and stability
Healthcare infrastructure	Hospitals, clinics, medical equipment	Provide access to healthcare services and support public health initiatives
Emergency services infrastructure	Police departments, fire departments, emergency medical services	Respond to emergencies and ensure public safety
Food and agriculture infrastructure	Farms, food processing facilities, distribution networks	Ensure a reliable and secure food supply
Chemical and hazardous materials infrastructure	Chemical production facilities, storage and transportation systems, disposal and remediation systems	Ensure safe and secure production, storage, transportation, and disposal of chemicals and hazardous materials
Government infrastructure	Government buildings, roads, bridges, public transit systems, communication networks, information technology systems	Support the functioning of government operations and public services
Information technology infrastructure	Hardware, software, data storage, networks, and cloud computing services	Provide a reliable and secure platform for communication, collaboration, and data processing
Defense industry infrastructure	Military installations and facilities, weapons and equipment, communications and command-and-control systems, research and development centers	Ensuring the readiness and effectiveness of military forces. Protecting national security interests

we have listed the specific components that make up the infrastructure, as well as the goals that these components aim to achieve.

One key takeaway from the table is that each sector has a unique set of components and goals, reflecting the specific needs and challenges of that sector. For example, the energy infrastructure is focused on providing a reliable and secure energy supply to homes, businesses, and essential services, while the telecommunications infrastructure supports the flow of information among individuals, government agencies, and public institutions. Another key takeaway is the interconnected nature of critical infrastructure. For example, the transportation infrastructure is essential for moving goods and people, which are necessary for supporting the functioning of the economy and ensuring access to essential services. Similarly, the information technology infrastructure supports communication and collaboration across different sectors, making it a critical component of many other sectors.

8 Challenges

The final step of the domain analysis process is to identify the key challenges. Many different challenges can be identified [1,52–55]. These challenges are not mutually exclusive and often overlap with one another. Addressing these challenges require a comprehensive and collaborative approach involving stakeholders from the public and private sectors, as well as academia and civil society. We describe the following key categories of challenges.

8.1 Design and construction

Design challenges for critical infrastructures are related to the initial planning, development, and construction of these systems. These challenges can be attributed to the complexity of these infrastructures, the evolving threat landscape, and the need to balance safety, security, and functionality.

One significant design challenge for critical infrastructures is the need to incorporate safety and security features into the design process. Critical infrastructures are essential for maintaining the functioning of modern society, and a failure in one of these systems can have severe consequences. Therefore, safety features need to be incorporated into the design process to ensure that these systems are as safe as possible.

Another design challenge is the need to incorporate security features into the design process. Critical infrastructures are a prime target for malicious actors, and these systems are vulnerable to various types of attacks. Designers must incorporate security features into the infrastructure design to ensure that these systems are as secure as possible.

A further challenge is the need to balance safety and security features with functionality. Critical infrastructures must be designed to perform their intended functions effectively while still incorporating safety and security features. This can be challenging, as safety and security features can often be seen as impeding functionality.

Moreover, critical infrastructure systems are often complex and interdependent, making it challenging to design and implement effective solutions. For instance, the integration of ICTs in these systems has created complex interdependencies and increased their vulnerability to cyber-attacks. The design process must account for these complexities and interdependencies to ensure that the system is resilient to various threats.

Lastly, the design process must consider the evolving threat landscape and anticipate potential threats that may arise in the future. This requires designers to engage in ongoing research and development to identify new threats and to incorporate new safety and security features into the infrastructure design. The design process must be agile and responsive to evolving threats to ensure the continued safety and security of critical infrastructures.

8.2 Evaluation

Evaluation challenges refer to the difficulties faced in assessing the effectiveness of critical infrastructure systems, processes, and procedures in mitigating threats and ensuring their continued functioning. These challenges can arise due to several reasons, including the lack of standardized evaluation methodologies, limited availability of data, and the dynamic nature of threats faced by critical infrastructures.

One of the primary challenges in evaluating critical infrastructure is the lack of a standardized evaluation framework. There is no universally accepted approach for evaluating critical infrastructure, and evaluations are often conducted on a case-by-case basis. This makes it difficult to compare the effectiveness of different critical infrastructure systems and processes.

Another challenge in evaluating critical infrastructure is the limited availability of data. Critical infrastructure owners and operators are often hesitant to share data about their systems due to concerns about security and confidentiality. This limits the ability of evaluators to assess the effectiveness of critical infrastructure systems and processes.

The dynamic nature of threats faced by critical infrastructures is another significant challenge in evaluation. Threats can evolve rapidly, and critical infrastructure systems and processes must adapt to address these changes. This requires frequent updates to evaluation methodologies to ensure that they remain relevant and effective.

Finally, evaluation challenges can also arise from the complexity of critical infrastructure systems. These systems often involve numerous interconnected components and processes, making it difficult to assess the effectiveness of individual components in isolation.

To address these challenges, there is a need for standardized evaluation methodologies that can be adapted to the specific needs of different critical infrastructure systems. There is also a need for greater collaboration among critical infrastructure owners and operators, government agencies, and academia to improve the availability of data for evaluation purposes. Additionally, evaluations should be conducted regularly and updated to reflect changes in the threat landscape and advances in technology.

8.3 Cybersecurity

Critical infrastructures are highly interconnected and dependent on digital systems, which makes them vulnerable to cyber threats such as hacking, ransomware, and other malicious attacks [12,56–58]. Cybersecurity challenges are one of the most significant challenges facing critical infrastructures today. A cyber-attack on critical infrastructure can have severe consequences, including disruption of services, loss of data, and compromise of sensitive information. As a result, cybersecurity is a critical challenge for critical infrastructure operators and requires continuous monitoring, assessment, and updating of security measures to mitigate the risks. Two of the significant challenges in cybersecurity include the constantly evolving nature of the threats and the growing volume of attack sources. Cyber attackers are constantly developing new methods and techniques to breach defenses, and critical infrastructure organizations must be proactive by continually adapting their defenses to keep up. This challenge is compounded by the increasing sophistication of cyber attackers, who are often well-funded and well-organized. Another challenge is the lack of skilled cybersecurity personnel. There is a significant shortage of cybersecurity professionals globally, and this shortage is particularly acute in the critical infrastructure sector. This shortage can make it challenging for organizations to implement and maintain effective cybersecurity measures, as they may not have the necessary expertise in-house. Finally, there is a challenge in balancing security with the need for accessibility and usability. Critical infrastructure systems must be accessible to authorized users, but this accessibility can create vulnerabilities that can be exploited by cyber attackers. Finding the right balance between security and accessibility is a significant challenge for critical infrastructure organizations.

8.4 Interdependence and complexity

Complexity is a significant challenge in critical infrastructure systems, given the interconnectedness of various components and subsystems that are often designed and operated by different organizations. With the increasing adoption of digital technologies, the complexity of these systems has only increased [59–62]. As a result, critical infrastructure systems are becoming more difficult to manage and maintain, leading to various operational and technical challenges.

One of the major complexity challenges is related to the integration of legacy systems with modern systems, which can lead to interoperability issues [5]. For example, many critical infrastructure systems were designed decades ago and were not initially intended to be connected to the internet or other modern communication networks. As a result, these systems may have different data formats, communication protocols, and security requirements that are incompatible with modern technologies. Furthermore, the use of different vendor-specific hardware and software components can create additional complexity challenges, such as compatibility issues, patch management, and system updates.

Another challenge related to complexity is the difficulty in understanding the dependencies between different components and subsystems within critical infrastructure systems [33]. Due to the complex nature of these systems, it is often challenging to identify all the interconnected components and their dependencies accurately. This lack of understanding can lead to unexpected consequences, such as cascading failures or unintended consequences when changes are made to the system.

Moreover, the large-scale nature of critical infrastructure systems also poses a challenge, as it can be difficult to maintain and manage these systems effectively. The large-scale nature of these systems means that they may have numerous subsystems, geographically dispersed locations, and a high number of users. These factors can make it challenging to implement effective monitoring and management strategies, which can increase the risk of failure or cyber-attacks.

8.5 Resilience and continuity

Resilience and continuity challenges refer to the ability of critical infrastructure systems to withstand and recover from disruptions or failures and to continue providing essential services to society. These challenges are becoming increasingly important due to the growing complexity and interdependence of critical infrastructure systems, as well as the increasing frequency and severity of natural disasters, cyber-attacks, and other threats.

One major challenge related to resilience and continuity is the need to design critical infrastructure systems that can withstand and recover from a wide range of disruptions and failures, including physical, cyber, and natural disasters. This requires the development of robust and redundant systems and processes, as well as the implementation of effective response and recovery plans.

Another challenge is the need to ensure the continuity of essential services in the face of disruptions or failures. This requires the development of backup and alternative systems and processes, as well as the implementation of effective communication and coordination mechanisms to ensure that all stakeholders are aware of the situation and can take appropriate actions.

Resilience and continuity challenges also highlight the importance of effective risk management and contingency planning. This involves identifying and assessing the various risks and threats facing critical infrastructure systems, and developing and implementing appropriate strategies and plans to mitigate these risks and ensure the continuity of essential services.

Finally, resilience and continuity challenges require the development of effective governance and regulatory frameworks that promote the resilience and continuity of critical infrastructure systems. This involves the establishment of clear roles and responsibilities for all stakeholders, as well as the development of effective mechanisms for monitoring and enforcing compliance with relevant regulations and standards.

8.6 Regulation and policy

Regulation and policy challenges for critical infrastructures involve the development and implementation of legal frameworks, standards, and guidelines that ensure the protection and resilience of critical infrastructure. There is a need for effective policies and regulations to provide a framework for addressing critical infrastructure vulnerabilities, identifying and assessing risks, and developing strategies to mitigate those risks. However, regulatory and policy challenges often arise due to the complex and dynamic nature of critical infrastructure, which can lead to difficulties in implementing policies and regulations that are effective and flexible enough to adapt to changing circumstances.

One challenge is the lack of coherence and harmonization in regulatory frameworks across different jurisdictions, which can create uncertainty for critical infrastructure operators and impede the sharing of best practices and information. This can lead to a fragmentation of regulatory frameworks and can be particularly problematic in cases where critical infrastructure is located across different countries or regions.

Another challenge is the need to balance security and resilience requirements with economic and operational considerations. Critical infrastructure operators need to balance the costs of implementing security measures with the benefits of mitigating risks, and regulators need to balance the need for security with the need to ensure that critical infrastructure remains affordable and accessible to all.

Additionally, regulatory and policy challenges can arise due to the rapid pace of technological change and the emergence of new threats, which can quickly make existing policies and regulations obsolete. This requires a proactive and adaptive approach to policy development and implementation, which can be challenging for regulatory bodies and policymakers.

Finally, there can also be challenges in ensuring compliance with regulatory frameworks, particularly in cases where there is a lack of awareness or understanding of the requirements or where critical infrastructure operators may be resistant to implementing costly security measures. This highlights the need for effective communication and engagement between regulators, policymakers, and critical infrastructure operators to ensure that regulatory frameworks are understood, accepted, and effectively implemented.

8.7 Aging infrastructure

Aging infrastructure is a critical challenge for critical infrastructure systems. Many critical infrastructure systems were built decades ago and have not been updated or replaced promptly, leading to a high risk of failure or disruption. Aging infrastructure is also exacerbated by the rapid pace of digitalization in recent years, as older systems may not be able to keep up with the latest technological advances or cybersecurity threats.

One key issue with aging infrastructure is the lack of funding and resources for maintenance and upgrades. Many critical infrastructure systems are owned and operated by government agencies or private companies that may not have the necessary budget or incentives to invest in modernizing their systems. This can lead to a vicious cycle where aging infrastructure becomes even more prone to failure or disruption, causing even more damage and disruption in the future.

Another challenge of aging infrastructure is the lack of interoperability with modern systems. As new technologies are developed and implemented, older infrastructure systems may not be able to communicate or integrate with these systems, leading to inefficiencies and potential vulnerabilities. For example, older power grids may not be able to integrate with renewable energy sources or smart grid technologies, leading to wasted energy and increased risk of power outages.

Digitalization also poses unique challenges for aging infrastructure. As more systems become connected and reliant on digital technologies, older systems may not be able to keep up with the demand or security requirements. This can lead to a higher risk of cyberattacks, as older systems may not have the necessary security protocols or updates to protect against modern threats. Additionally, as more data is generated and stored by critical infrastructure systems, older systems may struggle to handle the volume or complexity of this data, leading to potential performance issues or data breaches.

8.8 Public awareness and perception

Public awareness and perception are critical challenges for critical infrastructures. This challenge arises from the fact that critical infrastructure is often invisible to the general public, and its importance is not fully understood. This lack of awareness can lead to a lack of appreciation for the critical role that infrastructure plays in society and the need to invest in its maintenance and protection.

One factor that contributes to this challenge is the tendency for critical infrastructure to be viewed as a commodity rather than a service. In other words, people tend to take the services provided by critical infrastructure for granted, assuming that they will always be available. This can lead to a lack of investment in critical infrastructure, as people do not see the need to spend money on something that they do not fully understand or appreciate.

Another factor that contributes to the challenge of public awareness and perception is the complexity of critical infrastructure. Most critical infrastructure is made up of multiple systems and assets that are interconnected and interdependent. This complexity can make it difficult for people to understand the role that critical infrastructure plays in their daily lives and the potential consequences of infrastructure failure.

The rapid pace of technological change and digitalization is also contributing to the challenge of public awareness and perception. Many critical infrastructure systems are now highly digitized, making them more vulnerable to cyber threats and other forms of disruption. However, the public may not fully understand the risks associated with digitalization, leading to a lack of investment in cybersecurity and other protective measures.

To address the challenge of public awareness and perception, there is a need for increased education and outreach. This can involve working with schools and universities to teach people about the importance of critical infrastructure and the risks associated with infrastructure failure. It can also involve public awareness campaigns that highlight the critical role that infrastructure plays in society and the need to invest in its protection and maintenance.

Another important step is to increase transparency and communication around critical infrastructure. This can involve sharing information about infrastructure failures and the steps being taken to address them. It can also involve engaging with the public and stakeholders to understand their concerns and perceptions of critical infrastructure, and working to address these concerns through increased investment in resilience and continuity planning, as well as improved communication and outreach efforts.

8.9 Funding and resources

Funding and resources are significant challenges for critical infrastructure. Maintaining and upgrading critical infrastructure is costly, and ensuring that the necessary resources are available to address all potential threats and risks is essential.

One of the primary challenges with funding and resources is the competing demands for limited resources. Governments and private organizations must balance the need to invest in critical infrastructure with other priorities such as healthcare, education, and social programs. This can make it difficult to allocate sufficient resources to critical infrastructure, leading to delays in maintenance and upgrades.

Another challenge is the long-term nature of critical infrastructure investments. Many critical infrastructure systems have a lifespan of several decades or even centuries, making it challenging to plan for future funding and resources. Additionally, many critical infrastructure systems require ongoing maintenance and upgrades, which can add to the cost over time.

There is also a challenge in ensuring that funding and resources are distributed fairly and equitably. This is especially important in situations where critical infrastructure serves multiple communities or regions. Inadequate funding for critical infrastructure in one area can have ripple effects on other regions, leading to economic and social consequences.

Finally, a lack of funding and resources can lead to a lack of investment in research and development, making it difficult to innovate and develop new technologies to address emerging threats and risks. This can leave critical infrastructure vulnerable to new and evolving threats such as cyber-attacks.

8.10 Supply chain security

Supply chain security is a critical aspect of ensuring the reliability and resilience of critical infrastructure. It refers to the measures taken to safeguard the flow of goods, services, and information between suppliers and customers. Supply chains are complex and involve multiple stakeholders, including suppliers, manufacturers, distributors, and retailers. A disruption or compromise of any of these stakeholders can have significant impacts on the entire supply chain, leading to delayed or disrupted delivery of goods and services.

One of the major challenges in supply chain security is the lack of visibility and control over the entire supply chain. Critical infrastructure operators may not have complete knowledge of all the suppliers and subcontractors involved in the supply chain, making it difficult to assess and manage the risks associated with each. This can lead to vulnerabilities and blind spots that can be exploited by attackers.

Another challenge is the increasing globalization of supply chains, which can make it more difficult to ensure the security of critical infrastructure. Suppliers and subcontractors may be located in different countries with varying levels of security standards and regulations. This can make it challenging to enforce consistent security measures across the entire supply chain.

Furthermore, the increasing use of digital technologies in supply chains, such as the Internet of Things (IoT) devices, have introduced new vulnerabilities that can be exploited by cyber attackers. For example, an attacker could compromise an IoT device in the supply chain, leading to a chain reaction of compromise and disruption.

To address these challenges, critical infrastructure operators must take a risk-based approach to supply chain security. This includes conducting comprehensive risk assessments of the entire supply chain, identifying critical suppliers and subcontractors, and implementing security controls to reduce the risks associated with each stakeholder. Additionally, critical infrastructure operators should establish clear security policies and standards for all suppliers and subcontractors to follow. This can include requirements for cybersecurity certifications, security audits, and ongoing monitoring and reporting.

Finally, critical infrastructure operators should also establish contingency plans and backup suppliers to ensure continuity of operations in the event of a disruption or compromise in the supply chain.

8.11 Summary of the challenges

Table 4 summarizes the major challenges faced by critical infrastructures. These challenges are classified into eight categories: Cybersecurity, complexity, resilience and continuity, regulation and policy, aging infrastructure and digitalization, public awareness and perception, funding and resources, and supply chain security. For each challenge, a brief description is provided, along with some specific examples of issues that fall under that category. This table can serve as a reference for policymakers, infrastructure owners and operators, and researchers who seek to address these challenges and improve the resilience of critical infrastructures.

Please note that the challenges listed in the table are not exhaustive, and there may be other challenges that are specific to certain critical infrastructure sectors or regions. However, these challenges represent some of the most commonly recognized and significant issues facing critical infrastructure protection efforts.

Table 4 Summary of the selected key challenges.

Challenge category	Description
Cybersecurity	Threats from cyberattacks and data breaches, and the challenge of securing increasingly interconnected systems
Complexity	Managing the complexity of critical infrastructure systems and their interdependencies
Resilience and continuity	Ensuring continuity of critical infrastructure operations in the face of disruptive events
Regulation and policy	Balancing the need for regulation and standards with the need for innovation and flexibility
Aging infrastructure	Addressing the challenges associated with aging infrastructure, including maintenance and replacement
Digitalization	Managing the challenges of integrating digital technologies into critical infrastructure systems
Public awareness and perception	Raising public awareness of the importance of critical infrastructure and the risks associated with it
Funding and resources	Ensuring sufficient funding and resources to support critical infrastructure development and maintenance
Supply chain security	Managing risks to critical infrastructure posed by vulnerabilities in supply chains
Design, implementation, evaluation, and maintenance	Ensuring the effective design, implementation, evaluation and maintenance of critical infrastructure systems

9 Conclusion

Critical infrastructures are the backbone of modern society, providing essential services and systems that are crucial to the functioning of our economies, societies, and governments. These infrastructures include energy systems, transportation networks, water and sanitation systems, telecommunications, financial systems, and many others. However, these systems are also vulnerable to a wide range of threats, including natural disasters, cyber-attacks, terrorism, and human error. To ensure the resilience and sustainability of critical infrastructures, it is important to understand the key concepts and challenges that are involved in their design, operation, and maintenance.

The key objective of this chapter is to provide an overview of the key concepts and challenges that are involved in critical infrastructures. The chapter focused on identifying and defining key terms and concepts related to critical infrastructures, as well as exploring the main challenges that are involved in ensuring their resilience and sustainability. By providing this overview, the chapter aimed to provide a foundation for further research and discussion on critical infrastructures and inform policy and decision-making related to their design, operation, and maintenance.

To achieve this objective, a domain analysis approach was adopted. By applying domain analysis to the field of critical infrastructure, we identified the key components of the domain, such as the sectors and industries included, the terminology and definitions used, and the challenges and risks associated with critical infrastructure.

Based on the domain analysis process, the following key components of critical infrastructure were identified:
- Sectors and industries, such as energy, transportation, water, telecommunications, and financial systems
- Key stakeholders, including government agencies, regulators, industry associations, and private companies
- Key concepts and terms, such as resilience, sustainability, risk, and vulnerability
- Key challenges, including cybersecurity, complexity, resilience and continuity, regulation and policy, aging infrastructure, digitalization, public awareness and perception, funding and resources, and supply chain security.

Additionally, the feature diagram identified the common and variant features of critical infrastructures, including their physical components, operational characteristics, and management structures. Furthermore, examples of critical infrastructures include power grids, transportation networks, water treatment plants, communication networks, financial systems, and emergency response systems.

Future work should focus on exploring additional challenges and solutions for ensuring the resilience and sustainability of critical infrastructures. Additionally, the validation of the identified challenges and solutions should be further explored through empirical research and case studies, to ensure their relevance and effectiveness in different contexts and settings. Overall, critical infrastructure is a complex and evolving field that requires ongoing research and collaboration to address the challenges and risks associated with these essential systems.

Appendix. Glossary

Term	Definition
Critical infrastructure	The physical and cyber systems, assets, and networks are essential for the functioning of a society and economy
Interdependence	The state of mutual dependence between different systems, assets, and networks that make up critical infrastructure
Resilience	The ability of critical infrastructure to withstand and recover from disruptions or threats
Risk management	The process of identifying, assessing, and prioritizing risks to critical infrastructure and implementing strategies to mitigate those risks
Vulnerability	The state of being exposed to the possibility of harm or damage, especially in the context of critical infrastructure
Threat	Any potential event or action that could cause harm or damage to critical infrastructure
Cybersecurity	The protection of critical infrastructure from cyber threats such as hacking, malware, and other forms of cyber-attacks
Physical security	The protection of critical infrastructure from physical threats such as terrorism, sabotage, and natural disasters
Emergency response	The coordinated actions are taken in response to a disruption or threat to critical infrastructure
Business continuity	The planning and preparation to ensure the continued operation of critical infrastructure in the event of a disruption or threat
Disaster recovery	The process of recovering critical infrastructure after a disaster or major disruption
Risk assessment	The process of evaluating the likelihood and potential impact of different risks to critical infrastructure
Risk mitigation	The process of reducing or eliminating the likelihood and impact of different risks to critical infrastructure
Redundancy	The provision of duplicate or backup systems, assets, or networks to ensure the continuity of critical infrastructure in the event of a disruption
Emergency management	The process of coordinating emergency response and recovery efforts to protect critical infrastructure and minimize damage and loss
Continuity of operations	The ability of critical infrastructure to continue to operate in the face of disruptions or threats
Public-private partnerships	Collaborations between government agencies and private sector organizations to improve the security and resilience of critical infrastructure

Term	Definition
Situational awareness	The ability to monitor and understand the state of critical infrastructure in real-time to detect and respond to threats and disruptions
Information sharing	The exchange of information between different organizations and stakeholders involved in critical infrastructure to improve situational awareness and response efforts
Infrastructure as code	The practice of managing critical infrastructure as software code to improve agility, scalability, and security
Smart infrastructure	The use of technology such as sensors, automation, and data analytics to improve the efficiency and resilience of critical infrastructure
Geographic information systems (GIS)	The use of digital maps and geospatial data to better understand and manage critical infrastructure
Interoperability	The ability of different systems, assets, and networks to work together effectively to ensure the continuity of critical infrastructure
Cyber hygiene	The practices and procedures used to maintain the security and integrity of critical infrastructure from cyber threats
Digital resilience	The ability of critical infrastructure to adapt and recover from digital disruptions or cyber-attacks
Incident response	The process of responding to a security incident or cyber-attack on critical infrastructure
Threat intelligence	The process of gathering, analyzing, and sharing information on cyber threats and other risks to critical infrastructure
Business impact analysis	The process of assessing the potential impact of disruptions or threats to critical infrastructure on business operations
Operational technology (OT)	The hardware and software used to control and monitor physical systems such as manufacturing, transportation, and energy infrastructure
Critical infrastructure	The physical and cyber systems, assets, and networks are essential for the functioning of a society and economy

References

[1] Department of Homeland Security, Critical Infrastructure Sectors, 2020. https://www.dhs.gov/cisa/critical-infrastructure-sectors.
[2] V. Langer, R.R. Møller, O. Hjelmar, Integrating hazardous chemicals management into sustainable business models, J. Clean. Prod. 208 (2019) 1249–1257, https://doi.org/10.1016/j.jclepro.2018.10.040.
[3] National Infrastructure Advisory Council, Critical Infrastructure: A Framework for Understanding and Improving Resilience, 2018. https://www.dhs.gov/sites/default/files/publications/NIAC%20Critical%20Infrastructure%20Framework_FINAL%202018_508.pdf.
[4] P.A. Thiel, The case for the 21st-century infrastructure of government, Harv. Bus. Rev. (June 2019) (2019). https://hbr.org/2019/06/the-case-for-the-21st-century-infrastructure-of-government.
[5] B. Tekinerdogan, Obstacles of system-of-systems, in: 2022 IEEE International Symposium on Systems Engineering (ISSE), Vienna, Austria, 2022, pp. 1–7, https://doi.org/10.1109/ISSE54508.2022.10005491.

[6] U.S. Department of Homeland Security, National Infrastructure Protection Plan (NIPP) 2013: Partnering for Critical Infrastructure Security and Resilience, 2013, Retrieved from: https://www.dhs.gov/national-infrastructure-protection-plan.

[7] European Commission, Critical Infrastructure, 2018. https://ec.europa.eu/home-affairs/what-we-do/policies/crisis-and-terrorism/critical-infrastructure_en.

[8] The White House, Office of the Press Secretary, Presidential Policy Directive 21: Critical Infrastructure Security and Resilience, 2013, Retrieved from: https://obamawhitehouse.archives.gov/the-press-office/2013/02/12/presidential-policy-directive-critical-infrastructure-security-and-resil.

[9] International Atomic Energy Agency, Critical Infrastructure, 2019. https://www.iaea.org/topics/critical-infrastructure.

[10] Australian Government, Critical Infrastructure, 2020. https://www.homeaffairs.gov.au/about-us/our-portfolios/national-security/critical-infrastructure.

[11] Ministry of Internal Affairs and Communications, Japan, Cybersecurity Strategy Headquarters, Information Security Strategy for Protecting the Nation, 2011, Retrieved from: https://www.nisc.go.jp/eng/pdf/New_Strategy_English.pdf.

[12] DHS Cybersecurity and Infrastructure Security Agency (CISA), Cybersecurity Framework, 2018, Retrieved from: https://www.nist.gov/cyberframework.

[13] National Institute of Standards and Technology, Nist.gov, 2023. Accessed March 2023.

[14] Canadian Centre for Cyber Security, Critical Infrastructure, 2023, Retrieved from: https://cyber.gc.ca/en/glossary/critical-infrastructure.

[15] World Economic Forum, Global Risks 2012, seventh ed., 2012. Retrieved from: http://www3.weforum.org/docs/WEF_GlobalRisks_Report_2012.pdf.

[16] K. Kang, S. Cohen, J. Hess, W. Nowak, S. Peterson, Feature-Oriented Domain Analysis (FODA) Feasibility Study (PDF) (Report), Software Engineering Institute, Carnegie Mellon University, Pittsburgh, 1990.

[17] I. Eusgeld, C. Nan, S. Dietz, "System-of-systems" approach for interdependent critical infrastructures, Reliab. Eng. Syst. Saf. 96 (6) (2011) 679–686.

[18] E. Bompard, R. Napoli, F. Xue, Analysis of structural vulnerabilities in power transmission grids, Int. J. Crit. Infrastruct. Prot. 5 (1) (2012) 16–27.

[19] E. Bompard, R. Napoli, F. Xue, Power Grids and Critical Infrastructures: A Multidisciplinary Approach to Resilience, Security, and Vulnerability, Wiley-IEEE Press, 2019, https://doi.org/10.1002/9781119569405.

[20] A. Cherp, J. Jewell, A. Goldthau, The political economy of energy transitions: an interdisciplinary review of the literature, Energy Res. Soc. Sci. 37 (2018) 175–187, https://doi.org/10.1016/j.erss.2018.01.010.

[21] M. Panteli, P. Mancarella, The grid: stronger, bigger, smarter?: presenting a conceptual framework of power system resilience, IEEE Power Energy Mag. 13 (3) (2015) 58–66, https://doi.org/10.1109/MPE.2015.2397334.

[22] L. Zhang, G. Andresen, X. He, Assessing the security of electrical energy delivery systems using vulnerability and risk analysis, IEEE Trans. Smart Grid 5 (4) (2014) 2104–2113.

[23] D. Banister (Ed.), Critical Issues in Transportation Policy: Planning and Analysis, Edward Elgar Publishing, 2018, https://doi.org/10.4337/9781785367267.

[24] T. Litman, Evaluating transportation equity: guidance for incorporating distributional impacts in transportation planning, ITE J. 88 (2) (2018) 31–36.

[25] J.-P. Rodrigue, C. Comtois, B. Slack, The Geography of Transport Systems, Routledge, 2016, https://doi.org/10.4324/9781315618159.

[26] W. Fang, J. Zhuang, Telecommunications infrastructure, networked computing, and regional economic development: evidence from China, Telecommun. Policy 37 (10) (2013) 920–931.

[27] W.R. Neuman, D. Park, E. Panek, The Digital Difference: Media Technology and the Theory of Communication Effects, Harvard University Press, 2018, https://doi.org/10.4159/9780674977016.

[28] J.M. Bauer, M. Latzer (Eds.), Handbook on the Economics of the Internet, Edward Elgar Publishing, 2016, https://doi.org/10.4337/9780857939852.

[29] A.M. Madureira, S. Denazis, G. Pujolle, A survey of network virtualization, Comput. Netw. 51 (10) (2007) 2683–2700.

[30] A.K. Biswas, C. Tortajada, R. Izquierdo (Eds.), Water Management in 2020 and Beyond, Springer, 2019, https://doi.org/10.1007/978-3-030-11446-1.

[31] N.F. Gray, Water Technology: An Introduction for Environmental Scientists and Engineers, CRC Press, 2017, https://doi.org/10.1201/9781315371405.

[32] F.I. Khan, S.A. Abbasi, Techniques and methodologies for risk analysis in chemical process industries, J. Loss Prev. Process Ind. 11 (4) (1998) 261–277.

[33] I. Koutiva, C. Makropoulos, Assessing the vulnerability of water distribution systems: a multi-criteria decision analysis approach, Environ. Model. Softw. 26 (9) (2011) 1134–1142.

[34] N. Alexandratos, J. Bruinsma, World Agriculture Towards 2030/2050: The 2012 Revision, Food and Agriculture Organization of the United Nations 12 (3) 2012.

[35] F. Allen, E. Carletti, Systemic risk from global financial derivatives: a network analysis of contagion and its mitigation with super-spreader tax, J. Bank. Financ. 37 (10) (2013) 4055–4068, https://doi.org/10.1016/j.jbankfin.2013.07.002.

[36] S. Claessens, L. Kodres, The regulatory responses to the global financial crisis: some uncomfortable questions, IMF Working Paper WP/14/46, 2014, https://doi.org/10.5089/9781484309017.001.

[37] R. Haux, Health information systems—past, present, future, Int. J. Med. Inform. 79 (6) (2010) 383–395, https://doi.org/10.1016/j.ijmedinf.2010.01.003.

[38] C. Klinger, J. Landgren, E. Antonsen, The interdependencies between critical infrastructures: the impact of energy on healthcare systems, Energy Policy 128 (2019) 459–466, https://doi.org/10.1016/j.enpol.2019.01.012.

[39] J.G. Meara, A.J. Leather, L. Hagander, B.C. Alkire, N. Alonso, E.A. Ameh, et al., Global surgery 2030: evidence and solutions for achieving health, welfare, and economic development, Lancet 386 (9993) (2015) 569–624, https://doi.org/10.1016/S0140-6736(15)60160-X.

[40] A. Guha, M.J. Aziz, Environmental management of hazardous chemicals: a review, J. Hazard. Mater. 363 (2019) 98–115, https://doi.org/10.1016/j.jhazmat.2018.09.056.

[41] J. Tummers, B. Tekinerdogan, H. Tobi, C. Catal, B. Schalk, Obstacles and features of health information systems: a systematic literature review, Comput. Biol. Med. 137 (2021) 104785, https://doi.org/10.1016/j.compbiomed.2021.104785.

[42] J. Telfer, S. Smith, Measuring resilience to disasters: a systematic review of literature and approaches, Int. J. Disaster Risk Reduct. 38 (2019) 101200, https://doi.org/10.1016/j.ijdrr.2019.101200.

[43] J. Weichselgartner, P. Pigeon, The role of knowledge in disaster risk reduction, Int. J. Disaster Risk Sci. 6 (2) (2015) 107–116.

[44] International Organization for Standardization, ISO 22300:2018 Societal Security—Emergency Management—Guidelines for Colour-Coded Alerts, 2018. https://www.iso.org/standard/65735.html.

[45] H.C.J. Godfray, J.R. Beddington, I.R. Crute, L. Haddad, D. Lawrence, J.F. Muir, et al., Food security: the challenge of feeding 9 billion people, Science 327 (5967) (2010) 812–818.

[46] T.W. Hertel, J. Beckman, Modelling global food systems to support sustainable development goals, Nat. Food 1 (1) (2020) 2–4, https://doi.org/10.1038/s43016-019-0001-1.

[47] H. Kahiluoto, M. Kuisma, H. Lehtonen, Agriculture and food security in a changing climate: a review of research on adaptation and mitigation strategies, Wiley Interdiscip. Rev. Clim. Chang. 10 (6) (2019) e597, https://doi.org/10.1002/wcc.597.

[48] S. Das, R. Kanchan, Chemical process safety and hazard identification techniques: an overview, Process. Saf. Prog. 34 (2) (2015) 108–118.

[49] J.S. Gansler, W. Lucyshyn, Defense industry consolidation and the industrial base: prospects for future innovation and competition, Public Contract Law J. 35 (4) (2006) 733–748.

[50] R.A. Bitzinger, The Modern Defense Industry: Political, Economic, and Technological Issues, Praeger Security International, 2009.

[51] N. Shafqat, A. Masood, Cyber-security challenges in the defense industry: Protecting critical infrastructure and sensitive information, in: 2015 IEEE International Multi Topic Conference (INMIC), 2015, pp. 1–6.

[52] J. Johansson, H. Hassel, A. Cedergren, Quantitative risk assessment of infrastructure systems: a unifying approach, Int. J. Crit. Infrastruct. 9 (3) (2013) 251–270.

[53] T.G. Lewis, Critical Infrastructure Protection in Homeland Security: Defending a Networked Nation, John Wiley & Sons, 2018.

[54] E. Zio, Challenges in the vulnerability and risk analysis of critical infrastructures, Reliab. Eng. Syst. Saf. 152 (2016) 137–150.

[55] B. Tekinerdogan, Engineering Connected Intelligence: A Socio-Technical Perspective, Wageningen University & Research, The Netherlands, 2017.

[56] M. Dunn Cavelty, Cybersecurity in critical infrastructure, in: M. Dunn Cavelty, V. Mauer (Eds.), The Routledge Handbook of Security Studies, Routledge, 2012, pp. 203–214.

[57] M.J.G. Eeten, J.M. Bauer, Emerging threats to internet security: incentives, externalities and policy implications, J. Conting. Crisis Manag. 20 (4) (2012) 225–236.

[58] R. Setola, S. De Porcellinis, Critical Infrastructure Security and Resilience: Advances in Critical Infrastructure Protection: Information Infrastructure Models, Analysis, and Defense, Springer, 2018.

[59] S. Bouchon, The vulnerability of interdependent critical infrastructures systems: epistemological and conceptual state of the art, Saf. Sci. 82 (2016) 325–334.

[60] Center for Strategic and International Studies, Critical Infrastructure, 2014. https://www.csis.org/programs/technology-policy-program/cybersecurity-and-governance/critical-infrastructure.

[61] G. D'Agostino, A. Scala, Networks of Networks: The Last Frontier of Complexity, Springer, 2014.

[62] B. Tekinerdogan, M. Akşit, Classifying and evaluating architecture design methods, in: M. Akşit (Ed.), Software Architectures and Component Technology. The Springer International Series in Engineering and Computer Science, vol. 648, Springer, Boston, MA, 2002, https://doi.org/10.1007/978-1-4615-0883-0_1.

CHAPTER 3

Insider threats to critical infrastructure: A typology

Mathias Reveraert[a], Marlies Sas[b], Genserik Reniers[c], Wim Hardyns[d,e], and Tom Sauer[f]

[a]Advisor Security Management at Infrabel, Brussels, Belgium
[b]Attaché Noodplanning Nationaal Crisiscentrum, Brussels, Belgium
[c]Safety and Security Science Group, TUDelft, Delft, The Netherlands
[d]Faculty of Law and Criminology, Institute for International Research on Criminal Policy (IRCP), Ghent University, Ghent, Belgium
[e]Faculty of Social Sciences, Master of Safety Sciences, University of Antwerp, Antwerp, Belgium
[f]Department of Political Science, University of Antwerp, Antwerp, Belgium

1 Introduction

Whereas in the past organizations[a] often prioritized the protection against external threats over the mitigation of insider threats [6], lately attention to the latter has relatively increased due to high-level incidents in the United States [7], for instance, the Fort Hood shooting incident caused by US Army Psychiatrist Nidal Malik Hasan in 2009 [8] and the whistleblowing practices of private Chelsea Manning and Edward Snowden in, respectively, 2010 and 2012–13 [9]. Insider threats arise when a person who is authorized by the organization to perform certain activities decides to abuse the trust and cause damage to the organization.

Attacks committed by insiders are one of the most challenging threats organizations face today, especially in critical infrastructure where the damage resulting from insider threat incidents can have far-reaching consequences. In 2014, the nuclear reactor Doel 4 in Belgium was deliberately sabotaged by an individual who intentionally opened a valve in the steam turbine, thereby causing a deficiency of lubricating oil that severely damaged the reactor [10]. The costs of the sabotage amounted to millions of euros because the reactor was shut down for a few months, implying a loss of revenues, while hundreds of millions of euros were needed to repair the damage [11]. There is no need to explain that if the sabotage took place in the nuclear part of the reactor, the damage could have been much worse with, for instance, large-scale exposure to nuclear contamination and human casualties similar to the nuclear disaster in Chernobyl. Although the culprit

---

[a] In analogy with international relations realist theory that considers states as unitary actors [1, 2], "the organization" is interpreted in this study as a collective actor [3, 4] that is anthropomorphized [5], acting as if it were a single, united entity" should be added as a footnote by the word organizations.

Management and Engineering of Critical Infrastructures
https://doi.org/10.1016/B978-0-323-99330-2.00012-X

was never found [12], there is no doubt an insider was responsible for the incident, or at least deliberately assisted an external adversary [9,13].

Given that organizations first have to be aware of the insider threat problem before they can start thinking about insider threat mitigation [14,15], this chapter aims to increase insider threat awareness in critical infrastructure by illustrating the complexity of the problem. A typology of insider threats is outlined, discussing seven different insider threat domains in more detail. It is argued that organizations can build on the conclusions of the typology to develop a tailor-made, risk-based approach to insider threat mitigation.

In what follows, the chapter starts with a brief outline of the theoretical framework, elaborating on the meaning of the concepts of insider threat and critical infrastructure. After that, the methodology behind the typology is explained. Subsequently, the typology itself is illustrated by systematically referring to real examples of insider threat incidents in critical infrastructure sectors that are publicly available. Finally, the chapter concludes with a discussion of the shortcomings and recommendations for future research.

2 Theoretical framework

2.1 Insider threat

We draw upon Reveraert and Sauer's [16] interpretation of the insider threat problem, who refer to insider threats as the possibility that insiders cause harm to the organization because they intentionally misuse their privileged access to the organizational assets. The definition consists of five key concepts, namely, organizational assets, insiders, privileged access, harm, and intentional misuse, which will be briefly explained below.[b]

Organizational assets are valuable resources controlled by the organization that need to be protected and that are therefore situated within the proverbial security perimeter [9,17]. Critical assets are assets that are essential for the continuation of the organization's business [7,18]. The exact interpretation of the (critical) assets varies from one organization to another but commonly includes financial resources, intellectual property, equipment, or people.

Organizations have no other choice than "to allow people to get into positions where they can, if they choose, injure what [the organization] cares about, since those are the same positions that they must be to help [the organization]" [19]. Organizations have to provide access to the organizational assets to certain individuals and trust that they will handle the assets with care. These individuals are referred to as *insiders*. Insiders are not only employees that work permanently for their employer but also include individuals that are not part of the

---

[b] For a more detailed explanation of the conceptualization of insider threats, please see Reveraert and Sauer [15].

organization's permanent labor force, like contract employees [6]. Moreover, insiders are not only individuals currently belonging to the organization but are also individuals that used to be part of it and that were trusted by the organization in the past, at least the former confidants that still have access to the organizational assets [17,20,21].

Privileged access refers to the trust-based permission that the organization gives to insiders to penetrate the proverbial security perimeter that defends the organizational assets [22]. The access can be physical, for instance, the authorization to enter a building, or virtual, like the password to enter a network system [6,23].

Harm to the organizational assets is interpreted as negatively affecting the confidentiality, availability, or integrity of the organizational assets [17,24,25]. Harm to confidentiality implies that unauthorized individuals can access the proverbial security perimeter. Harm to integrity implies that the organizational asset is modified. Harm to availability implies that the organizational asset is destroyed or made inaccessible.

To conclude, harm caused by insiders results from misconduct, which can either be unintentional due to lack of proficiency (that is, insider hazards) or intentional due to lack of trustworthiness (that is, insider threats) [16]. Here, only *intentional misuse* of privileged access by insiders, whether or not to inflict harm, is interpreted as an insider threat. Unintentional misuse of privileged access, like employees who spread sensitive information to unauthorized individuals by accidentally hitting "reply all" instead of "reply" [15], is beyond the scope of this chapter.

2.2 Critical infrastructure

Critical infrastructure has become "the central nervous system of the economy in all countries" [26]. To achieve economic and social development or energy sustainability goals, guaranteeing that the infrastructure network is not at risk or vulnerable is indispensable. Over the past decades, the concept of "critical infrastructure" constantly evolved to respond to new emerging security challenges [27]. Events such as natural disasters, mechanical failures, or human actions (for example, theft or a terrorist attack) may disrupt or destruct their operations. The protection of critical infrastructure sectors has therefore become a high priority both at European and international levels [27].

Based on EC 2008/114 of the European Council, critical infrastructure is defined as "an asset, system or part thereof (…) which is essential for the maintenance of vital societal functions, health, safety, security, economic or social well-being of people, and the disruption or destruction of which would have a significant impact (…) as a result of the failure to maintain those functions." Countries have slightly different lists detailing their critical infrastructure sectors, but according to the European Council Directive, companies that belong to the following 12 sectors are considered as critical infrastructure: energy, information and communication technologies (ICTs), water, food, health, financial, public and legal order and safety, transport, chemical and nuclear industry, and space and research [28]. This list is indicative, which implies that European member states can identify their national critical sectors themselves according to some formal guidelines. In this chapter, we build upon the definition stated in the European Council Directive to illustrate our insider threat typology, referring to insider threat incidents that have taken place at different organizations within any of these 12 sectors.

3 Methodology

So far, we elaborated on our interpretation of the concepts of insider threat and critical infrastructure. In what follows, a typology of insider threats is outlined that breaks down the insider threat problem in a critical infrastructure context. To this end, the study relies upon a spin-off version of the who, what, where, when, why, and how (5W1H) methodology utilized by, for instance, Hart [29] and Homaliak et al. [30]. In contrast to these authors, we use the 4W3H methodology, asking the following seven questions: (1) **What** does the insider want to achieve? (2) **Who** suffers or benefits from the insider threat? (3) **Why** does the insider want to commit intentional misconduct? (4) **When** does the insider become untrustworthy? (5) **How** does the insider commit intentional misconduct? (6) **How serious** is the impact of the insider threat? and (7) **How many** insiders are involved with the insider threat? By asking these elementary questions, we tend to get a more complete picture of the characteristics of insider threats.

While the bedrock of the typology is based upon a theoretical analysis of existing literature, it was (and still is) challenged by real insider threat incidents that appear(ed) in (inter)national media sources[c] and that did (or do) not fit in any existing category. In other words, the answers to the four W and (derivates of) How came into existence via a constant interplay between insider threats literature that provides the foundation of the theoretical framework and publicly available examples of insider threat incidents that continuously reassess(ed) the theoretical framework to validate it. In the end, this enabled us to draw the mind map illustrated in Fig. 1 that displays the characteristics of the insider threat problem.

To illustrate the typology displayed in Fig. 1, we have opted to link each item of the typology with at least one insider threat incident that happened in one of the critical infrastructure sectors mentioned above, referring to examples that are publicly available in academic journals, (academic) books (nonfiction), or mainstream media. Although the dark or hidden number of insider threats remains high due to the tendency to avoid public announcements to safeguard the organization's reputation [25,31], we insisted on using real-case examples instead of hypothetical ones because they are functional both in a theoretical and a practical way. On the one hand, real cases help us to continually

[c] International media sources are BBC News, CNN, NOS, and The Guardian. National media sources De Morgen, De Standaard, De Tijd, Het Nieuwsblad, and VRT NWS.

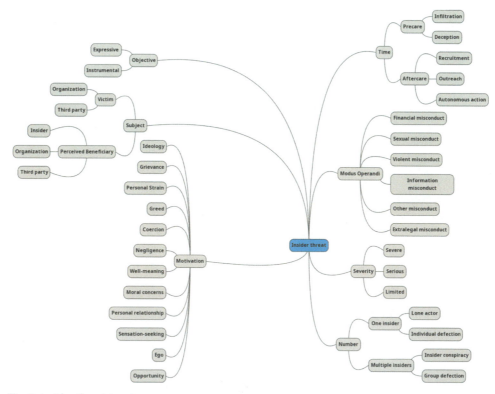

Fig. 1 Insider threat typology.

reassess and validate the theoretical framework. Moreover, discussing real insider threat incidents can help staff and organizational leaders to recognize their organization's vulnerability to insider threats [9] as they "bridge the gap between theoretical concepts and real-world problems" [32].

4 A typology of insider threats to critical infrastructure

4.1 Objective

The first domain of the typology refers to the objective of the insider [17], or **what** the insider wants to achieve. To answer this question, this study draws upon Willison and Warkentin's distinction between expressive and instrumental crimes. Expressive crimes are crimes where "the actual commission [of the crime] is considered an end in itself (…) [with] no additional goal to be met" [33], while instrumental crimes are crimes that "focus on achieving a goal where the criminal act is viewed as a means to an end" [33]. In analogy with Willison and Warkentin's expressive/instrumental division, expressive insider threats can be differentiated from instrumental insider threats in the sense that causing harm to the organization is an end in itself or a means to achieve a higher end.

4.1.1 Expressive insider threats

Insider threats that are principally aimed at causing harm to the organization are considered to be *expressive* insider threats. An example of an expressive insider threat is the case of David Burke, the man who was responsible for the crash of the Pacific Southwest Airlines Flight 1771. Being aware that the employer who recently dismissed him was on board the plane, Burke misused his access to the plane to smuggle a gun on board, kill his former employer and crash the plane [34,35]. Since Burke's mission was only accomplished when the organization (in a broad sense) got hurt, harming the organization was his main objective. Expressive insider threats are therefore not a pragmatic choice to harm an arbitrary organization, given that the identity of the organization (or organizational representative in Burke's case) is a decisive factor.

4.1.2 Instrumental insider threats

In contrast to expressive insider threats that primarily endeavor to cause harm to the organization, instrumental insider threats are mainly aimed at reaching a higher goal or personal gain through intentional misconduct. The potential damage to achieve this goal should thus not be regarded as an end in itself, but should rather be considered either unintentional or collateral damage.

On the one hand, the insider might have no intention at all to hurt the organization. Think, for instance, of a nurse at a hospital in Groningen, The Netherlands, who accidentally cut off a fingertip from a newborn baby in an attempt to remove a bandage from the baby girl's hand [36]. The nurse allegedly used a scissor, which was not according to the applicable procedures. Also, socially engineered insiders, referring to insiders who are manipulated by a third, unauthorized person (for example, the social engineer) into sharing their authorized access to the organizational assets with them [37], fit this description. Although the insider intentionally violates the organizational norm, there is no intention to hurt the organization. The insider does simply not take account of the fact that the witting misconduct might have counterproductive results.

However, the insider might perceive that the potential benefits of hurting the organization outweigh the potential costs. In other words, the insider can achieve an advantage or can reduce a disadvantage but has to make the organization suffer to achieve this advantage or reduction of the disadvantage, making the damage to the organization collateral. Instrumental insider threats, therefore, do not only arise from insiders that do not take account of possible counterproductive effects of witting misconduct but also from insiders that have a "the end justifies the means" mentality, implying that any means, even those that inflict

Fig. 2 Categorization of insider threat according to the insider's objective.

harm on the organization, can be used to reach the ultimate objective. The insider neutralizes his wrongdoing to overcome his potential moral barriers, convincing himself that he is allowed to harm the organization because it is for the greater good [33,38]. An example of such an insider threat is the case of Abdul-Majeed Marouf Ahmed Alani, a former American Airlines mechanic who was accused of trying to sabotage a commercial airliner. Although the former mechanic was upset over a contract dispute between the airlines and union workers, he claimed his intentions were purely financial as he allegedly explained that the sabotage enabled him to get overtime pay for repairing the plane, money he desperately needed to pay his children's study costs [39,40] (Fig. 2).

4.2 Subject

Second, insider threats can be divided according to the subject that is affected by them [9,41]. Two separate subquestions can be asked, namely, who[d] suffers from the insider threat (that is, victim) and **who** benefits from it (that is, perceived beneficiary)?

4.2.1 Victim

Concerning the victim, a distinction can be made between insider threats that solely cause harm to the organization, and insider threats that also cause harm to a third party outside the organization.[e] Harm to the *organization* is present when the insider threat only impacts the organization's assets. To illustrate, reference can be made to the case of Oswald "Ozzie" Bilotta, a former sales representative at a Swiss pharmaceutical company who acted as a whistleblower. In cooperation with the federal government, he used secret recording equipment to expose that the company had paid thousands of doctor's bribes to prescribe its own drugs [42].

In contrast, one can speak of harm to a *third party* when the impact of the insider threat exceeds the impact on the organizational assets and profoundly affects a third party negatively. The third-party may appear in many guises, ranging from customers and fellow

---

[d] The question "who poses the threat" is already addressed earlier in the chapter while discussing the definition of the insider.

[e] One could distinguish victims within the organization, separating individuals, groups, and the organization as such. However, given that we consider the organization as a unitary actor, we make an abstraction of this distinction.

organizations to ordinary citizens. To give an example, reference can be made to the case of Vitek Boden, who "worked for Hunter Watertech, a supplier of radio-controlled sewage control systems to the Maroochy Shire Council in Queensland, Australia. When Boden quit his job and was refused another by the council, he took his revenge by sabotaging the control systems, sending 800,000 L of raw sewage into local parks and rivers" [43]. Given that Boden's sabotage caused nuisances like contaminated water and stink, the harm caused by his actions was not confined to the organization's assets but also affected third parties.

4.2.2 Perceived beneficiary

Concerning the beneficiaries of the insider threat, a division can be made between insider threats that benefit the insider, benefit the organization (how strange it may sound), or benefit a third party. Insider threats aimed at benefitting the *insider* refer to insiders that are primarily interested in improving their own position, even at the expense of the organization. An example is the case of the employee of Bpost, the Belgian postal company, who stole over 300,000 euros by intentionally withholding certain letters over a period of 10 years [44].

Insider threats that perceive to cause benefit to the *organization* refer to insider threats where the ultimate goal is to aid the organization in the long run. It includes well-meaning insiders that wittingly circumvent organizational norms to help the organization. An example is the case of Oleg Savchuk, who, in an attempt to improve security, deliberately infected the computer system of the Ignalina Nuclear Power Plant in Lithuania with a computer virus as a wake-up call to the inadequate security measures [9]. Savchuk perceived that, as a cautionary tale, he had to hurt the organization in the short run, only to help the organization in the long run. Also, socially engineered insiders and employees who perceive that productivity is more important than security, and who therefore tend to ignore security protocols [45], fall under the scope of this category.

Next to the insiders that (want to) benefit either themselves or the organization, also insiders that benefit a *third party* should be considered. These insider threats refer to whistleblowers like Snowden who during his employment at the National Security Agency (NSA) leaked confidential information to newspapers [43,46], as well as to spies and moles like Robert Hanssen [9] and Ana Montes [47] who misused their access to classified US information to spy for, respectively, the Soviet Union (and later Russia) and Cuba. However, also the thwarted case of Auburn Calloway, whose "plan was to disable the DC-10's cockpit voice recorder, kill the crewmembers with hammers to simulate injuries consistent with an aircraft crash, and fly the aircraft into the ground so that his family would be able to collect on a $2.5 million life insurance policy provided by the company" [34] can be considered an insider threat that wants to benefit a third party (that is, his family) (Fig. 3).

Fig. 3 Categorization of insider threat according to the subject of the insider threat.

4.3 Motivation

A third domain concerns the motivation of the insider [17,25,31], or **why** the insider intentionally misuses his access to the organizational assets. The difference with domain one is that the latter questions the insider's intent, while the former questions the motivation behind the intent, whereby "multiple motivations may map into a single intent" [15]. This study provides a nonexhaustive list of 13 potential driving forces of insider threats.

4.3.1 Ideology

The first motivation is an *ideology* [31,46,48]. The ideological insider threat is posed by insiders that act out of ideological conviction. It refers to insiders that want to make a political, religious, or ideological statement or insiders that want to express their views on how (international) society should be managed. It includes, but does not exclusively consist of, cases of extremism and terrorism that can originate from different kinds of ideology.

The ideological insider threat can, for instance, be based on religious ideology, like the case of US Army psychiatrist Nidal Malik Hasan, who out of ideological convictions killed 13 people and wounded several others at Fort Hood [8]. During the lead-in time to his attack, his colleagues described him as "a ticking time bomb due to his radical views on Islam" [49]. Although this example refers to the Islamic religion, also other religious beliefs should be borne in mind when discussing insider threats based on religious ideology.

Apart from religious ideology, the insider threat can also originate from right-wing ideology. An example is the relatively recent insider threat incident in Belgium, where Jürgen Conings, a Belgian soldier with links to right extremism, misused his access to the army barracks to steal heavy weaponry while expressing intentions to kill known people

[50,51]. Similar examples of ideological insider threats based on right-wing ideology can be found in other countries as well. In the United States, reference can be made to the case of US soldier Ethan Melzer who has been charged for planning a terrorist attack on his unit by sending sensitive details to a neoNazi group [52] or to the members of the National Guard that were excluded from President Biden's inauguration because of their tie with right-extremist militants [53]. In Germany, the Parliamentary Oversight Panel (PKGr) stated in a report that "in the Bundeswehr and in several other security services on federal and state level (police and intelligence agencies)—despite a security screening—there are several public servants with an extreme far-right and violence-oriented mindset" (cited in Ref. [54]). In the United Kingdom, to conclude, one can think of the multiple investigations that were carried out among members of the UK military services in 2019 stemming from potential far-right concerns [55].

Next to religious and right-wing ideology, also left-wing ideology should be taken into account. Take, for instance, the case of Chinese double agent Larry Wu-Tai Chin who "joined the Communist party in 1942 and worked as an undercover agent while translating first for the US Army in China, then for the CIA in the United States, until his arrest in 1985" [56, p. 49]. In the present context, a possible breeding ground for left-wing ideological insider threats is the problem of climate change. Although the protests started with peaceful demonstrations, for instance, with the mobilization of over 70.000 people in Belgium [57], more radical climate organizations like Extinction Rebellion perceive that the reaction from the political authorities is unsatisfactory, requiring them to go a step further than simply marching the streets. It is possible that, if the legal measures do not have the desired effects, activists perceive that they will have to take the fate of the earth into their own hands. The insider threat is one possible alternative scenario that activists might take into consideration.

Additionally, also other ideological beliefs can be the subject of an insider threat incident. An example is the more recent debate about covid-19 vaccinations and other related topics. To illustrate this, reference can be made to the demand of Belgian hospitals to be able to fire staff who refuse to vaccinate themselves against covid-19 [58].

4.3.2 Grievance

A second motivation is a *grievance* [21,33,59]. It refers to insiders who believe that they are treated unfairly by the organization, which leads to a sense of disgruntlement. Disgruntled insiders often indicate that the organization did not hold its end of the deal, urging them to react to a perceived relative deprivation. To illustrate, one can refer to the case of Roger Duronio, the logic bomber of UBS PaineWebber that crippled the company's ability to exchange stocks by sabotaging the IT system of the company because "he didn't receive the large annual bonus he expected" [31]. Another example is the case of Ricky Joe Mitchell, who sabotaged the computer network of his employer EnerVest when he found out about his impending dismissal [43].

4.3.3 Personal strain

A third motivation is a *personal strain* [47,60], referring to negative personal experiences that are not directly related to the role of an insider. The difference between grievance and personal strain is that the former originates from the organization and is directed toward it, while the latter is not directly caused by the organization.[f] An example is the case of Helga Wauters, a Belgian anesthetist who has been hold responsible for the death of a pregnant woman in a French hospital in 2014. During the cesarean section, the woman's brain received too little oxygen which caused her death. An investigation revealed that Wauters had started her working day by drinking vodka with water, as she did every day for 10 years. She only worked in the hospital for less than 2 weeks after being discharged from a Belgian hospital for showing up drunk to work [61]. Another example is the case of a Dutch call center employee at an oil company who was suspected of ordering liquor and food worth more than a ton without paying for it by pretending to be the director of the company. He told in court that he resold the meals and drinks because he saw no other way to pay his debts [62].

4.3.4 Greed

A fourth motivation is a *greed*, applying to insiders that have a desire to have more of something [25,31,46]. An example of an insider threat based on greed is the case of an employee of Belgian bank KBC who ignored the investment mandate of a client and instead spent 200,000 euros on a new expensive car and other luxurious expenses [63]. Although greed is often interpreted as a striving for more money, we define the concept more broadly by also including other kinds of personal gain. Reference can be made to the case of a retired British major who misused his military rank to fraudulently collect 25 armored vehicles from several army museums to enlarge his personal collection of armored vehicles [64]. Moreover, greed can be expressed sexually, like former Norwegian politician Svein Ludvigsen who misused his political authority to sexually abuse asylum seekers that "believed their response to Ludvigsen's demands for sex could either result in being deported or securing permanent residency" [65]. In sum, greed refers to an overwhelming desire to have more of something.

4.3.5 Coercion

A fifth motivation is a *coercion*, which refers to insiders that are pressured by a third person to execute a certain action that hurts the organization [9,66]. To give an example, one can refer to the Northern Bank robbery where two bank officials were pressured by gang members to enable the bank robbery. The insiders had no other choice than to cooperate

---

[f] If the insider indicates insufficient support from the organization to mitigate his personal strain as primary motivation for the insider threat, it should be considered as an insider threat based on grievance rather than personal strain.

with the gang because their families were taken hostage and would have been murdered if the insiders did not collaborate [32,49].

4.3.6 Negligence

A sixth motivation is a *negligence*, which applies to "insiders who (…) take the path of least resistance (…) to make their working lives easier" [37]. In other words, it refers to insiders who deliberately deviate from the organization's behavioral guidelines just to make their own lives as easy as possible. An example is the case of an Italian man who did not show up for work at a hospital for 15 years but did receive his monthly salary. Also in Italy, 35 employees were caught in 2015 after camera images showed how they clocked in at their work at the Sanremo City Hall and afterward went shopping or canoeing [67].

4.3.7 Well-meaning

A seventh motivation concerns *well-meaning* insiders that have no intention to cause harm to the organization but "knowingly take risks to purposefully bypass bureaucratic security processes to be more effective in achieving what they think are organizational goals (…)" [37]. The difference between the well-meaning insiders and the negligent insiders is that the former act (or at least perceive to act) in the interest of their organization, while the latter act out of self-interest. An example is the case of "two control room trainees [who] poured caustic soda on fuel assemblies at a US nuclear power plant to draw attention to the lax security at the site" [68]. As in the previously mentioned case of Oleg Savchuck, the trainees perceived that, as a cautionary tale, they had to hurt the organization in the short run, only to help the organization in the long run. The well-meaning category also includes, for instance, insiders that bend the organization's behavioral guidelines if these are believed to slow down the achievement of the organization's objectives [6,45,69].

4.3.8 Moral concerns

An eighth motivation is *moral concerns*, which mainly corresponds to the category of whistleblowers[g] [37,49]. An example is the group of doctors in Brazil that shared a 10,000 paged document with investigators in which they accused Prevent Senior, a prominent healthcare provider in Brazil, of "covering up coronavirus deaths, pressuring doctors to prescribe ineffective treatments, and testing unproven drugs on elderly patients as part of ideologically charged efforts to help the Brazilian government resist a Covid lockdown" [70].

The principal objective of whistleblowers is to make a third party, like "competitive colleagues" or society as a whole, aware of the immorality of the organization. Applying a long-term perspective, it could be argued that whistleblowers help the organization

---

[g] For the sake of clarity, whistleblowing that originates from revenge or financial gain [41] is categorized under grievance and greed, respectively.

rather than harm it. This however does not alter the fact that the organization suffers in the short term [41].

4.3.9 Personal relationship (love and empathy)

A ninth motivation is a *personal relationship* ("*love and empathy*") [34], or insiders that misuse their insider privilege to benefit a third party they feel connected with. To illustrate this type of threat, reference can be made to the case of Joyce Mitchell who as a prisoner guard contributed to the escape of two prisoners, including one with whom she allegedly had a sexual affair [9], or to the paramedic in Pakistan who with the help of her colleagues stole a newborn to give to her childless aunt that desperately wanted to become a mother [71].

4.3.10 Personality disorder

A tenth motivation relates to insiders with a *personality disorder*[h] [17,66]. It concerns insider threats where the drive to deviate from the organization's behavioral guidelines primarily emerges from a certain inner drive that originates from personality traits, without a clear-cut alternative motive like financial gain or revenge. Given that personality traits play an important role in the thoughts and actions of an individual [17], personality disorders may be the main reason why certain individuals pose insider threats. An example is the case of Niels Högel [72] whose personality resulted in the death of several people. Working as a nurse, Högel deliberately injected patients with medication that resulted in cardiac arrest so that he could act as the hero that tried to save those patients. According to a psychiatric expert, Högel "displayed traits of noticeable personality disorders, such as a lack of shame, guilt, and empathy" [72].

4.3.11 Sensation-seeking

An eleventh motivation refers to insiders who are driven by the urge to seek sensation [17]. To illustrate this type of threat, reference can be made to the US military pilots who were posting selfies from their F–16 aircraft cockpits on social media, or to the one who was reading a book with his hands off the controls [73].

4.3.12 Ego

A 12th motivation concerns insider threats that are caused by insiders who cover up their mistakes to not lose face and save their egos [46,66]. An example is the case of the transplant surgeon responsible for the death of a 36-year-old transplant patient and the illness of two other patients who became ill of an infection from donated organs. Several organs became infected after the surgeon spilled stomach contents over other organs while

---

[h] Remember that to consider an incident an insider threat, the insider has to make the deliberate decision to misuse his access or knowledge. As a result, we assume that every insider with a personality disorder is able to make this conscious decision, and therefore make abstraction of the possible discussion on whether or not the insider "is of sound mind."

retrieving these from the donor. As the surgeon did not tell anyone about spilling contents, the organs were transplanted into the three patients with serious consequences [74].

4.3.13 Opportunity

A last motivation builds upon the "opportunity makes the thief" paradigm [31]. Sometimes, the insider threat is only caused because an opportunity presented itself to the insider, just at the right time. An example is the case of Britta Nielsen, who due to insufficient safeguards saw an opportunity to commit fraud. According to Nielsen, "it was a standing joke that you could easily add your account number and then be off to the Bahamas" (cited in Ref. [75]) (Fig. 4).

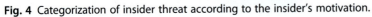

Fig. 4 Categorization of insider threat according to the insider's motivation.

4.4 Time

Next to the objective, subject, and motivation, the fourth domain of the insider threat typology refers to the time the insider slips through the trustworthiness net [9,30,76]. In other words, **when** does the insider become untrustworthy? A distinction is made between precare insider threats and aftercare insider threats.

4.4.1 Precare insider threats

Precare insider threats are insiders that are untrustworthy from the start of their employment. They should have been prevented from joining the organization. On the one hand, it can refer to insiders that infiltrate on behalf of an outside party and oppose the organization from the start. An example of such an insider threat is the case of Shannon Maureen Conley who infiltrated the US army to gain combat experience that she could subsequently take to Islamic State in Syria [49], or the case of Takuma Owuo-Hagood who started a career as a baggage handler at Delta Airlines to feed the Taliban sensitive information [20].

However, precare can refer to insiders that independently from an outside party deceive the organization, or circumvent the organization's screening procedures by falsifying their credentials. An example is the case of Zholia Alemi who "falsely claimed to have a medical degree from Auckland University when she registered in the UK in 1995" and who was able to legally exercise the medical profession for about 22 years without the necessary qualifications [77].

4.4.2 Aftercare insider threats

In contrast, *aftercare* insider threats are insiders who convert during or after their employment at the organization [6]. Aftercare insider threats can be a consequence of either recruitment, outreach, or autonomous actions [68]. An example of recruitment is the defection of Eugène Michiels, who worked for the Belgian Foreign Ministry since the 60s but who was contacted in 1978 by Romanian and Russian secret services with a request to sell classified information, which he eventually did until he was caught in 1983 [78].

While recruitment refers to insiders that are persuaded by an outside party to commit misconduct, outreach refers to insiders that on their own take the initiative to convert by reaching out to an outside party. An example is the case of Rajib Karim, a software engineer at British Airways who corresponded with Anwar al-Awlaqi, a senior Al-Qaeda figure. Karim exchanged information with al-Awlaqi about how he could attack the computer servers of British Airways to cause financial and operational disruption [32]. According to Hegghammer and Hoelstad Daehli, "it was Rajib Karim who first reached out to al-Awlaqi by e-mail" [68].

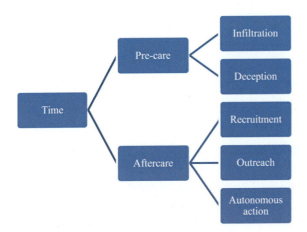

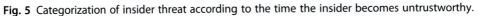

Fig. 5 Categorization of insider threat according to the time the insider becomes untrustworthy.

Insider threats based on autonomous actions, to conclude, refer to insiders that act individually without a link to an outside party. An example is the suicide of Andreas Lubitz, the co-pilot of German Wings who out of mental health problems deliberately crashed a plane, killing hundreds of innocent civilians [20] (Fig. 5).

4.5 Modus operandi

A fifth domain refers to the modus operandi of the insider, or **how** the insider misused his insider privilege. Inspired by Cools' [79] typology of employee crime, insider threats are divided into financial misconduct, sexual misconduct, violent misconduct, information misconduct, and a residual category of other forms of misconduct.

4.5.1 Financial misconduct

The first category refers to *financial misconduct*, such as the case of several leaders in the Russian space industry who were suspected of using false invoices and phantom firms in setting up a satellite navigation system Glonass [80], or the case of two managers of South Africa's ailing power firm Eskom who were suspected of manipulating contracts relating to the construction of two large power stations [81].

4.5.2 Sexual misconduct

Also, *sexual misconduct* can be committed by insiders. To give an example, reference can be made to the case of Teresa W., an employee of a German railway company who after work used her privileged access to enter one of the railway wagons to shoot porn videos and earn herself a little on the side [82]. A more serious case of sexual misconduct is the case of former British police officer Ian Naude, "who raped a young girl he met on duty" [83].

4.5.3 Violent misconduct

Violent misconduct can either be interpersonal or organizational [84]. With violent misconduct one spontaneously thinks of cases of interpersonal violence, like the shooting an (off-duty) fireman caused in his firehouse in Agua Dulce in the United States, thereby killing one colleague and wounding another [85]. However, violence can also be aimed at the property of the organization, referring to cases of sabotage [79]. The case of Rodney Wilkinson, who after exchanges with the African National Congress (ANC) used his privileged access to the Koeberg nuclear facility in South Africa to commit a terrorist attack on the nuclear plant [32,68], can therefore also serve as an example of *violent misconduct*.

4.5.4 Information misconduct

Insider threats also refer to *information misconduct*, such as the case of Xiang Haitao, a former employee of Monsanto, who pleaded guilty to stealing software developed by the agribusiness company. Xiang has admitted that "he attempted to take it to the People's Republic of China for the benefit of the Chinese government" [86]. Also, leakage of information to the press fits this category, like in the case of antiterror analyst Henry Kyle Frese who allegedly "leaked classified materials about a foreign country's weapons system to two journalists" [87].

4.5.5 Other misconduct

Finally, numerous *other types of misconduct* constitute a residual category. For instance, reference can be made to arms trafficking like the case of Eugene Harvey who as an airport employee misused his access to smuggle guns and ammunition in cooperation with an outside accomplice [34], human trafficking like the employees of a public hospital in Nairobi, Kenya, that misused their position in the hospital to steal and subsequently sell children for $400 [88] or drugs trafficking like the revelations in the context of 'Operation Sky' that discovered the assistance Belgian dockworkers provided to drug gangs [89].

Notice that so far, the examples cited above relate to illegal behavior. Indeed, when discussing the modus operandi of insiders, the center of attention seems to be on criminal actions like sabotage, fraud, theft (of intellectual property), and terrorism [31,49,90]. However, not every insider threat activity is equivalent to a criminal offense in the legal meaning of the word. In his work on employee crime, Cools [79] expands the legal meaning of the word "employee crime" to include behaviors that are criminalized by the organization itself, thereby relating it to sociological concepts like deviance. The broadening of the concept of employee crime, therefore, resembles our conceptualization of insider threat.

As a result, this study not only includes illegal misconduct where the insider commits a crime in a legislative sense but also includes *extralegal misconduct* where the deviation from the organizational norm does not correspond with a crime in a legal sense. An example is the malpractice of Belgian postmen who when delivering packages simply put a card in the client's post box that urges the client to pick it up at the post office, instead of ringing the doorbell to check whether the client is home as prescribed by the organizational norms [91]. What is considered to be misconduct however depends on the organizational culture and the applicable organizational norms. In other words, what is considered to be misconduct for one organization might be considered proper conduct for another. To give an example, some organizations allow love relationships among colleagues, while others prohibit it [92].

Likewise, whether or not misconduct is considered to be illegal or extralegal differs from organization to organization, as it depends on the legislation of the country of the organization, as well as on the employment contracts and code of conduct applicable within the particular organization [79] (Fig. 6).

4.6 Severity

A sixth insider threat domain refers to the severity of the insider threat, or **how serious** the impact of the insider threat is. The impact can appear in many guises, like damage to physical equipment, financial loss, litigation, reputational damage, or even loss of life. In this study, severe insider threats are distinguished from serious and limited insider threats [25].

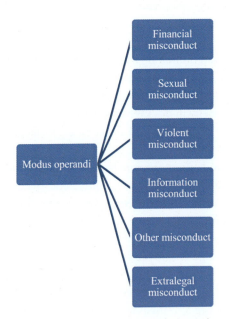

Fig. 6 Categorization of insider threat according to the insider's modus operandi.

4.6.1 Severe insider threats

Severe insider threats are threats that endanger the so-called critical assets of the organization [7], or the assets that are essential for the continuation of the organization's business [18]. In other words, severe insider threats put the survival of the organization at risk. An example is the case of the employee of American Superconductor who stole a crucial software program and passed it on to the organization's main consumer Sinovel Wind Group, who subsequently refrained from using American Superconductor's services. The theft of intellectual property and the resulting decrease in sales put the organization on the brink of insolvency [93].

4.6.2 Serious insider threats

Insider threats that do not pose a threat to the survival of the organization are either serious or limited, depending on the acceptable level of loss of the organization [24,94,95]. *Serious* insider threats refer to threats that do not necessarily endanger the survival of the organization, but that nevertheless inflict considerable harm to the organization that goes beyond the organization's level of acceptable loss.

4.6.3 Limited insider threats

Limited insider threats are threats that only slightly harm the organization and therefore fall within the scope of the organization's acceptable level of loss. Bunn and Glynn [94], for instance, indicate that both the casino and the pharmaceutical industries "accept that in some cases the expense of preventing small thefts may not be worth the cost of prevention." In other words, although the threat causes some harm to the organization, the benefits of mitigating the threat do not outweigh the costs of mitigation, making it a risk the organization is willing to take.

Whether an insider threat is considered to be severe, serious, or limited again differs from organization to organization, depending on the interpretation of the organization's critical assets and the organization's risk appetite (Fig. 7).

4.7 Number

Seventhly, a distinction can be made according to the number of insiders that participate in the insider threat. In other words, **how many** insiders are involved in the insider threat? Insider threats may either be the work of one single insider or multiple insiders, with or without outside accomplices [9].

4.7.1 One insider

First, the insider threat might be posed by only *one single insider*. Here, a subdivision can be made between lone actors and individual defectors. On the one hand, lone actor insider threats are insider threats where the insider acts completely on his own, without outsider involvement. An example of a lone actor is Jan Karbaat, a Dutch doctor who against the

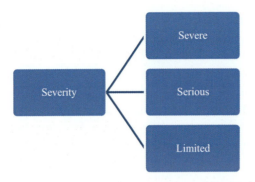

Fig. 7 Categorization of insider threat according to the severity of the insider threat.

rules donated his sperm on multiple occasions which resulted in the parenthood of dozens of children [96].

However, a combination between one insider and one or multiple outsiders is possible, whereby the insider is significantly linked[i] to the outsider(s). The insider defects[j] by shifting allegiance to the outside party. Reference can be made to the case of A.Q. Khan who stole confidential information during his employment at the Physics Dynamics Research Laboratory in the Netherlands and shared it with Pakistani researchers that were in charge of the Pakistani nuclear weapons program [32,97]. Khan misused his access to the classified information of the nuclear facility to give his country of birth a competitive advantage by acquiring nuclear weapons.

4.7.2 Multiple insiders

In contrast to insider threats posed by one single insider, insider threats can also be posed by *multiple insiders*. Again, a subdivision can be made between insiders that operate independently of outsider involvement, referred to as insider conspiracies, and insiders that defect to an outside party in a group, referred to as group defection. To demonstrate the insider conspiracies, one can refer to the case of Indira Gandhi. As prime minister of India, Gandhi was murdered by two of her Sikh personal guards as retaliation for her military action against the Sikh population at the Golden Temple in Amritsar [9,49]. Regarding group defection, reference can be made to the case of the employees of a security company who helped a drug gang to smuggle thousands of kilos of cocaine through the port of Rotterdam into the Netherlands [98] (Fig. 8).

[i] This should be interpreted as at least one direct contact (physically or virtually) between the insider(s) and the outsider(s).

[j] Premeditated in case of insiders that infiltrate on behalf of an outside party.

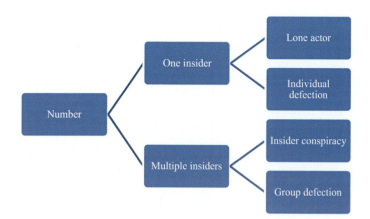

Fig. 8 Categorization of insider threat according to the number of insiders that are involved.

4.8 Summary

In sum, a typology of seven insider threat domains (that is, objective, subject, motivation, time, modus operandi, severity, and number) was outlined and illustrated by referring to relevant situations in critical infrastructure. Table 1 provides a schematic overview of the typology of insider threats, demonstrating the diversity and complexity of the issue and showing the different ways in which the insider threat can be expressed.

Table 1 An overview of seven insider threat domains.

1. Objective: *What* does the insider want to achieve?	
To cause harm to the organization To reach a higher goal or personal gain through harming the organization	Expressive Instrumental

2. Subject: *Who* suffers or benefits from the insider threat?	
The party that suffers • Only the organization • The organization and a third party The subject that benefits • The insider • The organization • A third party	The victim The perceived beneficiary

3. Motivation: *Why* does the insider want to misuse his access or knowledge?	
Because of the belief in an ideology • Religious ideology • Right-wing ideology • Left-wing ideology • Other	Ideology

Continued

Table 1 An overview of seven insider threat domains—cont'd

Because of a sense of disgruntlement	Grievance
Because of negative personal experiences	Personal strain
Because of the desire to have more of something	Greed
Because of the pressure of a third party	Coercion
To make things easier	Negligence
To act in the interest of the organization	Well-meaning
To make a third party aware of the immorality of the organization	Moral concerns
To benefit a third party, they feel connected with	Love and empathy
Because of a personality disorder	Personality disorder
Because of the urge to seek sensation	Sensation-seeking
Because of the fear to lose face	Ego
Because of an interesting opportunity	Opportunity

4. Time: *When* does the insider become untrustworthy?

Before employment • Infiltration = with outsider involvement • Deception = without insider involvement	Precare
During or after employment • Outreach = insider → outsider • Recruitment = outsider → insider • Autonomous action = without outsider involvement	Aftercare

5. Modus operandi: *How* does the insider misuses his access/knowledge?

Illegal financial behavior (for example, fraud, bribery, theft, …)	Financial misconduct
Illegal sexual behavior (for example, sexual assault, sexual harassment, …)	Sexual misconduct
Illegal violent behavior (for example, physical abuse, murder, sabotage, …)	Violent misconduct
Illegal information behavior (for example, espionage, cybercrime, …)	Information misconduct
Other types of illegal behavior (for example, drugs, human trafficking, …)	Other misconduct
Extralegal deviant behavior (for example, bullying, …)	Extralegal misconduct

6. Severity: *How serious* is the impact of the insider threat?

The insider threat affects the survival of the organization	Severe
Insider threat > acceptable level of loss	Serious
Insider threat < acceptable level of loss	Limited

7. Number: *How many* insiders are involved with the insider threat?

The insider acts on his own • Lone actor = without outsider involvement • Individual defection = with outsider involvement	One insider
At least two insiders are involved • Insider conspiracy = without outsider involvement • Group defection = with outsider involvement	Multiple insiders

5 Limitations

Our conceptualization consists of some limitations [16]. First of all, separating acceptable from unacceptable behavior is far more complicated than assumed in our study [99,100]. While the organization can provide guidelines to its members to behave appropriately, insiders are daily confronted with several unforeseen events. This implies that the insider's behavior cannot be exactly prescribed and that a certain discretion is needed. However, "this does not give them carte blanche to do as they wish" [101] as the discretionary power is limited and acceptable as long as "any fluctuations between perceived intentions and actions do not exceed the [organization's] expectations" [102]. Trying to provide insiders with directions on how to handle organizational assets beats the alternative situation of anarchy in which insiders are allowed to do whatever they want.

Secondly, despite the (sometimes) ambiguous character of the norms provided by the organization, which might lead to a discrepancy between the interpretation of the norm by the insider and the interpretation of the norm by the organization, the study is based on organizational norms rather than organizational rules. While there always apply some formal rules and regulations within the organization, we believe it is unwieldy for an organization to formalize all expected behavior in rules and regulations. Instead, it should be the organization's objective to create a strong organizational culture to guide the behavior of its employees [103]. This implies that the organization should aim to persuade its members with argumentation and justification into accepting the organizational norms, rather than to hegemonically enforce a regulatory framework without the general acceptance of its members [104,105].

Thirdly, it should be mentioned that insiders always have the opportunity (or even the propensity) to claim that an observed deviation of the organizational norms was unwitting because they were not aware of the norm, because they lack the skills to comply with the norm or because they argue it was an accident. Although this can be the case, we argue that in most instances the organization can accurately evaluate whether the insider's misuse of privilege is witting or not, as well as whether the claim of unawareness or incompetence is credible or not [15,106]. By any means, it should be the organization's first objective to communicate to its insiders in a clear way which behavior is expected to make sure they are aware of the applicable organizational norms and to train them to acquire the necessary skills that enable norm compliance [107].

Fourthly, many of the answers to the W's and How questions on which the typology is grounded are based on subjective interpretations rather than objective observations [100]. Given that one single story consists of several aspects, insider threat cases can be categorized according to the interpretation of the reader. In other words, each reader will have his or her way of looking at the story, reading it from their perspective. The distinction between negligent and well-meaning insiders is, for instance, a thin and ambiguous line, as illustrated by the case of John Deutch who as a CIA director "handled highly sensitive classified information on an insecure computer connected to the internet" [9]. Here, it can be argued that the insider had well-meaning intentions and handled in the interest of the organization or that the insider was negligent and handled in his self-interest. The categorization of such ambiguous cases as negligent or well-meaning insider threats, therefore, depends on the persuasiveness of the insider.

An interrogation of the insider can indicate why the insider deviated from the organizational norm, but precaution is still recommended as the insider has the propensity to lie and control the truth. Referring back to the case of Oleg Savchuk who allegedly sabotaged the computer system of the Ignalina Nuclear Power Plant, Bunn and Sagan indicate that he did not only do it "to call attention to the need for improved security" [9], but also "to be rewarded for his diligence" [9]. It can be assumed that during an investigation, the insider has the propensity to put more emphasis on the former (that is, well-meaning) than on the latter (that is, ego). Although each insider threat indeed has alternative readings, the quest for the veracious interpretation of the insider threat resembles our judicial system, where a breach of law is investigated and each party involved has the opportunity to give its interpretation of the facts. Ultimately, the interpretation that is most likely to be true determines the alleged guilt of the perpetrator. It is therefore believed that, in similarity with the judicial system, a thorough investigation of the insider threat, collecting evidence à charge and à décharge by taking into account the perspective of the different stakeholders that are involved and giving them the possibility to explain their version of events, should make it possible to properly judge the situation.

6 Conclusion and discussion

Based on our findings, the following conclusions can be made. First, the division of the insider threat into different domains and categories "does not mean that these categories are readily formed and should be hunted down and exposed" [99]. Instead, the domains and categories are just indicative of the different characteristics of the insider threat problem and the possible scenarios in which the insider threat can take place. It is not argued that, for instance, the examples that were given to illustrate the different motivations only apply to that specific category, as motivation is difficult to observe and subject to interpretation [15]. The goal of the study was therefore not to claim that insider threats should be siloed into these categories. Instead, the study simply wants to illustrate the variation of insider threats, like the variety of motivations that might urge an insider to misuse his access. In similarity with Eoyang's interpretation of espionage, who indicates that "espionage is not unitary and simple behavior; we must treat it as a class of criminal behaviors and not as a single distinct crime" [108], insider threat should be interpreted as a class of threats and not as a single distinct threat.

In this chapter, we took the first step to analyze this class of threats to critical infrastructure. Insiders who constitute a threat to critical infrastructure can have a significant impact on the vulnerable assets of the organization and ultimately cause irreparable damage to its activities and reputation. To tackle this problem properly, organizations in critical infrastructure need to have a good understanding of the threats that they face. By gaining a proper understanding of the different types of insider threats, critical infrastructure organizations can implement adequate policies to mitigate intentional misconduct among their members. The current study can serve as a framework that can guide security professionals into getting a better understanding of the phenomenon.

Nevertheless, more empirical studies are needed to consider the validity of the typology presented in this chapter and to develop possible mitigation strategies. In this light, it should be clear that each organization belonging to critical infrastructure should make an organization-specific threat analysis to examine which of the threats apply to their organization, and a corresponding risk analysis to evaluate the odds that correspond with those threats. We urge each organization to interpret the insider threat within its organizational context. We believe that our (nonexhaustive) typology might help organizations in this difficult exercise. The diversity of insider threats implies that there is not one holy grail that can mitigate the insider threat problem, but that a tailor-made approach is needed [109]. The typology might be a good starting point in the development of such a tailor-made approach, both to create awareness of the topic within the organization as well as to brainstorm about the different types of insider threats applicable to the organization. Subsequently, the organization can use risk assessments build on the conclusions of the insider threat typology to determine the probabilities and impact that are associated with these different scenarios.

With our study, we paved the way for future research that might apply our typology to less-security-minded sectors. Although some organizations dismiss the applicability of insider threats to their organization, referred to by Bunn and Sagan [9] as the NIMO (Not in My Organization) bias, the insider threat is generally considered to be a threat that is universal and therefore applicable to every organization [24,25]. Or as Gelles [7] indicates, "[i]nsider threats exist within every organization because employees, or insiders, comprise the core of an organization's operational plan and are the key drivers of its business objectives". As illustrated in this chapter, the insider threat is a nondiscriminatory threat that applies to organizations across different countries and sectors, including the financial sector [21], the nuclear sector [68,110], the energy sector [111,112], the aviation sector [20,34,35,109], the military sector [8,113,114], the intelligence and security services [108,115,116], the pharmaceutical sector [94] and so on. We believe our typology outlined, in this chapter, might help to understand the complexity of the insider threat to these (and other) sectors and organizations.

Funding

Funding was provided by Bel-V, Brussels Airport Company, Elia, Engie-Electrabel, the Federal Agency of Nuclear Control (FANC), and G4S.

References

[1] K. Waltz, Theory of International Politics, Addison-Wesley Publishing Company, London, 1979.

[2] P.R. Viotti, M.V. Kauppi, International Relations Theory, Longman, 2012.

[3] N. Isaeva, C. Hughes, M. Saunders, Trust, distrust and human resource management, in: K. Townsend, K.D. Cafferekey, A. McDermott (Eds.), Elgar Introduction to Theories of Human Resources and Employment Relations, Edward Elgar, Cheltenham, 2019, pp. 247–263.

[4] R.H. Searle, HRM and trust, or trust and HRM? An underdeveloped context for trust research, in: R. Bachmann, A. Zaheer (Eds.), Handbook of Advances in Trust Research, Edward Elgar Publishing Limited, 2013, pp. 9–28.

[5] S.C. Monahan, B.A. Quinn, Beyond 'bad apples' and 'weak leaders' toward a neo-institutional explanation of organizational deviance, Theor. Criminol. 10 (3) (2006) 361–385, https://doi.org/10.1177/1362480606065911.

[6] C. Colwill, Human factors in information security: the insider threat—who can you trust these days? Inf. Secur. Tech. Rep. 14 (2009) 186–196.

[7] M. Gelles, Insider Threat: Detection, Mitigation, Deterrence and Prevention, Elsevier-Health Science Division, 2016.

[8] A.B. Zegart, The Fort Hood terrorist attack: an organizational postmortem of army and FBI deficiencies, in: M.S. Bunn (Ed.), Insider Threats, Cornell University Press, 2016, pp. 42–73.

[9] M. Bunn, S. Sagan, Insider Threats, Cornell University Press, Ithaca, 2016.

[10] Federal Agency for Nuclear Control, Sabotage van de stoomturbine van Doel 4, 2021. https://fanc.fgov.be/nl/dossiers/kerncentrales-belgie/actualiteit/sabotage-van-de-stoomturbine-van-doel-4.

[11] K. Dewey, C. Hobbs, G. Foster, D. Salisbury, S. Tzinieris, Reconceptualising nuclear security as a business enabler: opportunities and challenges, in: IAEA International Conference on Nuclear Security (ICONS 2020), International Atomic Energy Agency (IAEA), Vienna, 2020, pp. 1–8. Retrieved from: https://conferences.iaea.org/event/181/contributions/15290/.

[12] L. Bové, Onderzoek naar sabotage kerncentrale Doel na 7 jaar afgerond, De Tijd, 2021, August 13. https://www.tijd.be/politiek-economie/belgie/algemeen/onderzoek-naar-sabotage-kerncentrale-doel-na-7-jaar-afgerond/10325602.html.

[13] Federal Police, Sabotage Doel 4, 2019. https://www.politie.be/5998/nl/opsporingen/gezocht/onbekende-verdachten/sabotage-doel-4.

[14] T. Dinev, Q. Hu, The centrality of awareness in the formation of user behavioral intention toward protective information technologies, J. Assoc. Inf. Syst. 8 (7) (2007) 386–408.

[15] C.W. Probst, J. Hunker, D. Gollmann, M. Bishop, Aspects of insider threats, in: C.W. Probst, J. Hunker, D. Gollmann, M. Bishop (Eds.), Insider Threats in Cyber Security, Springer, 2010, pp. 1–15.

[16] M. Reveraert, T. Sauer, Redefining insider threats: a distinction between insider hazards and insider threats, Secur. J. 34 (2021) 1–21.

[17] J.R. Nurse, O. Buckley, P.A. Legg, M. Goldsmith, S. Creese, G.R. Wright, M. Whitty, Understanding insider threat: a framework for characterising attacks, IEEE Security and Privacy Workshops, 2014, pp. 214–228.

[18] M. Bishop, C. Gates, D. Frincke, F.L. Greitzer, AZALIA: an A to Z assessment of the likelihood of insider attack, in: IEEE Conference on Technologies for Homeland Security, IEEE, Boston, 2009, pp. 385–392.

[19] A. Baier, Trust and antitrust, Ethics 96 (2) (1986) 231–260.

[20] K. Krull, The Threat Among Us: Insiders Intensify Aviation Terrorism, Pacific Northwest National Laboratory - Prepared for the US Department of Energy, 2016.

[21] M.R. Randazzo, M. Keeney, E. Kowalski, D. Cappelli, A. Moore, Insider Threat Study: Illicit Cyber Activity in the Banking and Finance Sector, Carnegie Mellon Software Engineering Institute, 2005.

[22] M. Bishop, S. Engle, D.A. Frincke, C. Gates, F.L. Greitzer, S. Peisert, S. Whalen, A risk management approach to the 'Insider Threat', in: C.W. Probst, J. Hunker, D. Gollmann, M. Bishop (Eds.), Insider Threats in Cyber Security, Springer, 2010, pp. 115–137.

[23] A. Munshi, P. Dell, H. Armstrong, Insider threat behavior factors: a comparison of theory with reported, in: 45th Hawaii International Conference on System Sciences, IEEE Computer Society, Hawaii, 2012, pp. 2402–2411, https://doi.org/10.1109/HICSS.2012.326.

[24] E. Cole, S. Ring, What is there to worry about? in: E. Cole, S. Ring (Eds.), Insider Threat: Protecting the Enterprise From Sabotage, Spying, and Theft, Syngress Publishing, 2006, pp. 3–48.

[25] K.R. Sarkar, Assessing insider threats to information security using technical, behavioural and organisational measures, Inf. Secur. Tech. Rep. 15 (2010) 112–133.

[26] J.M. Yusta, G.J. Correa, R. Lacal-Arántegui, Methodologies and applications for critical infrastructure protection: state-of-the-art, Energy Policy 39 (10) (2011) 6100–6119.

[27] D. Markopoulou, V. Papakonstantinou, The regulatory framework for the protection of critical infrastructures against cyberthreats: the example of the health sector in particular, Comput. Law Secur. Rev. 41 (2021) 105502.

[28] European Council, Council directive 2008/114/EC on the identification and designation of European critical infrastructures and the assessment of the need to improve their protection, Off. J. Eur. Union 3 (2008) 0075–0082.

[29] G. Hart, The five W's: an old tool for the new task of audience analysis, Tech. Commun. 43 (1996) 139–145.

[30] I. Homaliak, F. Toffalini, J. Guarnizo, Y. Elovici, M. Ochoa, Insights into insiders and IT: a survey of insider threat taxonomies, analysis, modeling, and countermeasures, ACM Comput. Surv. 52 (2) (2018) 1–40, https://doi.org/10.1145/3303771.

[31] J.E. Mehan, Insider Threat: A Guide to Understanding, Detecting, and Defending Against the Enemy From Within, IT Governance Publishing, Cambridgeshire, 2016.

[32] C. Hobbs, M. Moran, Insider Threats: An Educational Handbook of Nuclear and Non-Nuclear Case Studies, King's College London, 2015.

[33] R. Willison, M. Warkentin, Beyond deterrence: an expanded view of employee computer abuse, MIS Q. 37 (1) (2013) 1–20.

[34] P.J. Greco, Insider threat: the unseen dangers posed by badged airport employees and how to mitigate them, J. Air Law Comm. (2017) 717–742.

[35] J.M. Loffi, R.J. Wallace, The unmitigated insider threat to aviation (part 1): a qualitative analysis of risks, J. Transp. Secur. 7 (2014) 289–305.

[36] Het Nieuwsblad, Verpleegkundige knipt per ongeluk vingertopje van baby af: "Mijn dochter is verminkt voor het leven", December 4, Het Nieuwsblad (2021). https://www.nieuwsblad.be/cnt/dmf20211204_94835758.

[37] D.S. Wall, Enemies within: redefining the insider threat in organizational security policy, Secur. J. 26 (2013) 107–124.

[38] M. Siponen, A. Vance, Neutralization: new insights into the problem of employee information systems security policy violations, MIS Q. 34 (3) (2010) 487–502.

[39] N. Chavez, D. Royal, A former american airlines mechanic admitted he tried to sabotage a plane at the miami airport, December 19, CNN (2019). https://edition.cnn.com/2019/12/18/us/former-american-airlines-mechanic-plea/index.html.

[40] NOS, Saboterende monteur American Airlines moet drie jaar de cel in, NOS, 2020, March 4. Retrieved from: https://nos.nl/artikel/2325836-saboterende-monteur-american-airlines-moet-drie-jaar-de-cel-in.

[41] D.J. Moberg, On employee vice, Bus. Ethics Q. 7 (4) (1997) 41–60.

[42] J. De Schamphelaere, Klokkenluider van Novartis krijgt 109 miljoen dollar, De Tijd (2020). July 23 https://www.tijd.be/ondernemen/farma-biotech/klokkenluider-van-novartis-krijgt-109-miljoen-dollar/10240450.html.

[43] T. Ring, The enemy within, Comput. Fraud Secur. 12 (12) (2015) 9–14.

[44] Het Nieuwsblad, Postsorteerster die jarenlang brieven achteroverdrukte maakte 150.000 tot 300.000 euro buit, Het Nieuwsblad (2019). July 1 https://www.nieuwsblad.be/cnt/dmf20190701_04488256.

[45] K. Roemer, Treating employees as a threat, Netw. Secur. 2008 (2008) 9–11.

[46] D. Fischbascher-Smith, The enemy has passed through the gate: insider threats, the dark triad, and the challenges around security, J. Organ. Eff. 2 (2) (2015) 134–156.

[47] E. Shaw, L. Sellers, Application of the critical-path method to evaluate insider risks, Stud. Intell. 59 (2) (2015) 1–8.

[48] M. Maasberg, J. Warren, N.L. Beebe, The dark side of the insider: detecting the insider threat through examination of dark triad personality traits, in: 48th Hawaii International Conference on System Sciences, IEEE Computer Society, Hawaii, 2015, pp. 3518–3526.

[49] D. BaMaung, D. McIlhatton, M. MacDonald, R. Beattie, The enemy within? The connection between insider threat and terrorism, Stud. Confl. Terror. 41 (2) (2018) 133–150.

[50] K. Heylen, 11 militairen die Defensie volgt voor connecties met extreemrechts, verliezen rechten, nog veel vragen over Conings, VRT NWS (2021). May 23 https://www.vrt.be/vrtnws/nl/2021/05/23/zoektocht-juergen-conings-zondag/.

[51] Vast Comité van toezicht op de inlichtingen- en veiligheidsdiensten, Toezichtonderzoek naar het opsporen en het opvolgen – door de twee inlichtingendiensten – van de radicalisering van een militair werkzaam bij Defensie, en anderzijds naar hun samenwerking met hun partnerdiensten, waaronder Defensie, onder meer wat betreft hun informatie-uitwisseling, Comité I, 2021. https://www.comiteri.be/images/pdf/publicaties/RAPP%20UNCLASS%20JC%20NL%2001%2007%202021.pdf.

[52] BBC News, US soldier ethan melzer accused of planning attack on own unit, June 23, BBC News (2020). https://www.bbc.com/news/world-us-canada-53145806.

[53] De Morgen, Twaalf leden National Guard geweerd van inauguratie Biden, onder meer om banden met extreemrechts, De Morgen (2021). January 19. Retrieved from: https://www.demorgen.be/nieuws/twaalf-leden-national-guard-geweerd-van-inauguratie-biden-onder-meer-om-banden-met-extreemrechts~b0442b4d/.

[54] F. Flade, The insider threat: far-right extremism in the German military and police, CTC Sentinel 14 (5) (2021) 1–10. Retrieved from: https://ctc.usma.edu/the-insider-threat-far-right-extremism-in-the-german-military-and-police/.

[55] B. Quinn, Attractiveness of British military for far right continues to be a threat, The Guardian (2021). May 31 https://www.theguardian.com/world/2021/may/31/attraction-of-british-military-to-the-far-right-continues-to-be-a-threat-prevent.

[56] Katherine L. Herbig, A history of recent American espionage, in: T.R. Sarbin, R.M. Carney, C. Eoyang (Eds.), Citizen Espionage: Studies in Trust and Betrayal, Greenwood Publishing Group, 1994, pp. 39–68.

[57] De Morgen, Tot 75.000 deelnemers op grootste Belgische klimaatmars ooit, December 2, De Morgen (2018). https://www.demorgen.be/nieuws/tot-75-000-deelnemers-op-grootste-belgische-klimaatmars-ooit~bbaafbca/.

[58] D. Baert, Ziekenhuizen willen personeel dat zich niet laat vaccineren kunnen ontslaan, VRT NWS (2021). October 20 https://www.vrt.be/vrtnws/nl/2021/10/20/vaccinatieplicht-ziekenhuizen/.

[59] F.L. Greitzer, L.J. Kangas, C.F. Noonan, A.C. Dalton, R.E. Hohimer, Identifying at-risk employees: modeling psychosocial precursors of potential insider threats, in: Hawaii International Conference on System Sciences, IEEE Computer Society, Hawaii, 2012, pp. 2392–2401, https://doi.org/10.1109/HICSS.2012.309.

[60] R. Agnew, Foundation for a generalism strain theory, Criminology 30 (1) (1992) 47–87.

[61] L. Hodge, Belgische anesthesiste krijgt drie jaar cel wegens dood zwangere vrouw, VRT NWS (2020). November 12 https://www.vrt.be/vrtnws/nl/2020/11/12/belgische-anesthesiste-krijgt-drie-jaar-cel-wegens-dood-zwangere/.

[62] NOS, Nepdirecteur bestelt voor ruim een ton aan sterke drank en eten, NOS, 2020, July 3. https://nos.nl/artikel/2339451-nepdirecteur-bestelt-voor-ruim-een-ton-aan-sterke-drank-en-eten.

[63] L. Milio, Voormalig bankier uit Heist-op-den-Berg veroordeeld omdat hij geld van klant in eigen zak stak, VRT NWS (2020). October 10. Retrieved from: https://www.vrt.be/vrtnws/nl/2020/10/08/voormalig-bankier-veroordeeld-omdat-hij-geld-van-klant-in-eigen/.

[64] Y. Barbieux, Britse majoor veroordeeld voor stelen tanks uit Belgisch legermuseum: "U zou uit pure schaamte de rechtszaal met neerhangend hoofd moeten verlaten", Het Nieuwsblad (2021). September 11 https://www.nieuwsblad.be/cnt/dmf20210911_93847854.

[65] BBC News, Norway ex-minister svein ludvigsen guilty of sexually abusing asylum seekers, July 5, BBC News (2019). https://www.bbc.com/news/world-europe-48880551.

[66] C. Noonan, Spy the Lie: Detecting Malicious Insiders, Pacific Northwest National Laboratory Prepared for the US Department of Energy, Richland, Washington, 2018.

[67] De Standaard, Italië vervolgt man die niet ging werken maar wel half miljoen euro aan salaris opstreek, De Standaard (2021). April 22 https://www.standaard.be/cnt/dmf20210422_97040986.

[68] T. Hegghammer, A. Hoelstad Daehli, Insiders and outsiders: a survey of terrorist threats to nuclear facilities, in: M. Bunn, S. Sagan (Eds.), Insider Threats, Cornell University Press, 2016, pp. 10–41.

[69] E.E. Thompson, Introduction, in: E.E. Thompson (Ed.), The Insider Threat: Assessment and Mitigation of Risks, CRC Press Taylor and Francis Group, New York, 2018, pp. 1–34.

[70] T. Phillips, F. Milhorance, Brazil hospital chain accused of hiding covid deaths and giving unproven drugs, The Guardian (2021). September 29. Retrieved from: https://www.theguardian.com/global-development/2021/sep/29/brazil-prevent-senior-hospital-chain-covid-accusations.

[71] M. Kazim, Pakistan paramedic 'stole baby to give to childless aunt', December 9, BBC News (2019). Retrieved from: https://www.bbc.com/news/world-asia-50713433.

[72] P. Oltermann, German nurse given second life sentence for murder of 85 patients, The Guardian (2019). June 6 https://www.theguardian.com/world/2019/jun/06/german-nurse-niels-hogel-second-life-sentence-murder-of-85-patients.

[73] J. McCurry, Sky-high selfies: japan warns us over 'outrageous' antics of military pilots, The Guardian (2019). November 8 https://www.theguardian.com/world/2019/nov/08/sky-high-selfies-japan-warns-us-over-outrageous-antics-of-military-pilots.

[74] BBC News, Patient died after 'transplant surgeon error', November 21, BBC News (2019). https://www.bbc.com/news/uk-wales-50493248.

[75] P. Oltermann, Danish social worker jailed for stealing £13m of government funds, The Guardian (2020). February 18. Retrieved from: https://www.theguardian.com/world/2020/feb/18/danish-social-worker-jailed-britta-nielsen.

[76] N. Catrantzos, No dark corners: a different answer to insider threats, Homel. Secur. Aff. 6 (2) (2010) 1–20.

[77] BBC News, Zholia Alemi: foreign doctor checks after fake psychiatrist case, November 19, BBC News (2018). https://www.bbc.com/news/health-46258687.

[78] K. Lasoen, Geheim België: De geschiedenis van de inlichtingendiensten 1830-2020, Lannoo Meulenhoff-Belgium, 2020.

[79] M. Cools, Werknemerscriminaliteit, VUB Press, 1994.

[80] VRT NWS, Schandaal bij Russisch satelliet-programma, 2012 November 9. https://www.vrt.be/vrtnws/nl/2012/11/09/schandaal_bij_russischsatelliet programma-1-1478076.

[81] BBC News, Eskom crisis: arrests over $50m South Africa power station 'fraud', December 12, BBC News (2019). https://www.bbc.com/news/world-africa-50854186.

[82] Het Nieuwsblad, Treinbegeleidster ontslagen omdat ze pornovideo's maakte in wagon, Het Nieuwsblad (2019). May 22 https://www.nieuwsblad.be/cnt/dmf20190522_04417883.

[83] BBC News, Ian Naude: how predatory paedophile joined police, BBC News (2018). November 15. Retrieved from: https://www.bbc.com/news/uk-england-shropshire-46220586.

[84] S.L. Robinson, R.J. Bennett, A typology of deviant workplace behaviors: a multidimensional scaling study, Acad. Manag. J. 38 (2) (1995) 555–572.

[85] De Standaard, Brandweerman gedood door collega bij schietpartij in Amerikaanse kazerne, 2021 June 2. https://www.standaard.be/cnt/dmf20210602_95259747.

[86] The Guardian, China scientist pleads guilty to stealing trade secret from monsanto, The Guardian (2022). January 6 https://amp-theguardian-com.cdn.ampproject.org/c/s/amp.theguardian.com/us-news/2022/jan/07/china-scientist-pleads-guilty-to-stealing-trade-secret-from-monsanto.

[87] The Guardian, US counter-terrorism analyst charged with leaking classified materials, 2019, October 9. Retrieved from:, The Guardian. https://www.theguardian.com/us-news/2019/oct/09/henry-kyle-frese-defense-intelligence-arrest-leaker.

[88] BBC News, Kenya arrests four more after BBC Africa eye baby stealers exposé, BBC News (2020). November 19. Retrieved from: https://www.bbc.com/news/world-africa-54986993.

[89] S. Sanen, Opnieuw arrestaties in Antwerpse drugsmilieu na grote telefoonkraak "Operatie Sky": 10 verdachten opgepakt, VRT NWS (2021). October 25. Retrieved from: https://www.vrt.be/vrtnws/nl/2021/10/25/opnieuw-arrestaties-in-antwerpse-drugsmilieu-na-grote-telefoonkr/.

[90] M. Maasberg, N.L. Beebe, The enemy within the insider: detecting the insider threat through addiction theory, J. Inf. Priv. Secur. 10 (2) (2014) 59–70.

[91] VRT NWS, Postbode die niet aanbelt en enkel kaartje in de bus steekt, terwijl je thuis op pakje wacht? "Kan niet", zegt CEO Bpost, VRT NWS (2020). October 28. Retrieved from: https://www.vrt.be/vrtnws/nl/2020/10/28/ceo-bpost-postbodes-moeten-aanbellen-als-ze-een-pakje-leveren/.

[92] M. Biaggio, T.L. Paget, M.S. Chenoweth, A model for ethical management of faculty-student dual relationships, Prof. Psychol. Res. Pract. 28 (2) (1997) 184–189.

[93] National Insider Threat Task Force, Protect Your Organization From the Inside Out: Government Best Practices, NITTF, 2016. https://www.dni.gov/files/NCSC/documents/products/Govt_Best_Practices_Guide_Insider_Threat.pdf.

[94] M. Bunn, K.M. Glynn, Preventing insider theft: lessons from the casino and pharmaceutical industries, in: M. Bunn, S. Sagan (Eds.), Insider Threats, Cornell University Press, 2016, pp. 121–144.

[95] S. Kumar, A. Deskmukh, J. Liu, K.E. Stecke, An analysis of trust, employee trustworthiness, fraud, and internal controls, Int. J. Strateg. Decis. Sci. 4 (3) (2013) 66–89.

[96] The Guardian, Dutch fertility doctor 'secretly fathered at least 49 children', April 12, The Guardian (2019). https://www.theguardian.com/world/2019/apr/12/dutch-fertility-doctor-secretly-fathered-at-least-49-children.

[97] L. Barbé, Hoofdstuk 5: Pakistan, België en de bom: de rol van België in de proliferatie van kernwapens, Own Publication, 2012, pp. 69–122. http://www.lucbarbe.be/sites/default/files/boek/files/Belgie-en-de-bom.pdf.

[98] NOS, Securitas-medewerkers helpen drugsbende bij cocaïnesmokkel via haven R'dam, NOS, 2021, January 13. https://nos.nl/artikel/2364165-securitas-medewerkers-helpen-drugsbende-bij-cocainesmokkel-via-haven-r-dam.

[99] S. Dekker, Just Culture: Restoring Trust and Accountability in Your Organization, CRC Press, 2017.

[100] S.W. Dekker, Just culture: who gets to draw the line? Cogn. Tech. Work 11 (3) (2009) 177–185.

[101] T. Bailey, On Trust and Philosophy, The Philosophy of Trust, 2002.

[102] S.M. Ho, R.R. Katukoori, Agent-based modelling to visualise trustworthiness: a socio-technical framework, Int. J. Mob. Netw. Des. Innov. 5 (1) (2013) 17–27.

[103] R. Von Solms, B. Von Solms, From policies to culture, Comput. Secur. 23 (2004) 275–279.

[104] M. Siponen, A conceptual foundation for organizational information security, Inf. Manag. Comput. Secur. 8 (1) (2000) 31–41.

[105] M. Siponen, J. Kajava, Ontology of organizational IT security awareness-from theoretical foundations to practical framework, in: Proceedings Seventh IEEE International Workshop on Enabling Technologies: Infrastucture for Collaborative Enterprises, IEEE, Stanford, CA, USA, 1998, pp. 327–331.

[106] S.M. Ho, M. Kaarst-Brown, I. Benbasat, Trustworthiness attribution: inquiry into insider threat detection, J. Assoc. Inf. Sci. Technol. 69 (2) (2018) 271–280.

[107] T. Gundu, S.V. Flowerday, The enemy within: a behavioural intention model and an information security awareness process, in: Information Security for South Africa (ISSA), IEEE, South Africa, 2012, pp. 1–8.

[108] C. Eoyang, Models of espionage, in: T.R. Sarbin, R.M. Carney, C. Eoyang (Eds.), Citizen Espionage: Studies in Trust and Betrayal, Greenwood Publishing Group, United States of America, 1994, pp. 69–91.

[109] D. BaMaung, The hidden threat, Int. Airpt. Rev. 22 (4) (2018) 22–25.

[110] International Atomic Energy Agency, Preventive and Protective Measures Against Insider Threats, IAEA Nuclear Security Series No. 8, 2008.

[111] A.J. Bell, M.B. Rogers, J.M. Pearce, The insider threat: behavioral indicators and factors influencing likelihood of intervention, Int. J. Crit. Infrastruct. Prot. 24 (2019) 166–176.

[112] D. Rehak, M. Hromada, T. Lovecek, Personnel threats in the electric power critical infrastructure sector and their effect on dependent sectors: overview in the Czech Republic, Saf. Sci. 127 (2020) 1–10.

[113] N.J. Armstrong, With an eye open and a round chambered: explaining the Afghan insider threat and its implications for sustained partnership, J. Interv. Statebuilding 7 (3) (2013) 223–240.

[114] A. Long, Green-on-blue violence: a first look at lessons from the insider threat in Afghanistan, in: I.M. Bunn, S. Sagan (Eds.), Insider Threats, Cornell University Press, 2016, pp. 103–120.

[115] M. Hershkowitz, The "insider" threat, J. Police Crisis Negot. 7 (1) (2007) 103–111.

[116] L.A. Kramer, R.J. Heuer Jr., America's increased vulnerability to insider espionage, Int. J. Intell. Counterintell. 20 (1) (2007) 50–64.

PART II

Management of critical infrastructures

CHAPTER 4

Health management of critical digital business ecosystems: A system dynamics approach

Abide Coskun-Setirek[a], William Hurst[b], Maria Carmela Annosi[c], Bedir Tekinerdogan[b], and Wilfred Dolfsma[c]
[a]Department of Social Sciences, Wageningen University & Research, Wageningen, The Netherlands
[b]Information Technology Group, Wageningen University & Research, Wageningen, The Netherlands
[c]Business Management & Organisation Group, Wageningen University & Research, Wageningen, The Netherlands

1 Introduction

The term "critical infrastructure" (CI) refers to systems and assets that are vital to a nation's physical security, economic security, or public health [1,2]. Critical infrastructure systems (CIS) provide essential services for modern society [3]. Such separate but interdependent systems coexist in modern societies to form a socio-technical system of systems with complex behavior and provide critical economic and societal services in critical domains such as water and energy supply, telecommunication, transportation, power systems, agriculture, banking, health, education, and public administration [4,5]. In these domains, business ecosystems have started to be developed around digital platforms for efficiency, scalability, smartness, and security. According to Argyroudis et al. [6], digital technologies supported by collaborative partnerships are essential for CI. Public-private partnership (PPP) for CI, for example, can be viewed as business ecosystem, which is a network of interrelated companies, such as governing bodies/regulators, private owners or financiers, operators, service providers, lobbies, and users/customers [7].

However, not all CIS are successful because of invulnerability to disruptive events such as cyber or physical attacks, natural disasters, epidemics, wars, and faulty operations or accidents that may cause cascading effects and threaten business continuity [8]. Yet, resilience towards such threats is essential for the health and continuity of digital business ecosystems (DBEs), especially in CI sectors, to ensure public safety, economic success, and human well-being [9]. Resilience, both technical and organizational, and business continuity are key to CI and they can benefit from collaboration, which enables co-innovations and co-creation in business ecosystems [10,11]. While the likelihood of co-creation in a DBE can be increased through member acquisition if an improvement in ecosystem health. As such, there is a circular causality among co-creation, DBE health,

Management and Engineering of Critical Infrastructures
https://doi.org/10.1016/B978-0-323-99330-2.00004-0

and resilience, meaning it is necessary to analyze the relations between the health and resilience of such complex, interconnected, non–linear adaptive systems using the system dynamics methodology. Traditional analytical approaches do not scale or are not sufficient to analyze the system. A holistic, system thinking and modeling approach identifies the key variables and their interactions that impact the health of the critical DBEs. System dynamics is a useful approach to understand the complicated relations in complex dynamic systems and a powerful tool to analyze system behavior [12].

To the best of our knowledge, at the time of writing this manuscript, no attempt has been made to construct a dynamic conceptual model focusing on the DBE health and resilience for CI. Thereby, this study aims to identify the key variables of the health and resilience of critical DBEs, examine the cause-and-effect relationships among them, and construct a dynamic conceptual model based on existing literature and the views of an interdisciplinary team. To build the model, the relevant theoretical knowledge from the DBE, CI, and resilience literature is reviewed, and all elements that might play a role are specified. An initial causal loop diagram is created after the key causal effects, and feedback loops between these variables have been discovered. An interdisciplinary team of authors examines the diagram before presenting the final conceptual model to validate the model. In the end, the importance of elements that impact DBE health is highlighted by examining feedback loops. The study may benefit researchers by presenting a holistic view of the elements, relations, and dynamics of DBE health and resilience, while it has implications for practitioners by improving their understanding of DBE administration in CI sectors.

The rest of this chapter is organized as follows. The study's background is presented in Section 2. The selected research approach is explained in Section 3. The developed dynamic conceptual model is described and a causal loop diagram, which is modeled using the simulation tool Vensim is presented in Section 4. Section 5 explains the validation of the model. The results and contributions to the research are discussed in Section 6. Finally, Section 7 brings the paper to a close.

2 Background

This section provides the background of the study, which includes the resilience and health of DBEs for CI, and system dynamics subjects.

2.1 Digital business ecosystems

Business ecosystems are economic communities of interacting species like corporations and individuals in the business environment [13,p. 26]. Digital innovations have enabled the development of DBEs [14]. Business ecosystems have been usually evolved around digital platforms as DBEs, which are defined by Nachira [15] as *"a digital environment populated by digital species."* Multi-sided platforms, collaboration platforms, or platform–based ecosystems present a socio-technical environment based on collaboration among actors and provide technological infrastructures to create value [14]. They allow collaborations among heterogeneous members from different geographical regions via digital opportunities for a common purpose [16]. Value co-creation is achieved by combining the value chain and platform logic [17]. Therefore, they support collaborative value creation [17] and enhance product innovation [18].

With the deployment of new digital technologies, digital systems within CI sectors have also increased [19]. CI is defined as *"an asset, a system, or a part of it necessary to maintain society's vital functions, health, safety, security, or economic or social welfare"* [5]. Although it varies by nation, common CI sectors identified include energy, transportation, water, telecommunication, health, financing, food supply, and public administration. Digital technologies supported by collaborative partnerships like PPP, are essential for CI [6]. The actors of such a business ecosystem can interact with each other for knowledge sharing and value co-creation through a digital platform and this DBE can ensure efficiency, quality, and cost-effectiveness in value creation and delivery [7].

However, not all DBEs are successful, and, for business continuity, DBE health is essential. Productivity, robustness, and the ability to create niches are three factors for healthy business ecosystems identified by Iansiti and Levien [20]. According to Iansiti et al., keystones, dominators, hub landlords, and niche players have roles in structuring DBEs. Keystones improve the health of the business ecosystem, which also affects their own development. Dominators enable vertical or horizontal integration and have the rights to manage the network. Hub landlords, for instance, have a low physical presence with few network nodes and provide little value to ecosystems. Furthermore, niche players constitute a large part of the ecosystem both in mass and variety. This structure leads to network effects and there will be an exponential increase in the benefits provided by a network node as the total number of nodes increases considering the effect of one actor's actions on the well-being of the other [16]. Therefore, business ecosystem health is impacted by the collaboration of individual members [21].

Another concept that supports managers when protecting their systems against both foreseen and unpredicted events is resilience [5]. Former disruptive events, such as COVID-19, have demonstrated that CI organizations need to improve CI's safety and continuous operation against foreseen and unpredicted events [22]. Resilience and business continuity are key to CI and risk identification and problem-solving by collaborative networks enhance them [11]. System resilience is described by the National Research Council [23] as the system's preparedness, absorbability, recoverability, and adaptability against disruptive events. Bruneau et al. [24], for example, stated four properties of both physical and social systems' resilience, these are (i) robustness, (ii) rapidity, (iii) redundancy, and (iv) resourcefulness. In addition to the preventive focus of risk management, other scientists have introduced the concept of resilience against unexpected events, considering contingencies and these properties of resilience [25]. Technical, organizational, economic, and social resilience are also studied as resilience dimensions or domains [26]. Scholars stated that the most relevant domains for CI sectors are technological and organizational resilience [27].

As the aforementioned literature demonstrates, there is a circular causality between DBE health and resilience. A holistic, system thinking and modeling approach are necessary to identify the key variables and their dynamic interactions that impact the health of the DBEs in CI sectors. Therefore, the system dynamics methodology can be used to analyze such complex, interconnected, non-linear adaptive systems.

2.2 System dynamics

System dynamics methodology is covered by early systems theory, such as in the works by Von Bertalanffy [28], and based on system thinking, which is a causality-driven and holistic approach to understanding the interactions among elements within complex systems [29]. Direct causality, circular-feedback causality, and that the system's dynamic behavior is caused by the internal structure of it are some assumptions for system thinking [30]. The methodology was developed in the late 1950s by researchers at the Massachusetts Institute of Technology under the leadership of Jay W. Forrester [30–33]. System dynamics often use simulations to generate the dynamic behavior of models since finding analytical solutions to most non-linear and complex feedback models are difficult or impossible [34]. System dynamics methodology can be used to model and analyze the structure and dynamics of complex and non-linear systems using a feedback perspective. Feedback loops are employed to determine the dynamics that occur from these interactions. As such, they are the engine of system dynamics modeling, and a feedback loop emerges when the outcome of one action in a system influences its place of origin [35].

System dynamics methodology can be applied to business issues to develop corporate strategies and effective policies, and to design better organizations, in addition to engineering problems [33]. In literature, for example, system dynamics has been successfully applied to many management subfields such as operations, organizational behavior, marketing, behavioral decision-making, and strategy [36]. The technique has contributed to strategic management in terms of many theoretical perspectives such as strategic planning, organizational learning, stakeholder theory, knowledge elicitation, strategy formulation, knowledge and resource management, project management, and performance management [37].

In system dynamics literature, DBEs have been studied to analyze the performance indicators [38], the success drivers [39], the evolution mechanism [40], the operational mechanism [41], the development and competition [42], technological innovation and value creation [43], and innovation diffusion [44]. The concept of resilience has also been investigated using a system dynamic approach [45–48]; however, there are few studies on DBE health [21].

In summary, the literature indicates that there has been no clear attempt to develop a dynamic conceptual model that focuses on both DBE health and resilience, although they are essential together for the DBE administration, especially in the CI sector.

3 Methodology

In this study, a system dynamic approach is used to analyze the cause-and-effect relationships between the variables of the resilience and health of DBEs, especially for CI sectors, to analyze the feedback loops formed by the relationships, and to construct the conceptual model.

A system dynamics methodology consists of problem definition, model conceptualization, model formulation, testing, policy analysis, and implementation stages, which are defined similarly by system dynamic scholars [30,31,33]. In this study, the first two stages of the system dynamics methodology, which are the problem definition and the model conceptualization, are applied in line with the purpose of the study (Fig. 1). The model conceptualization phase of the system dynamics modeling is applied by examining the real problem, listing all of the possible variables, identifying the major causal relations among these variables, identifying the feedback loops, and constructing an initial causal loop diagram [30]. Therefore, first, the relevant theoretical information in the DBE literature is examined to establish the model. For the literature review, Scopus and Web of Science databases and Google Scholar are searched to access the relevant studies in the resilience and health of DBEs and CI literature. A careful review of the literature involved the key variables that have a potential role and the causal relations among them are identified. Only the concepts which have scientifically proven relationships are considered. After the feedback loops are identified, an initial causal loop diagram is developed. Vensim® PLE Plus simulation software is used for the construction of the causal loop diagram. The initial causal loop diagram is examined by the interdisciplinary team, which consists of four senior researchers in the fields of management and engineering and the final conceptual model is obtained.

4 Model conceptualization

The aforementioned disruption events may pose a danger to business continuity. Resilience, both technical and organizational, is essential for the health and financial performance of critical DBEs. On the other hand, resilience can benefit from collaboration, co-creation, and co-innovations in DBEs. DBEs are so complex and thus system dynamics can be used to make sense of the cause-effect relations that are important for preserving DBE health and supporting resilience. For this reason, the dynamic conceptual model is established through a holistic combination of related concepts to understand how to ensure a healthy environment in DBEs. Combining these three lines of research resulted

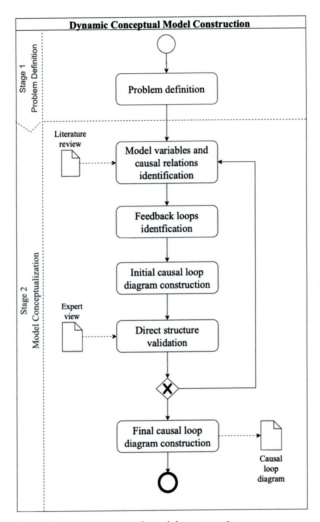

Fig. 1 The process of the dynamic conceptual model construction.

in three sub-structures for the model, which are collaboration and value creation, DBE health, and disruption and resilience.

In this section, the sub-structures, their elements, the interactions between them, and the feedback loops, which form the conceptual model, are explained and the final causal feedback loop diagram is created. At the end of the section, the analysis of how the whole structure might be affected by disruption events is provided.

In system dynamics, a causal feedback loop diagram is used to visualize the structure of a system. The arrows in a causal loop diagram demonstrate the causal relationships among the elements. Arrows with a plus (+) indicate that both elements change in the same direction whereas a minus (−) means that the change in the opposite direction. The reinforcing loops are shown by "R," whereas the balancing loops are denoted by "B."

The first sub-structure is the co-creation sub-structure, which consists of platform development, value creation to customers, customers, platform revenue, total DBE members, collaboration level, and innovation capability elements. The definitions of these elements for this study are provided in Table 1.

The evidence in the literature to the relations between them are outlined in Table 2. Platform development enhances value creation by the developments in service concepts and technical performance [42]. While value creation attracts more customers, revenues from them can be used in platform development, which can be described with the creation of a service idea and technical performance improvement. Moreover, the total number of ecosystem members changes with the changes in market opportunity and the number of customers. More ecosystem members mean a higher collaboration level

Table 1 The co-creation sub-structure elements.

Sub-structure elements	Definitions
Platform development	Service concept development and development of the technical performance of a DBE
Value creation to customers	Creating value for DBE customers to aid in the sale of goods and services
Customer	The person who buys goods or a service offered by collaborators in a DBE
Platform revenue	The revenue derived from customers utilizing a platform
Total DBE members	Total number of customers and collaborators in a DBE
Collaboration level	The level of collaboration among the ecosystem members in a DBE
Innovation capability	The capacity to successfully produce, utilize, and develop new ideas in a platform

Table 2 The relations among the co-creation sub-structure elements.

Element relations	References for the relations
Platform development ➜ Value creation	[42]
Value creation ➜ Customer	
Customer ➜ Platform revenue	
Platform revenue ➜ Platform development	
Customer ➜ Total DBE members	A general rule
Total DBE members ➜ Collaboration level	A general rule
Collaboration level ➜ Innovation capability	[49]
Innovation capability ➜ Value creation	[50]
Innovation capability ➜ Learning and growth	[21]
Platform revenue ➜ Financial performance	A general rule

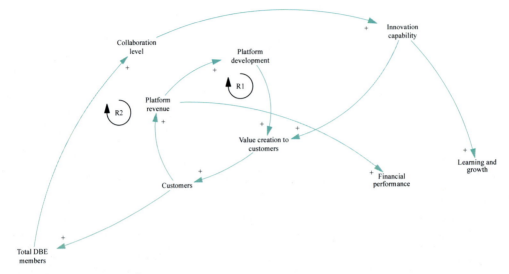

Fig. 2 The co-creation sub-structure.

among the members. Collaboration with other members may be extremely beneficial to a company's ability to innovate [49]. The capacity to successfully produce, utilize, and develop new ideas is known as innovation capability [51]. The ability of organizations to innovate may be viewed as a prerequisite for the creation of new value [50]. Moreover, it enhances the learning and growth perspective of a business ecosystem [21]. Eventually, a change in platform development impacts the DBE health and keystone performance through value creation, platform revenue, and innovation capability [21,52].

As indicated in Fig. 2, the relationships between the elements formed the following two reinforcing loops (they are examined together with other loops at the end of this section):

The second sub-structure is the DBE health sub-structure, which includes the DBE health, value creation to customers, customers, robustness, productivity, keystone performance, financial performance, learning and growth, market opportunity, and total DBE members elements (Table 3).

The causal effect relations between the elements of the DBE sub-structure are given in Table 4. For sustainable business ecosystems, it is crucial to maintain and improve their health [53]. Iansiti and Levien [20] identified three success factors for business ecosystems including (i) productivity, (ii) robustness, which means the capabilities of surviving across threats and changes, and (iii) the ability to create niches and opportunities. Keystones have the most influential roles in structuring a successful business ecosystem by contributing the productivity. According to Yim et al. [21], there is a circular feedback causality among business ecosystem health, keystone performance, and productivity. Keystone performance improves with the increase in learning and growth and financial

Table 3 The DBE health sub-structure elements.

Sub-structure elements	Definitions
DBE health	A DBE's financial well-being and strength to continue to create opportunities for its domains
Value creation to customers	Creating value for customers to aid in the sale of goods and services in a DBE
Robustness	The capability of a DBE to absorb the impacts of a disruptive event and surviving
Productivity	A network's ability to transform resources into goods and services
Keystone performance	The financial performance of DBE leaders
Financial performance	The capacity of a platform to generate profit from its economic activities, after subtracting all related costs
Learning and growth	Changes and improvements that a platform needs to make to realize its vision
Market opportunity	The chance to respond to a market need or interest in a DBE
Total DBE members	Total number of customers and collaborators in a DBE

Table 4 The relations among the DBE health elements.

Element relations	References for the relations
Value creation ➜ DBE health	[20,21]
Robustness ➜ DBE health	
Productivity ➜ DBE health	
DBE health ➜ Keystone performance	
Keystone performance ➜ Productivity	
Financial performance ➜ Keystone performance	[52]
Learning and growth ➜ Keystone performance	
Keystone performance ➜ Market opportunity	[20,21,53]
Market opportunity ➜ Total DBE members	
Total DBE members ➜ Productivity	
Total DBE members ➜ Robustness	
Robustness ➜ DBE health	

performance [52]. There is another circular feedback causality among DBE health, keystone performance, market opportunity, total DBE members, and productivity [21]. An increase in DBE health will increase the keystone performance, and then create market opportunity, in the end, improve business ecosystem health by increasing productivity and robustness [20,21,53].

The DBE health sub-structure is combined with the co-creation substructure (Fig. 3). The links of each sub-structure are shown in different colors. An additional four

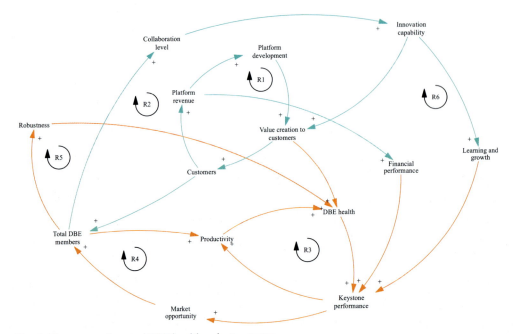

Fig. 3 The co-creation and DBE health sub-structures.

reinforcing loops are formed by combining the relationships between the DBE health sub-structure elements and co-creation substructure elements. These loops are

The last sub-structure is the disruption and resilience sub-structure, which consists of the disruptive events, disruption, resilience, technical resilience, robustness, recoverability, organizational resilience, risk management, learning and growth, platform development, cost, and financial performance elements (Table 5).

When disruptions adversely affect the development of a platform, a change in platform development results in a circular feedback causality [42]. Resilience was defined by Berkeley Iii et al. [54] as the capacity to absorb, adapt to, and recover from disruptions. Rehak et al. [27] studied CI resilience in terms of technical resilience and organizational resilience. The "robustness" and "recoverability" of CI determine the technical resilience [55], whereas organizational resilience is influenced by the knowledge gathered from prior disaster [27]. Therefore, in the model, technical resilience is strongly influenced by robustness, and learning and growth are the basis for organizational resilience. Resilient subsystems cause less disruption during disruptive events, and the recovery time is demonstrably shorter [27]. Therefore, resilience hurts disruption. On the other hand, disruption cause recovery cost [46] and affect platform development negatively whereas learning and growth from disruption take place [27]. The causal effect relations between the elements of the disruption and resilience sub-structure are given in Table 6.

Table 5 The disruption and resilience sub-structure elements.

Sub-structure elements	Definitions
Disruptive events	Unforeseeable events that threaten a DBE
Disruption	The harmful effects of disruptive events on a DBE
Resilience	The capacity of a DBE to absorb, adapt to, and recover from disruptions
Technical resilience	The robustness and recoverability of a DBE
Recoverability	The capacity of a DBE to recover its function after a disruptive event
Robustness	The capability of a DBE to absorb the impacts of a disruptive event and surviving
Organizational resilience	The ability and capacity of a DBE to adapt to and learn from disruptive events
Risk management	A DBE's process of ensuring security/safety and strengthening resilience at the prevention stage
Learning and growth	Changes and improvements that a platform needs to make to realize its vision
Platform development	Service concept development and development of the technical performance of a platform
Cost	Recovery cost caused by a disruptive event to return the original performance of a DBE
Financial performance	The capacity of a platform to generate profit from its economic activities, after subtracting all related costs

Table 6 The relations between the disruption and resilience sub-structure elements.

Element relations	References for the relations
Robustness ➜ Technical resilience	[55]
Recoverability ➜ Technical resilience	[27]
Learning and growth ➜ Organizational resilience	
Risk management ➜ Organizational resilience	
Technical resilience ➜ Resilience	
Organizational resilience ➜ Resilience	
Disruption ➜ Learning and growth	
Disruption ➜ Platform development	
Disruptive events ➜ Disruption	[46]
Resilience ➜ Disruption	[55]
Disruption ➜ Cost	
Cost ➜ Financial performance	

After the elements and causal effects of disruption and resilience sub-structure are identified, an additional four reinforcing loops and a balancing loop are formed (Fig. 4) as follows:

Therefore, the dynamic conceptual model of critical DBE health consists of ten reinforcing feedback loops and balancing feedback loop. The feedback loops tangent to each

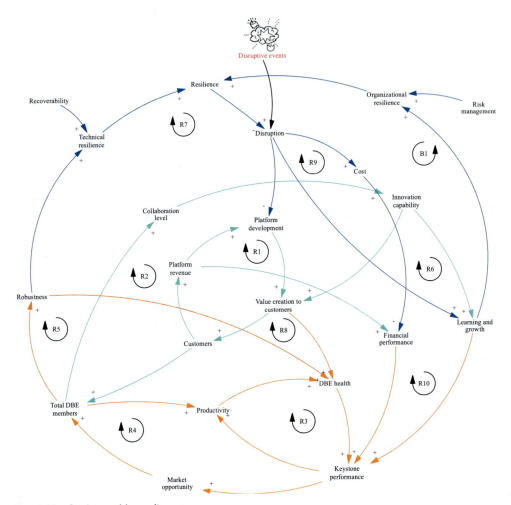

Fig. 4 The final causal loop diagram.

other should be examined together. In the case of disruptive events, the feedback loop R1 is affected adversely by disruption, the extent of its impact is dependent on resilience, and platform development and value creation are interrupted. This interruption in value creation causes customer loss and a decrease in revenues, which may decrease the investment for platform development. In the end, the value creation to customers is affected negatively. R1 triggers the feedback loop R2 by the customer variable. Customer loss leads to a decline in the collaborative level, which impacts innovation capability and then value creation.

Therefore, value creation to customers is a notable variable of the platform-development sub-structure and a driver of DBE health. A change in R1 or R2 triggers all other feedback loops. A decrease in value creation and financial performance that stems from a disruption event triggers the R3, R4, R5, and R6 and makes worse the DBE health and keystone performance; however, an increase in learning and growth after disruption improves them. The result of these loops determines the changes in robustness that trigger the feedback loops R7, R8, R9, R10, and B1. A change in robustness causes a change in technical resilience depending on the recoverability of the infrastructure, while an increase in learning and growth improves organizational resilience depending on risk management. Technical resilience and organizational resilience determine overall resilience, which has a significant role in the impact of disruptive events. As time progresses, recovery can occur, or disruption can cause irreversible damage. Therefore, the strengthening of resilience and thus DBE health are based on the continual enhancement of the level of robustness, and learning and growth in addition to current recoverability and risk management levels. It is necessary to develop a formal model to better analyze the impacts of disruption events on DBE health.

5 Structural validation

One of the notable steps of system dynamics methodology is model validation [56]. Rather than trying to fully test the model, the aim is to increase the level of trust in the model's credibility in system dynamics [57]. Structure validity and behavior validity are the two validation methods for system dynamics models. In this study, structural validity is assessed since the only conceptual model is constructed. The structural validity of a model refers to whether the model's structure gives a meaningful explanation of real-world relationships [30]. The direct structured tests are used to assess structural validity [56]. In this study, the structure-confirmation and boundary adequacy (structure) tests, which are suggested by several researchers [56–59] are performed for direct structure validity, and the process is facilitated using the cause-effect graphing features of Vensim.

With the structure-confirmation test, the consistency of a model structure with the real system structure is tested. It might be theoretical, based on generalized information from the literature, or empirical, based on real-world correlations [56]. The structural validity of the model is justified in this study by using existing models in the literature. Moreover, the major relationships in the model are examined and validated by the four experts, two information technology and two business, management, and organization academicians. The interdisciplinary team also performs the boundary adequacy test to test whether the model includes all important concepts for addressing the problem [56]. Cause-effect graphing aids model correctness evaluation by identifying causes and effects and determining if they are correctly portrayed in the model [57]. Vensim's "causal tracing" and "documentation" functions are employed to aid this procedure [56]. Causal tracing is a technique for determining the link between elements by displaying a tree of causes and the uses of each variable [60].

6 Discussion

In this study, a dynamic conceptual model is developed to describe the feedback relationships between the resilience and health of DBEs, especially for CI sectors. The model conceptualization phase of the system dynamics methodology is applied to construct the model. This phase includes the following steps: (i) Listing all of the possible variables related to the problem, (ii) identifying the major causal relations among these variables, (iii) identifying the feedback loops, and (iv) constructing the causal loop diagram [30]. After constructing the causal loop diagram, interactions among the feedback loops are examined and key variables for critical DBE health administration are identified.

The scope of the problem revealed three sub-structures, which are DBE health, resilience, and platform development and collaboration. The variables of these sub-structures and the relationship between them are identified by reviewing the related literature. DBEs are collaborative platforms and the collaboration and its relations with the platform development should be included in the model if the subject is DBE health. On the other hand, resilience is required for protecting the DBEs in CI sectors against both foreseen and unpredicted events. Therefore, both resilience and co-creation variables are linked with the variables of DBE health.

The diagram (Fig. 4) shows the importance of resilience in the impact of disruptive events on DBEs health and how the system feedback loops affect each other in a complex way. During disruptive events, a decrease in DBE health is shown depending on robustness, and it can return to the original level depending on organizational resilience and recoverability. In the worst-case scenario, the disruption irreversibly affects the DBE health and performance due to the low-level robustness. In the best-case scenario, the DBE health and performance level raises to a higher level eventually than the original level through higher robustness, recoverability, and adaptability values. It is necessary to test the interaction of feedback loops with different disruption, recoverability, and risk management values to see how the result is.

The causal loop diagram shows only the main effects and links in the model and there may be other factors that need to be included in a final quantitative model. For example, collaboration with external actors facilitates innovation capability depending on the level of absorptive capacity [49]. This variable is not included in the developed causal loop diagram since it can be included in the stock-flow diagram as an auxiliary variable, that is, a constant. On the other hand, some variables (such as recoverability and risk management) that are not impacted by the other variables of the model, can be included as auxiliary variables or linked with some auxiliary variables. In addition, some pertinent variables, for example, other variables covered by the studies in Section 4, can be omitted from the causal loop diagrams as they are outside the scope analysis, as well as the need to ensure brevity and simplicity [61]. A balance needs to be struck between the creation of a model that is too detailed or too simplified. Uncomplicated models also can be explanatory by focusing on only necessary details. A model that is too detailed will lose its clarity and become harder to control and understand.

To date, the dynamic relationships between the resilience and health of DBEs have not been investigated to any extent. The advantage of a causal loop diagram is the unique potential to depict explicitly the structure of dynamic systems and processes. Therefore, the developed model can provide an opportunity to explore the dynamic structure of the resilience and health of DBEs. This study has implications for researchers by providing a holistic view of the variables, relations, and dynamics of the DBE health, especially for CI sectors. It provides a meta-structure that can frame case studies, especially in CI sectors to contribute to the understanding of the resilience and health of DBEs dynamics. It has implications for practitioners such as managers at keystone firms, by enhancing their visualization of the decision-making on DBE administration. For example, they may utilize the structure to test the different disruption and resilience scenarios and understand the interactions between CI resilience and DBE health, that the scenarios triggered.

This study has some limitations that could be considered in future studies. Model conceptualization is the first step of system dynamics modeling and this study presents a causal loop diagram, which provides a holistic view of the variables, relations, and dynamics of the DBE health and resilience. These issues are important for further research, but until there is sufficient scientific evidence, reasonable estimations of the relationships based on the current literature are the first step in the modeling process. Although the feedback loops tangent to each other are discussed together, it is important to examine the mutual interactions of all six feedback loops and the evolvement of them together over time. For this reason, a formal model must be developed to test the relationships among the variables by quantifying them, and the model should be validated and simulated to analyze the behaviors of the key variables.

The scope of the developed model can be found simple. When this model has been scrutinized and understood by those who are interested it could be extended to a more complex model and examined in detail. The variable of technical resilience and organizational resilience, for example, are discussed very superficially and can be described in detail in future studies. It is also worth adding the elements of DBE architectural approaches and governance mechanisms and exploring their impacts on the dynamics of DBE health. Moreover, different types of business ecosystems and collaborations with their different types of actors can be considered in future dynamic models.

This study can be improved in the future by following methodologies: Constructing a stock-flow diagram, testing it with case studies in various CI sectors; adopting a statistical approach to understand the causality within and between the variables and developing a conceptual model; and integrating ecosystem actors as agents in the model and linking it with an agent-based modeling approach.

7 Conclusion

This study makes the first attempt to develop a dynamic conceptual model by considering interactions and feedback loops between the variables of the resilience and health of critical DBEs. Key variables and relations are defined, and feedback loops are identified. A causal loop diagram is constructed using the feedback loops. It consists of three sub-structures, which are co-creation, DBE health, and resilience sub-structures. Interactions between feedback loops and potential changes in the sub-structures are described. The conceptual model provides an overview of the processes that should be considered when developing policy to administrate DBE health. The study reveals that the effects of disruptive events on DBE health depend on robustness while returning to the original level of DBE health depends on organizational resilience and technical resilience. Moreover, collaboration, which is one of the main characteristics of a DBE, improves DBE health by enabling co-innovation and co-creation among DBE members. For future studies, it is recommended to construct a formal model by quantifying relationships among the variables and simulate the model to analyze the behaviors of the key variables.

References

[1] M.S. Gordon, Economic and National Security Effects of Cyber Attacks against Small Business Communities, 2018. ProQuest dissertations and theses (August).

[2] P.W. Parfomak, Vulnerability of concentrated critical infrastructure: background and policy options, Development (2008).

[3] S. Chiaradonna, F. Di Giandomenico, P. Lollini, Evaluation of critical infrastructures: challenges and viable approaches, in: Lecture Notes in Computer Science (Including Subseries Lecture Notes in Artificial Intelligence and Lecture Notes in Bioinformatics), Vol. 5135, Springer, 2008. https://doi.org/10.1007/978-3-540-85571-2_3.

[4] K. Gopalakrishnan, S. Peeta, Sustainable and resilient critical infrastructure systems: simulation, modeling, and intelligent engineering, in: Sustainable and Resilient Critical Infrastructure Systems: Simulation, Modeling, and Intelligent Engineering, 2010. https://doi.org/10.1007/978-3-642-11405-2.

[5] A. Mottahedi, F. Sereshki, M. Ataei, A.N. Qarahasanlou, A. Barabadi, The resilience of critical infrastructure systems: a systematic literature review, Energies 14 (6) (2021). https://doi.org/10.3390/en14061571.

[6] S.A. Argyroudis, S.A. Mitoulis, E. Chatzi, J.W. Baker, I. Brilakis, K. Gkoumas, et al., Digital technologies can enhance climate resilience of critical infrastructure, Clim. Risk Manag. 35 (2022). https://doi.org/10.1016/j.crm.2021.100387.

[7] P. Leviäkangas, Y. Ye, O.A. Olatunji, Sustainable public–private partnerships: balancing the multi-actor ecosystem and societal requirements, Front. Eng. Manag. 5 (3) (2018) 347–356. https://doi.org/10.15302/j-fem-2018020.

[8] P.F. Cleary, A.D. Banasiewicz, Toward resilience of business ecosystems: the internet as a critical infrastructure, Aust. Acad. Account. Finance Rev. 4 (1) (2018) 1–10.

[9] M. Ouyang, Review on modeling and simulation of interdependent critical infrastructure systems, Reliab. Eng. Syst. Saf. 121 (2014). https://doi.org/10.1016/j.ress.2013.06.040.

[10] S. Muegge, D. Craigen, A design science approach to constructing critical infrastructure and communicating cybersecurity risks, Technol. Innov. Manag. Rev. 5 (6) (2015). https://doi.org/10.22215/timreview902.

[11] H. Ruoslahti, Business continuity for critical infrastructure operators, Ann. Disaster Risk Sci. 3 (1) (2020). https://doi.org/10.51381/adrs.v3i1.46.

[12] J.W. Forrester, System dynamics, systems thinking, and soft OR, Syst. Dyn. Rev. (1994). https://doi.org/10.1002/sdr.4260100211.

[13] J.F. Moore, The Death of Competition: Leadership and Strategy in the Age of Business Ecosystems, HarperBusiness, New York, NY, 1996.

[14] P.K. Senyo, K. Liu, J. Effah, Digital business ecosystem: literature review and a framework for future research, Int. J. Inf. Manag. 47 (2019). https://doi.org/10.1016/j.ijinfomgt.2019.01.002.

[15] F. Nachira, Toward a Network of Digital Business Ecosystems Fostering the Local Development, 2002, Retrieved from https://web.archive.org/web/20130903235803/http://digital-ecosystems.org/doc/discussionpaper.pdf.

[16] F. Nachira, P. Dini, A. Nicolai, A Network of Digital Business Ecosystems for Europe: Roots, Processes and Perspectives, European Commission, Information Society and Media (Com), 2007.

[17] A. Nucciarelli, F. Li, K.J. Fernandes, N. Goumagias, I. Cabras, S. Devlin, et al., From value chains to technological platforms: the effects of crowdfunding in the digital game industry, J. Bus. Res. 78 (2017). https://doi.org/10.1016/j.jbusres.2016.12.030.

[18] W. Ben Arfi, L. Hikkerova, Corporate entrepreneurship, product innovation, and knowledge conversion: the role of digital platforms, Small Bus. Econ. 56 (3) (2021). https://doi.org/10.1007/s11187-019-00262-6.

[19] A.V. Gheorghe, M. Schläpfer, Ubiquity of digitalization and risks of interdependent critical infrastructures, in: Conference Proceedings—IEEE International Conference on Systems, Man and Cybernetics, 1, 2006. https://doi.org/10.1109/ICSMC.2006.384447.

[20] M. Iansiti, R. Levien, The Keystone Advantage: What the New Dynamics of Business Ecosystems Mean for Strategy, Innovation, and Sustainability, Harvard Business Press, 2004.

[21] J. Yim, J. Joo, A. Marakhimov, Customer participation in business ecosystems: an integrated approach of system dynamics and fuzzy sets, Int. J. Bus. Syst. Res. 12 (3) (2018). https://doi.org/10.1504/ijbsr.2018.10011349.

[22] T. Carvalhaes, S. Markolf, A. Helmrich, Y. Kim, R. Li, M. Natarajan, et al., COVID-19 as a harbinger of transforming infrastructure resilience, Front. Built Environ. 6 (2020). https://doi.org/10.3389/fbuil.2020.00148.

[23] National Research Council, Disaster resilience: a national imperative, in: Disaster Resilience: A National Imperative, National Academy Press, Washington, DC, 2012.

[24] M. Bruneau, S.E. Chang, R.T. Eguchi, G.C. Lee, T.D. O'Rourke, A.M. Reinhorn, et al., A framework to quantitatively assess and enhance the seismic resilience of communities, Earthquake Spectra 19 (2003). https://doi.org/10.1193/1.1623497.

[25] B. Rød, D. Lange, M. Theocharidou, C. Pursiainen, From risk management to resilience management in critical infrastructure, J. Manag. Eng. 36 (4) (2020). https://doi.org/10.1061/(asce)me.1943-5479.0000795.

[26] L. Labaka, J. Hernantes, J.M. Sarriegi, Resilience framework for critical infrastructures: an empirical study in a nuclear plant, Reliab. Eng. Syst. Saf. 141 (2015). https://doi.org/10.1016/j.ress.2015.03.009.

[27] D. Rehak, P. Senovsky, S. Slivkova, Resilience of critical infrastructure elements and its main factors, Systems 6 (2) (2018). https://doi.org/10.3390/systems6020021.

[28] L. Von Bertalanffy, An outline of general system theory, Br. J. Philos. Sci. (1950). https://doi.org/10.1093/bjps/I.2.134.

[29] D.V. Behl, S. Ferreira, Systems thinking: an analysis of key factors and relationships, Procedia Comput. Sci. (2014). https://doi.org/10.1016/j.procs.2014.09.045.

[30] Y. Barlas, System dynamics: systemic feedback modeling for policy analysis, in: Knowledge for Sustainable Development: An Insight into the Encyclopedia of Life Support Systems, 2002, pp. 1131–1175. Encyclopedia of Life Support Systems (EOLSS).

[31] J.W. Forrester, Industrial dynamics. A major breakthrough for decision makers, Harv. Bus. Rev. 36 (4) (1958) 37–66.

[32] J.D.W. Morecroft, The feedback view of business policy and strategy, Syst. Dyn. Rev. (1985). https://doi.org/10.1002/sdr.4260010103.

[33] J. Sterman, Business Dynamics: Systems Thinking and Modeling for a Complex World, McGraw-Hill Higher Education, 2000.

[34] H. Yaşarcan, Feedback, Delays and Non-linearities in Decision Structures, 2003, (Doctoral dissertation), Retrieved from http://www.ie.boun.edu.tr/~dynamics/wp-content/uploads/2019/publications/theses/yasarcan2003.pdf.

[35] J.W. Forrester, The Beginning of System Dynamics, The System Dynamics Group, 1995. Retrieved from http://static.clexchange.org/ftp/documents/system-dynamics/SD1989-07BeginningofSD.pdf.

[36] M.S. Gary, M. Kunc, J.D.W. Morecroft, S.F. Rockart, System dynamics and strategy, Syst. Dyn. Rev. (2008). https://doi.org/10.1002/sdr.402.

[37] F. Cosenz, G. Noto, Applying system dynamics modelling to strategic management: a literature review, Syst. Res. Behav. Sci. (2016). https://doi.org/10.1002/sres.2386.

[38] P. Graça, L.M. Camarinha-Matos, Evolution of a collaborative business ecosystem in response to performance indicators, in: IFIP Advances in Information and Communication Technology, 506, Springer, 2017. https://doi.org/10.1007/978-3-319-65151-4_55.

[39] V. Vignieri, Crowdsourcing as a mode of open innovation: exploring drivers of success of a multisided platform through system dynamics modelling, Syst. Res. Behav. Sci. 38 (1) (2021). https://doi.org/10.1002/sres.2636.

[40] X. Zhang, J. Yuan, B. Dan, R. Sui, W. Li, Simulation analysis on the evolution mechanism of the multi-value chain network ecosystem, in: Proceedings—2nd International Conference on E-Commerce and Internet Technology, ECIT 2021, 2021. https://doi.org/10.1109/ECIT52743.2021.00085.

[41] Y. Chen, H. Hu, Internet of intelligent things and robot as a service, Simul. Model. Pract. Theory 34 (2013) 159–171. https://doi.org/10.1016/j.simpat.2012.03.006.

[42] S. Ruutu, T. Casey, V. Kotovirta, Development and competition of digital service platforms: a system dynamics approach, Technol. Forecast. Soc. Chang. 117 (2017). https://doi.org/10.1016/j.techfore.2016.12.011.

[43] P. Zhang, E. Zhou, Y. Lei, J. Bian, Technological innovation and value creation of enterprise innovation ecosystem based on system dynamics modeling, Math. Probl. Eng. 2021 (2021). https://doi.org/10.1155/2021/5510346.

[44] J. Wang, J.-Y. Lai, Exploring innovation diffusion of two-sided mobile payment platforms: a system dynamics approach, Technol. Forecast. Soc. Chang. 157 (2020). https://doi.org/10.1016/j.techfore.2020.120088.

[45] J.F. Carías, L. Labaka, J.M. Sarriegi, A. Tapia, J. Hernantes, The dynamics of cyber resilience management, in: Proceedings of the International ISCRAM Conference, 2019-May, 2019.

[46] H. Cho, H. Park, Constructing resilience model of port infrastructure based on system dynamics, Int. J. Saf. Secur. Eng. 7 (3) (2017). https://doi.org/10.2495/SAFE-V7-N3-352-360.

[47] S.V. Croope, S. McNeil, Improving resilience of critical infrastructure systems postdisaster: recovery and mitigation, Transp. Res. Rec. 2234 (2011). https://doi.org/10.3141/2234-01.

[48] P. Trucco, B. Petrenj, S.E. Birkie, Assessing supply chain resilience upon critical infrastructure disruptions: a multilevel simulation modelling approach, in: Supply Chain Risk Management: Advanced Tools, Models, and Developments, Springer, 2018. https://doi.org/10.1007/978-981-10-4106-8_18.

[49] S. Najafi-Tavani, Z. Najafi-Tavani, P. Naudé, P. Oghazi, E. Zeynaloo, How collaborative innovation networks affect new product performance: product innovation capability, process innovation capability, and absorptive capacity, Ind. Mark. Manag. 73 (2018). https://doi.org/10.1016/j.indmarman.2018.02.009.

[50] S. Konsti-Laakso, T. Pihkala, S. Kraus, Facilitating SME innovation capability through business networking, Creat. Innov. Manag. 21 (1) (2012). https://doi.org/10.1111/j.1467-8691.2011.00623.x.

[51] D. Francis, J. Bessant, Targeting innovation and implications for capability development, Technovation 25 (3) (2005). https://doi.org/10.1016/j.technovation.2004.03.004.

[52] R.S. Kaplan, D.P. Norton, The balanced scorecard: measures that drive performance, Harv. Bus. Rev. 83 (2005).

[53] E. Den Hartigh, M. Tol, W. Visscher, The health measurement of a business ecosystem, in: ECCON 2006 Annual Meeting, 2783565 (Secretary 2781150), 2006.

[54] A.R. Berkeley Iii, M. Wallace, NIAC, A framework for establishing critical infrastructure resilience goals: final report and recommendations, in: Final Report and Recommendations by the Council, 2010.

[55] C. Nan, G. Sansavini, A quantitative method for assessing resilience of interdependent infrastructures, Reliab. Eng. Syst. Saf. 157 (2017). https://doi.org/10.1016/j.ress.2016.08.013.

[56] Y. Barlas, Formal aspects of model validity and validation in system dynamics, Syst. Dyn. Rev. 12 (3) (1996) 183–210. https://doi.org/10.1002/(SICI)1099-1727(199623)12:3<183::AID-SDR103>3.0.CO;2-4.

[57] O. Balci, Verification, validation, and testing, in: Handbook of Simulation, John Wiley & Sons, Inc, 2007. https://doi.org/10.1002/9780470172445.ch10.

[58] J.W. Forrester, P.M. Senge, Tests for Building Confidence in System Dynamics Models, The System Dynamics Group, 1980. Retrieved from http://static.clexchange.org/ftp/documents/roadmaps/RM10/D-2926-7.pdf.

[59] G.P. Richardson, A.L. Pugh, Introduction to system dynamics modeling with dynamo, J. Oper. Res. Soc. (1981). https://doi.org/10.1038/sj.jors.2600961.

[60] R.L. Eberlein, D.W. Peterson, Understanding models with Vensim™, Eur. J. Oper. Res. (1992). https://doi.org/10.1016/0377-2217(92)90018-5.

[61] M. Davies, J.K. Musango, A.C. Brent, A systems approach to understanding the effect of Facebook use on the quality of interpersonal communication, Technol. Soc. 44 (2016). https://doi.org/10.1016/j.techsoc.2015.10.003.

CHAPTER 5

Key performance indicators of emergency management systems

Mehmet Akşit[a], Mehmet Arda Eren[b], Hanne Say[b], and Umur Togay Yazar[b]
[a]Department of Computer Science, University of Twente, Enschede, The Netherlands
[b]Department of Engineering, TOBB University of Economics and Technology, Ankara, Turkey

1 Introduction

Emergency management systems are critical systems that are established to steer the necessary resources to minimize the negative effects of disasters, such as earthquakes [1]. Disasters that can negatively influence critical infrastructures such as communication, electricity, gas and water networks, roads, bridges, and centers for task forces such as police departments, fire stations, and hospitals, must be handled with high efficiency. Efficiency in emergency management is also important to restrict the number of casualties.

Emergency services are defined as aid services to deal with accidents and urgent problems [2]. Naturally, there may be many possible ways to plan, establish, and run emergency services [3]. The location of the logistic centers, the number of assigned aid workers, and the availability of task forces and their equipment, all influence the efficiency of emergency management. Moreover, there may be several kinds of alternative strategies for prioritizing and compromising resources.

Most importantly, to decide on the efficiency and effectiveness factors, one needs to agree on criteria to qualify, compare, and select the better-performing alternatives. In the literature, key performance indicators (KPIs) [4,5] are defined as a means of measuring a company's progress toward the desired goals. KPIs are defined for many kinds of businesses and domains. They are generally used as a reference framework for improving the quality of businesses. Unfortunately, to the best of our knowledge, there are no significant publications on the definition and use of KPIs for emergency management. We consider it essential to define rigorous KPIs so that they can be computed based on the measured and/or simulated parameters of the emergency management centers.

The novel contribution of this chapter is twofold. It introduces 14 KPIs and defines them as algebraic specifications. The KPIs are organized into three categories: Throughput, time performance, and supply and demand performance. Second, an intuitive explanation of a selected set of KPIs is given based on four disaster examples and emergency management center configurations.

This chapter is organized as follows. The following section introduces the background work and the research questions. The definition of KPIs is given in Section 3. Section 4 presents examples where a set of KPIs are implemented and intuitively evaluated. Section 5 includes a discussion about the threats to the validity of the presented solutions. A proof-of-concept implementation is presented in Section 6. Finally, Section 7 gives conclusions.

2 Background work and research questions

2.1 Background work

Although the introduced formulas and techniques are general in nature for any disaster management, for illustration purposes, this chapter takes an earthquake as an example of a disaster. An earthquake can be defined as the sudden movement of the Earth's surface [1], which may cause considerable damage in case of high intensity. Depending on the mechanics of the soil, the construction of the buildings, and the intensity of the earthquake, disasters of the kind "collapse," "flood," "fire," and "landslide" can be experienced. Consequently, casualties of persons and animals, and destruction of properties are likely to happen.

To ease the negative effects of disasters and to carry out aid operations efficiently and effectively, emergency control centers and systems are established [2,3]. When an earthquake occurred, emergency control centers receive emergency reports from various kinds of sources such as victimized persons, telephone calls, authority reports, sensors, drones, and satellite images [6,7]. Control centers analyze the characteristics of earthquakes and activate the necessary task forces such as rescue teams, firefighters, security forces, medical teams, and repair teams.

To increase the efficiency of the overall process, emergency management centers may adopt different kinds of computer-aided systems such as sensors and IoT networks, communication systems, decision support systems, and logistic systems [8,9] (Chapter 10).

To determine the efficiency of business operations, KPIs are defined. In the Cambridge dictionary, a KPI is defined as "a way of measuring a company's progress toward the goals it is trying to achieve" [4].

KPIs are defined for a large category of businesses [5,10]. Within the context of emergency management, KPIs are studied mainly from three perspectives: Disaster metrics, emergency preparedness, and emergency handling. Along this line, various studies have been published for humanitarian aid [11], tsunami preparedness [12], accident hazards preparedness [13], hospital emergency handling [14,15], and traffic incident handling [16].

To measure KPIs, some researchers propose simulation-based methods. Discrete event simulation is one of the commonly used methods, especially in virtual time-based simulations [17]. Simulation techniques are applied in a large category of applications [18,19], and various libraries are developed for software programmers [20,21].

Unfortunately, to the best of our knowledge, KPIs for emergency management centers and systems for disasters such as earthquakes have not been studied before.

2.2 Research questions and research method

To define the relevant KPIs for emergency management centers, it is considered important to address the following two research questions:

q1. What are the relevant KPIs for emergency management, and how to mathematically formulate them?

q2. How to design and implement a proof-of-concept realization of a software system to measure KPIs?

As for the research method, the following two approaches are combined:

m1. Algebraic formulation: KPIs are expressed as algebraic formulas.

m2. Emergency generation and discrete event simulation: Emergencies are simulated based on probabilistic distribution functions and discrete event simulation is used for emergency handling processes in which KPIs are measured.

3 Definition of key performance indicators for emergency management

In the UML activity diagram, Fig. 1 depicts a set of typical emergency management procedures which are considered relevant in determining the KPIs. The activity "Occurred" represents the existence of an emergency condition that is reported to the control center through various communication channels such as telephone calls and authority reports. Reporting is represented by the activity "Transferring." The registered emergencies are processed by the activities "Registering" and "Processing," respectively. Depending on the types and size of emergencies, various task forces are activated which are shown as "Task Force 1" to "Task Force N." The activity "Handled" corresponds to the final state of operations. The dashed and dotted vertical lines in the figure correspond to the time points and the activities where certain KPIs are defined and applied, respectively. The actor "Emergency Manager" is responsible to determine which KPIs are to be evaluated.

3.1 Throughput

Throughput is considered one of the fundamental performance indicators of a system. It indicates the number of elements that are processed by the system in a unit of time. The kind of an element depends on the process. For instance, in computer networks, it is a bit of information, in factories, it is the product that is manufactured. In our case, it is an emergency instance that is handled. Throughput can be computed at different levels and for different element kinds.

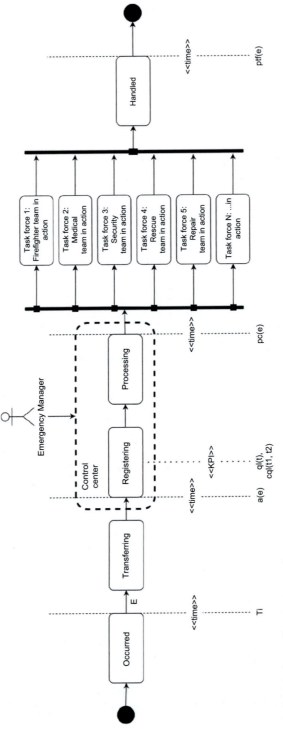

Fig. 1 A typical set of procedures of an emergency management system.

3.1.1 Throughput of the control center

$$tcr(t1, t2) = |\{e \in E \mid t1 \leq pc(e) \leq t2\}| \qquad (1)$$

Here,
- E is the set of emergencies.
- $pc(e)$ is the completion time of the processing by the control center.
- $t1$ is the start of the throughput measurement time.
- $t2$ is the end of the throughput measurement time.

An emergency instance is defined as a septette:

$$e_i = (K_i, I_i, P_i, T_i, L_i, O_i, M_i) \qquad (2)$$

where
- e_i is the ith emergency instance.
- K_i is the kind. K is the set of names {Fire, Collapse, Flood, Landslide, Unhealth}.
- I_i is the intensity. I is the set of names {Low, Medium Low, Medium, Medium High, High}.
- P_i is the kind of physical object. K is a set of names {Factory, Government Building, Historical Building, Hospital, Museum, Residence, Office Building, Religious Building, Restaurant, School, Shopping Center, Shop}.
- T_i is the time when the ith emergency instance occurs.
- L_i are the coordinates of the location.
- O_i is the occupation of the physical object. It is measured by the number of people.
- M_i is the state. M is the set of names {Occurring, Transferring, Queueing, Allocating, Dispatching, Acting, Finalizing}. If necessary, the state Acting can be further decomposed into Firefighter Team Acting, Medical Team Acting, etc.

3.1.2 Throughput of the control center per kind

$$tcrk(t1, t2, k) = |\{e \in E \mid k(e) = k, t1 \leq pc(e) \leq t2\}| \qquad (3)$$

Here,
- E is the set of emergencies.
- $pc(e)$ is the completion time of the processing by the control center.
- $t1$ is the start of the throughput measurement time.
- $t2$ is the end of the throughput measurement time.
- k is the emergency kind concerned.
- $k(e)$ is the kind of emergency e.

3.1.3 Overall throughput

$$to(t1, t2) = |\{e \in E \mid t1 \leq ptf(e) \leq t2\}| \tag{4}$$

Here,
- E is the set of emergencies.
- $ptf(e)$ is the completion time of the processing by the task forces.
- $t1$ is the start of the throughput measurement time.
- $t2$ is the end of the throughput measurement time.

3.1.4 Overall throughput per kind

$$tok(t1, t2, k) = |\{e \in E \mid k(e) = k, t1 \leq ptf(e) \leq t2\}| \tag{5}$$

Here,
- E is the set of emergencies.
- $ptf(e)$ is the completion time of the processing by the task forces.
- $t1$ is the start of the throughput measurement time.
- $t2$ is the end of the throughput measurement time.
- k is the emergency kind concerned.
- $k(e)$ is the kind of emergency e.

3.2 Time performance

Time performance is generally considered an important factor in emergency handling since delays in the processing of emergency conditions may lead to higher damages and casualties. As shown in the following, time performance can be measured from various perspectives.

3.2.1 The average duration of processing by a control center

$$adp = \frac{1}{|E|} \sum_{i=1}^{|E|} pc(e_i) - a(e_i) \tag{6}$$

Here,
- E is the set of emergencies.
- $pc(e)$ is the completion time of the processing by the control center.
- $a(e)$ is the arrival time.

3.2.2 The average duration of intervention

$$adi = \frac{1}{|E|} \sum_{i=1}^{|E|} ptf(e_i) - pc(e_i) \tag{7}$$

Here,
- E is the set of emergencies.
- $ptf(e)$ is the completion time of the processing by the task forces.
- $pc(e)$ is the completion time of the processing by the control center.

3.2.3 The average overall duration
It is the sum of the average duration of processing time and the average duration of intervention time.

$$ado = \frac{1}{|E|} \sum_{i=1}^{|E|} ptf(e_i) - a(e_i) \tag{8}$$

Here,
- E is the set of emergencies.
- $ptf(e)$ is the completion time of the processing by the task forces.
- $a(e)$ is the arrival time.

3.2.4 The average duration of awareness time
It is the average reporting time of emergencies.

$$ada = \frac{1}{|E|} \sum_{i=1}^{|E|} a(e_i) - T_i \tag{9}$$

Here,
- E is the set of emergencies.
- $a(e)$ is the arrival time.
- T_i is the time when the i_{th} emergency instance has occurred.

3.3 Supply and demand performance
Supply and demand models are useful to determine the KPIs for emergency management systems. From this perspective, emergency conditions demand a set of aid operations, which are modeled as a supply. A supply is a set of resources that are necessary for detecting, transferring, and processing emergency conditions, and for dispatching and allocating tasks forces for aid operations. If supply is sufficient for the demand,

emergency conditions can be handled without unnecessary delay. Otherwise, queues are likely to occur for the resources. Depending on supply and demand conditions, queues may emerge for each resource.

3.3.1 Agility factor of control center (bandwidth)
This KPI determines the agility of a control center concerning the change in the number of the emergency conditions.

$$\frac{1}{ada \times ado} \tag{10}$$

Here,
- $1/_{ada}$ is the frequency of emergency arrival.
- ado is the average overall duration.

3.3.2 Queue length

$$ql(t) = \left|\left\{e \in E \mid M(e, t) = \text{``Queueing''}\right\}\right| \tag{11}$$

Here,
- E is the set of emergencies.
- t is the measurement time.
- $M(e, t)$ is the state of the emergency instance e at the time t.

3.3.3 Change of queue length

$$cql(t1, t2) = ql(t2) - ql(t1) \tag{12}$$

Here,
- $t1$ is the start of the queue length measurement time.
- $t2$ is the end of the queue length measurement time.

3.3.4 The absolute emergency complexity factor
Emergency complexity factors give a measure of the complexity of emergency conditions that an emergency management system can handle. An emergency management center that is planned and established for a relatively low-populated area cannot be compared with a center of a large city. The following KPIs can be used to give a better evaluation on the supply and demand basis.

$$aecf(t1, t2) = \sum_{k \in K} w_k(d_k - r_k), \text{ where } \sum_{k \in K} w_k = 1 \tag{13}$$

Here,
- w_k is the weighted factor of the kind k.

- r_k is the total number of available resources for the kind k.
- d_k is the total number of demands for the kind k.

3.3.5 The relative emergency complexity factor

A relative emergency complexity factor is a Sigmoid function. If supply is much higher, equal, or much less than the demand, this function approaches to zero, equals to 0.5, and approaches to 1, respectively.

$$recf(t1, t2) = \frac{1}{1 + e^{-aecf(t1,t2)}} \tag{14}$$

Here,

- e is the natural number.
- $t1$ is the start of the complexity measurement time.
- $t2$ is the end of the complexity measurement time.

4 Examples

To intuitively explain the KPIs presented in the previous section, we will present four examples.

1. **Medium intensity earthquake—low in resources**: There is an emergency management center processing 200 emergency conditions with a low number of resources, 10 for processing the demands and 20 for the task forces in their aid operations. The computed KPIs are shown in Fig. 2.
2. **High intensity earthquake—low in resources**: There is an emergency management center managing 400 emergency conditions with the same number of resources as Example 1. The computed KPIs are shown in Fig. 3.
3. **Medium intensity earthquake—medium in resources**: There is an emergency management center processing 200 emergency conditions with a relatively higher number of resources than the previous example, 20 for processing the demands and 40 for the task forces in the aid operations. The computed KPIs are shown in Fig. 4.
4. **High intensity earthquake—medium in resources**: There is an emergency management center processing 400 emergency conditions with the same number of resources as Example 3. The computed KPIs are shown in Fig. 5.

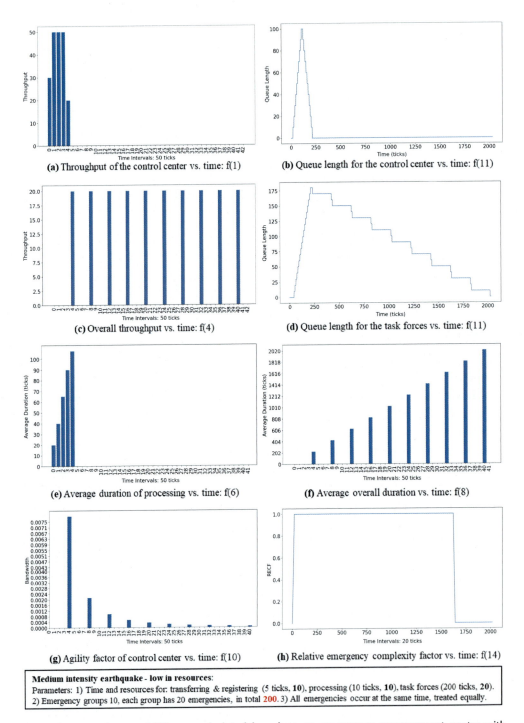

(a) Throughput of the control center vs. time: f(1)

(b) Queue length for the control center vs. time: f(11)

(c) Overall throughput vs. time: f(4)

(d) Queue length for the task forces vs. time: f(11)

(e) Average duration of processing vs. time: f(6)

(f) Average overall duration vs. time: f(8)

(g) Agility factor of control center vs. time: f(10)

(h) Relative emergency complexity factor vs. time: f(14)

Medium intensity earthquake - low in resources:
Parameters: 1) Time and resources for: transferring & registering (5 ticks, **10**), processing (10 ticks, **10**), task forces (200 ticks, **20**).
2) Emergency groups 10, each group has 20 emergencies, in total **200**. 3) All emergencies occur at the same time, treated equally.

Fig. 2 An example set of KPIs are calculated based on an emergency management center with relatively low available resources (10–20) and managing 200 emergency conditions.

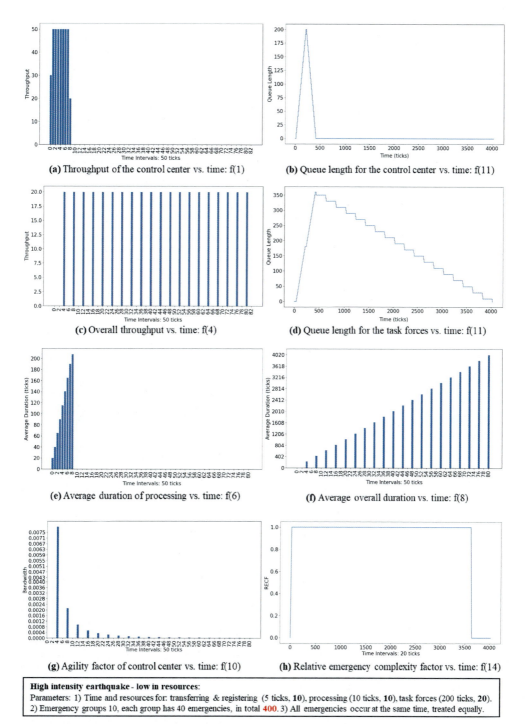

(a) Throughput of the control center vs. time: f(1)

(b) Queue length for the control center vs. time: f(11)

(c) Overall throughput vs. time: f(4)

(d) Queue length for the task forces vs. time: f(11)

(e) Average duration of processing vs. time: f(6)

(f) Average overall duration vs. time: f(8)

(g) Agility factor of control center vs. time: f(10)

(h) Relative emergency complexity factor vs. time: f(14)

High intensity earthquake - low in resources:
Parameters: 1) Time and resources for: transferring & registering (5 ticks, **10**), processing (10 ticks, **10**), task forces (200 ticks, **20**).
2) Emergency groups 10, each group has 40 emergencies, in total **400**. 3) All emergencies occur at the same time, treated equally.

Fig. 3 An example set of KPIs are calculated based on an emergency management center with relatively low available resources (10–20) and managing 400 emergency conditions.

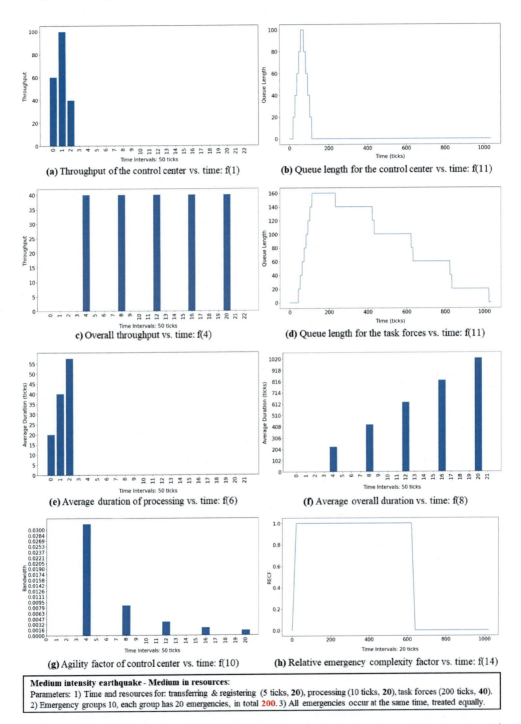

(a) Throughput of the control center vs. time: f(1)

(b) Queue length for the control center vs. time: f(11)

c) Overall throughput vs. time: f(4)

(d) Queue length for the task forces vs. time: f(11)

(e) Average duration of processing vs. time: f(6)

(f) Average overall duration vs. time: f(8)

(g) Agility factor of control center vs. time: f(10)

(h) Relative emergency complexity factor vs. time: f(14)

Medium intensity earthquake - Medium in resources:
Parameters: 1) Time and resources for: transferring & registering (5 ticks, **20**), processing (10 ticks, **20**), task forces (200 ticks, **40**). 2) Emergency groups 10, each group has 20 emergencies, in total **200**. 3) All emergencies occur at the same time, treated equally.

Fig. 4 An example set of KPIs are calculated based on an emergency management center with relatively high available resources (20–40) and managing 200 emergency conditions.

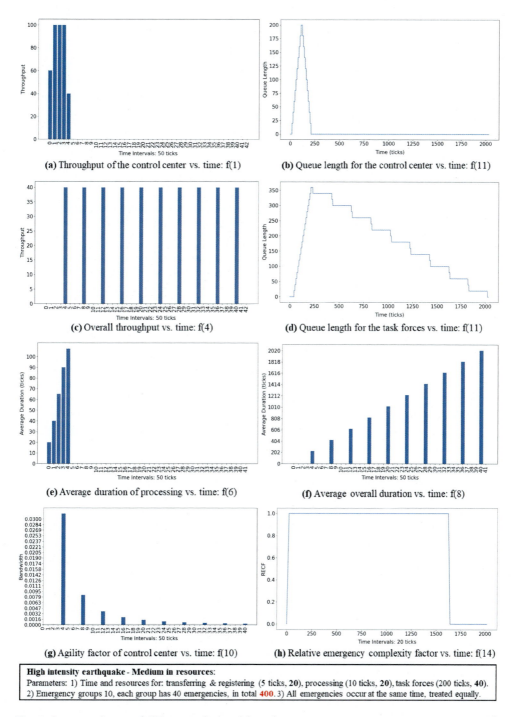

(a) Throughput of the control center vs. time: f(1)

(b) Queue length for the control center vs. time: f(11)

(c) Overall throughput vs. time: f(4)

(d) Queue length for the task forces vs. time: f(11)

(e) Average duration of processing vs. time: f(6)

(f) Average overall duration vs. time: f(8)

(g) Agility factor of control center vs. time: f(10)

(h) Relative emergency complexity factor vs. time: f(14)

High intensity earthquake - Medium in resources:
Parameters: 1) Time and resources for: transferring & registering (5 ticks, **20**), processing (10 ticks, **20**), task forces (200 ticks, **40**).
2) Emergency groups 10, each group has 40 emergencies, in total **400**. 3) All emergencies occur at the same time, treated equally.

Fig. 5 An example set of KPIs are calculated based on an emergency management center with relatively high available resources (20–40) and managing 400 emergency conditions.

Example 1. Medium intensity earthquake—low in resources

In Fig. 2A, the throughput of the control center is depicted against time which is calculated by the formula (1). The throughput value increases to 50 with the availability of resources, and when all the emergency conditions are dispatched to the desired task forces, it naturally drops to zero. Here, the X-axis shows the simulation time.

Fig. 2B shows the emergency conditions which are queued for processing by the authorities in the control center. When the authorities decide on the way how the emergency conditions must be handled, the orders are prepared and dispatched to the task forces. For simplicity, we assume that each emergency condition requires the same number of resources, and each task force takes the same amount of time for the aid operations.

As shown by Fig. 2C, this results in a throughput performance with a periodic character; when the first group of tasks is delivered to the task forces, the throughput value reaches to its peak, but soon the available resources are exhausted, and the throughput value drops to zero until a new set of resources become available again. This process continues until all the emergency conditions are handled.

Fig. 2D depicts that at first, the queue length of tasks waiting for the available tasks forces reaches its peak value since the task forces demand more time to complete than the processing time by the control center. The queue length gradually drops to zero, when the task forces pick up all the queued tasks. This plot shows a staircase like behavior since it is assumed that each emergency condition requires the same number of resources, and each task force takes the same amount of time for the aid operations.

Fig. 2E shows the calculation of the processing time by the control center. Since the number of emergency conditions are much higher than the available resources for processing, the emergency conditions which are handled at a later stage, must wait for a longer period for completion. This graph linearly increases because it is assumed that each emergency condition demands the same number of resources.

The overall completion time is shown in Fig. 2F. Similar to Fig. 2D, this graph shows a periodic character, with a difference in that the completion time naturally increases for the lately handled emergency conditions. The increase in value is linear since by assumption, each emergency condition demands the same number of task forces.

Fig. 2G plots the agility factor (a band with) of the control center. At first, the agility factor is high because the control center can respond to the increase in emergency conditions rapidly due to the availability of resources. When all the emergency conditions are received and the available resources are exhausted, the agility factor gradually drops to zero.

Since the number of emergency conditions are much higher than the available resources, in Fig. 2H, the relative emergency complexity factor approaches to one immediately at the beginning, meaning that the experienced disaster is much more complex than the capabilities of the emergency management center. Here, it is assumed that all disaster kinds have an equal weighting factor.

Example 2. High intensity earthquake—low in resources

The difference between this example and the previous one is that the number of emergency conditions are doubled to 400, while the number of resources remain the same. We compare this example with the previous case.

Fig. 3A shows that although the number of emergency conditions waiting to be handled are increased, the resources remained the same, and therefore the throughput value remained at 50.

Due to the doubling of emergency conditions, the queue length shown in Fig. 3B is also doubled.

In Fig. 3C, task forces must take more rounds to handle all the emergency conditions. The overall throughput value remains the same.

Fig. 3D shows a staircase function, where more steps are necessary to reduce the queue toward zero.

As shown in Fig. 3E and F, both the average duration of processing time and the overall time are worsened.

The plot of the agility factor in Fig. 3G shows that at first the value is the same but drops to zero in a steeper path.

Due to the increased number of emergency conditions, in Fig. 3H, the relative emergency complexity factor has a slightly higher rate of rise than the previous example.

Example 3. Medium intensity earthquake—medium in resources

The difference between this example and the previous two is that the number of resources is doubled. The number of emergency conditions is 200 as in the first example.

Compared to the previous two cases, Fig. 4A and B show that due to an increase in resources, the throughput of the control center is increased, and the queue length is reduced to zero in a shorter time. The same observation can be made also in Fig. 4C and D.

The processing time and overall processing time decrease as well (Fig. 4E and F).

Fig. 4G shows that the agility factor of the control center is relatively higher than in the previous cases.

Due to the increased number of resources, in Fig. 3H, the relative emergency complexity factor has a slightly lower rate of rise than the previous example.

Example 4. High intensity earthquake—medium in resources

The difference between this example and the previous one is that the number of emergency conditions is doubled to 400, while the available resources remain the same as in the previous example. As expected, Fig. 5 shows that all KPIs are worsened concerning the third example due to the increase in emergency conditions.

5 Threats to validity

Concerning the research questions as presented in this article, we observe the following threats to the validity:

1. **KPIs are not validated with realistic test cases**: It is considered important to conceptualize the meaning and importance of the presented KPIs. There is a need for examples that are simple and where intuitive explanation of KPIs is possible. To this aim, for example, it is assumed that each emergency condition demands the same amount of task forces, and all task forces take the same amount of time to complete their duties. As a result, KPI plots are largely structural in character. If the task forces would complete their duties at random times, for example, it would be much more difficult to give an intuitive explanation of these figures. The novelty of this chapter is to introduce a new set of KPIs for emergency management centers but not the application of KPIs for realistic case studies.

2. **Parameters are difficult to measure in realistic cases**: One may claim that the parameters that are used in the formulas are difficult to obtain. With the use of modern monitoring and tracking systems, most of the parameters can be measured online and registered on a server. The average duration of awareness time (formula 9) here is a special case, where the exact time of the occurrence of the earthquake can be taken from the seismic monitoring centers.

3. **Parameters values may contain errors**: In case data about the emergency conditions are informally gathered, information about the coordinates, the intensity of the disaster, the number of casualties, the impact on the location, etc., may contain errors. These errors may be reduced by using multiple information sources, but the correction process can take some time and may not eliminate the errors completely. One way to minimize such errors is to adopt an IoT-based information system infrastructure to gather data about emergency conditions more accurately (Chapter 10).

4. **Irrelevance of the KPIs**: One may question the relevance of the selected KPIs. The KPIs defined in this chapter are consistent with the KPIs defined within other application domains [5,10–16].

5. **The KPIs definitions are incomplete**: Of course, new KPIs can be introduced when necessary. For example, the degree of lesser errors in the detection of emergency conditions can be statistically formulated as a KPI. Measure such a KPI parameter online, however, can be difficult.

6 A proof-of-concept implementation

To understand the intuition behind the KPIs and to validate their utility, a software system is developed in Python consisting of the following parts:

1. **Emergency condition generator**: The main purpose of this part is to feed the simulator component with emergency instances, and it is implemented as a Python generator. Two different strategies are defined for generating emergencies: Parametric strategy and fixed strategy. Various perspectives of the emergency creation process such as randomized initialization of emergency attributes are parametrized in parametric strategy. In fixed strategy, all the emergencies and their attributes are prepared beforehand and are passed to the next component before the simulation starts. Fixed strategy, in a sense, can be considered a scenario writing tool. The latter strategy used during the experiment.

2. **The simulator**: This component is built upon SimPy [21], a discrete event simulation framework. SimPy works with Python generators and regards them as processes. Time consuming tasks are considered as processes and each process requires a resource to be executed. These resources should be shared across incoming requests. In the proof-of-concept implementation, there are four different tasks regarded as processes: Transferring, Registering, Processing, and Dispatching. Incoming emergencies pass through the processes in the specified order. For each of these processes, several resources are assigned to handle related tasks in their dedicated time interval.

3. **Evaluator**: This component contains the calculations to determine the predefined KPIs. Matplotlib library is used to plot the KPI values. Intervals for each KPI are listed below:

    Throughputs: 50 ticks.

    Average duration and agility factor: 50 ticks.

    Relative emergency complexity factor: 20 ticks.

7 Conclusions

This chapter aims at defining and formalizing the relevant KPIs for emergency management. In Chapter 3, 14 KPIs are introduced and described for this purpose. In Section 4, a selected set of KPIs are implemented, tested, and the results are intuitively explained. In addition, a list of threats to the validity of assumptions are discussed. We conclude that the research questions as presented in this chapter are adequately addressed.

Acknowledgment

We appreciate the contribution of TOBB ETU MSc. Student S. C. Altunsoy in the formulation of KPIs. We thank the Disaster and Emergency Presidency and, in particular, Nihan Karacameydan İncemehmetoğlu for her firm support. This work has been generously supported by the TÜBİTAK BİDEB program 2232 International Fellowship for Outstanding Researchers.

References

[1] Cambridge Dictionary: "Earthquake", Referred to on December 28, 2022.

[2] M.J. Fagel, R.C. Mathews, J.H. Murphy (Eds.), Principles of Emergency Management and Emergency Operations Centers (EOC), CRC Press, 2021.

[3] M. Ryan, Planning in the emergency operations center, in: Technological Forecasting and Social Change, 2013, pp. 1725–1731. 80(9).

[4] Cambridge Dictionary: "Key Performance Indicator", Referred to on December 28, 2022.

[5] A. Coppola, et al., An integrated simulation environment to test the effectiveness of GLOSA services under different working conditions, Transp. Res. Part C Emerg. Technol. 134 (2022) 1–20.

[6] S.M.S.M. Daud, et al., Applications of drone in disaster management: a scoping review, Sci. Justice 62 (1) (2022) 30–42.

[7] K. Kaku, Satellite remote sensing for disaster management support: a holistic and staged approach based on case studies in sentinel Asia, Int. J. Disaster Risk Reduct. 33 (2019) 417–432.

[8] M. Akşit, A system architecture for IoT-based earthquake management, in: S. Baş, et al. (Eds.), Workshop Proceedings on Software and Earthquake Engineering in Seismic Risk Management, Istanbul Technical University, 2022, pp. 14–17.

[9] P. Toth, D. Vigo (Eds.), Vehicle Routing Problems, Methods, and Applications, second ed., Society for Industrial and Applied Mathematics, 2014.

[10] T. Kasaeva, et al., Business performance assessment simulation of enterprises engaged in production textiles, clothing, AIP Conf. Proc. 2430 (1) (2022) 1–6.

[11] D.F. Wong, et al., Disaster metrics: a comprehensive framework for disaster evaluation typologies, Prehosp. Disaster Med. 32 (5) (2017) 501–514.

[12] R. Patrisina, et al., Key performance indicators of disaster preparedness: a case study of a tsunami disaster, in: MATEC Web of Conferences, vol. 229, EDP Sciences, 2018.

[13] J. Larken, et al., Performance Indicators for the Assessment of Emergency Preparedness in Major Accident Hazards, Great Britain, Health and Safety Executive, 2001.

[14] M. Khalifa, I. Zabani, Developing emergency room key performance indicators: what to measure and why should we measure it? ICIMTH 226 (2016) 179–182.

[15] N. Li, et al., Optimising key performance indicator adherence with application to emergency department congestion, Eur. J. Oper. Res. 272 (1) (2019) 313–323.

[16] M. Won, Outlier analysis to improve the performance of an incident duration estimation and incident management system, Transp. Res. Rec. 2674 (5) (2020) 486–497.

[17] G.S. Fishman, Discrete-Event Simulation: Modeling, Programming, and Analysis, vol. 537, Springer, New York, 2001.

[18] A. Atalan, A cost analysis with the discrete-event simulation application in nurse and doctor employment management, J. Nurs. Manag. 30 (2022) 733–741.

[19] F. Peña-Graf, et al., Discrete event simulation for machine-learning enabled mine production control with application to gold processing, Metals 12 (2) (2022) 225.

[20] https://simpy.readthedocs.io.

[21] A. Varga, OMNeT++, in: Modeling and Tools for Network Simulation, Springer, Berlin, Heidelberg, 2010, pp. 35–59.

CHAPTER 6

Responding cyber-attacks and managing cyber security crises in critical infrastructures: A sociotechnical perspective

Salih Bıçakcı and Ayhan Gücüyener Evren
Department of International Relations, Kadir Has University, Istanbul, Türkiye

1 Introduction

Our modern societies and urban life routines heavily rely on continuous, systematic, and proper operation of the Critical Infrastructures (CIs). We do not consider the presence of such services in our daily lives until their services are disrupted. Electricity is the primary source of energy for our several devices and controls. Several telecommunication systems are not visible as services, but they are essential for the communication of various platforms and components in our daily lives. The introduction of digitalization and hyper-connectivity have altered CI management systems. The use of Industrial (Digital) Control Systems (ICS) simplified CI functionality and control, but this paradigm also resulted in enormous interdependency and complexity among various CI systems (Fig. 1). Industry 4.0 has already introduced the Industrial Internet of Things (IIoT), which changed our conventional approaches to network management in ICS. Machine-to-machine communication complicates our cyber security playbooks and configurations even further.

Indeed, the content of the CIs varies by country, and its criticality is determined by the country. The primary concern is the function of CIs. By definition, CIs serve the public and ensure that the state appropriately functions to maintain its role. Any interruption in CI services jeopardizes the state's authority and capacity. This correlation is also associated with the idea that CIs are a country's most vulnerable points. To this end, any disorder in CI systems might paralyze several interdependent sectors' daily routines and functionality. The massive blackout on August 14, 2003 was one of the well-known examples to understand the possible outcomes of such crises. The outage affected eight Northeast states in the United States and approximately 50 million people. During the blackout, the traffic lights stopped working, and passengers were trapped in subway

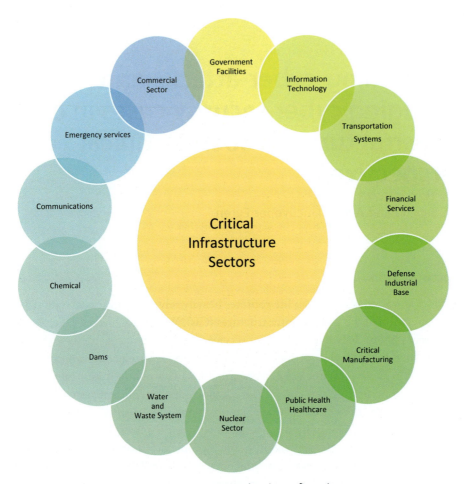

Fig. 1 Examples of critical infrastructure sectors. *(Authors' own figure.)*

cars. The crowds took the Brooklyn Bridge to commute home on foot. The 2003 massive blackout demonstrates the catastrophic consequences of such crises in the literature.

The economic and social criticality of the CIs makes them a soft spot and strategic target in case of cyber and terrorist attacks. Recently, the digitalization of the CIs also presented opportunities to the cyber threat actors, generating new attack surfaces, which raised the number of cyber-attacks. While cyber-attacks targeting CIs are on the rise, most existing studies are devoted to developing complex mathematical/engineering models to comprehend the limits of machinery and relevant infrastructures that affect CI's functionality. However, all CIs comprise several components and machinery encompassing multiple layers, including the physical, digital, and human layers. The physical layer comprises the production sublayer, machinery, PCs, servers, sensors, generators, and hardware. Network, software, and SCADA/EMS systems are in the digital layer. Above all, the human layer, including technicians, engineers, and management, are in attendance and are not just passive users [1]. However, they shape and define the overall CI ecosystem and interconnections among different layers. While cyber-attacks can affect all these layers, attackers increasingly use semantic attacks that mislead the operators and abuse the human-machine interaction interface.

The digitalization of CI automation systems speeds up and simplifies production, but the process builds up several dependencies within a production line. The introduction Industrial of the Internet of Things (IIoT) would also be increasing in the CIs as a new component. The increasing interdependence and interaction between various CIs complicate their management. Currently, the CIs are operated by public and private operators. This significant division also requires detailed planning to coordinate various CI operators critical to each other's operationality. The private sector owns approximately 85% of the CIs in the market. However, these private companies must follow the regulatory agencies and coordinate with public services. In addition to national-level cooperation, some CI categories, such as Eurocontrol (the European Aviation Organization), Galileo (the European Global Satellite Navigation System), the European Electricity Transmission Grid network, and the European Gas Transmission Network, require regional and global level coordination. Concerning this high level of interdependency, the tremendously evolving complexity and deepening uncertainty endanger the management and functionality of CIs.

The increasing number of actors, both in the public and private sector, to handle the crisis also entangles the process and jeopardizes information collection and coordination phases. Since the CIs are critical for national security, any crisis in these facilities ends with the involvement of several actors. During the Senate Homeland Security Hearing of Colonial Pipeline, CEO of the Colonial Pipeline, Joseph Blount, underlined that his management team worked closely with different institutions, including the FBI, Department of Energy, and Department of Homeland Security, during the ransomware crisis [2]. Any cyber-attack or cyber security crisis tests the limits of these costly systems' technical capacity and leadership, organizational structure, communication, and management. These critical assets are technical and involve complex Sociotechnical Systems (STSs), where people increasingly interact with technology and represent a single organism to achieve a specific purpose. In that sense, it is crucial to highlight those human operators running CIs are affected by physical conditions and psychological factors.

To sum up, responding to cyber-attacks and crises from a sociotechnical perspective is one of the critical questions waiting for us on the table. While the management of CIs should address evolving technological, economic, societal, and cultural needs, they are not one-size-fits-all systems. Furthermore, their operating model's paradigm shift makes them more vulnerable to man-made (that is, cyber-attacks) and natural threats (that is, climate change). CIs are also operated under an inherent and deep uncertainty encompassing various factors, including unpredictable threats/hazards and ambiguous/

insufficient data. These factors complicate robust planning and management. The rising embeddedness of digital technologies in CIs introduces additional and uninvited uncertainties in managing CIs since cyberspace is subjected to embracing ambiguities related to the source, scope, and consequences of a potential cyber incident.

Furthermore, STSs are "too complex to fully imagine, describe, understand, or analyze within human cognitive abilities" [3]. This approach advocates that failures and flaws in the functioning of CIs can be rooted in the organizational and individual levels and not necessarily located at the engineering or design level. Yet, as Malatji, Von Solms, and Marnewick also emphasize, "the application of sociotechnical systems theory to information and cyber security domain has not received much attention" [4], p. 233. In other words, social perspectives are missing, which can improve the cyber security crisis management perspective.

This chapter aims to "call the social back" and develop a "human-centric" perspective for responding to cyber-attacks and managing cyber security crises for CIs. To this end, the chapter is structured as follows. First, we aimed to conceptualize CIs as STSs, identify their principal characteristics, and elaborate on why these properties are significant for cyber security. Second, we aim to revisit the nature of the crisis and investigate what makes cyber security crises more complex. Third, while this study takes humans as the essential components of managing CIs, we will investigate how cognitive factors and uncertainty surrounding complex CI systems can be problematic in cyber security crisis management. Also, in terms of organizational level, we will appraise the role of organizational culture for successful cyber security crisis management in complex critical assets.

2 Critical infrastructures as complex sociotechnical systems

Any nation's wealth, prosperity, and continuity rely on the uninterrupted and secure working of CIs. Although Critical Infrastructure Protection (CIP) is regarded as a modern concept, concerns regarding the security of these infrastructures can be traced back to ancient Rome, where road systems and food stores were considered "indispensable" and kept "hidden to protect from external threats" [5], p. 765. Notwithstanding, the modern world's CIs are much more complex than their historical twins. Industry 4.0 creates a different landscape about how the social and technical interact and co-exist [6]. Even though infrastructure is designed simply, its evolution toward complexity is unavoidable since technology, user requirements, and consumption styles constantly change [7].

Modern electrical systems, railroads, air transportation, and water management facilities demonstrate a complex landscape where various agents act simultaneously, different components of social, environmental, and technological systems interact, and an action-reaction chain is present. Above all, a distributed/decentralized control takes place. In addition to these, Setola et al. suggest six main factors that increase CI-related risks, including (i) liberalization and privatization of CIs, (ii) increased use of Information Communication Technologies (ICT), (iii) 24/7 working principle, (iv) urbanization and aging infrastructures, (v) complex supply chain relations and interdependencies and, and (vi) recognition of catastrophic consequences of a successful attack by adversaries [8].

CIs and CIP are the products of complex engineering processes and techniques, including probabilistic risk assessments, hazard modeling, and network visualization. Nevertheless, besides the physicality and materiality of these assets, it shall be highlighted that CIP is "ultimately about social/political action and human life" [9]. Moreover, even though the CI industry and operators quickly embrace new materials and technologies, current practices see "social elements as contextual" [10], p. 133, which demonstrates the necessity to develop a novel approach emphasizing CIs; as "sociotechnical systems." In other words, the prevailing mindset that puts "technology is independent" and "mental and social conditions of human work had to follow the given technical structures" [11] shall be revised.

The sociotechnical design thinking can be traced back to the end of the 1940s when Tavistock Institute initiated several projects for the British Coal Industry to enhance work performance by focusing on the behaviors of human actors. More specifically, as Pasmore et al. mention, the claim of STS design is empowering workers so that they can handle "technological uncertainty, variation, and adaptation." Compared to the 1950s, the current gap between the rapid pace of technology and human capabilities has become so massive that STS-focused thinking is much more relevant [12]. While it is open to generate various definitions, the STSs can be generally defined as "networks of people and technological components which interacts in the course of usage or deployment of designed processes or products" and "cluster of elements including technology, regulation, user practices, and markets" [13] or "systems that involve the collaboration of a decision-maker with an autonomous system to fulfill their objectives" [14].

Despite the variation in definitions, any STS puts forward complex human and technology interaction and interdependency. From this dimension, CIs can be considered unique examples to elaborate on STS in which people have different roles and involvement in various ways. To better grasp, Ottens, Franssen, and Kroes's categorization are also relevant as they suggest three different types of systems. While the first category of methods (that is, landing gear) can function without actors of social institutions, the second category (that is, airplanes) needs the involvement of actors but not social institutions. In contrast, the third category, which can also be considered in line with CIs, is more hybrid (that is, civil aviation system). They need actors and social institutions to function appropriately [10].

Such unique orchestration between technical and social components also presents itself in subsystems in an STS proposed by Kelly as "goals and values subsystem, technical subsystem, psychosocial subsystem, structural subsystem, and managerial subsystem" [15], p. 6. According to this categorization, while technical subsystems within an STS encompasses the equipment and facilities, psychosocial subsystems include human resources, leadership, attitude, and perception of employees. Besides, each subsystem has dynamic interactions and inevitable intersections with other subsystems and the organization. Baxter and Sommerville mention five key elements that STS (i) should have including interdependent parts, (ii) should pursue goals in external environments, (iii) should have an internal environment comprising separate but interdependent technical and social subsystems, (iv) system goals can be achieved by more than one means, and (v) system performance relies on the joint optimization of technical and social subsystems [16]. Saurin and Gonzalez suggest similar characteristics for these systems, such as several interacting elements with diversity and unanticipated variability strengthened by uncertainty and openness [17].

For these characteristics, CIs can be considered unique examples of STS. Because most of the CIs are combinations of technical and nontechnical functionalities containing specific systemic and spatial features. For instance, when natural gas transportation is considered, we can identify its technical components, such as compressor stations, storage facilities, and city gate stations [18]. Besides these visible components, as Feleke discusses, such systems also have less visible facets, such as managing/maintaining human staff organizational processes [19]. In a natural gas transportation company, besides the technical assets on the ground, the maintenance staff in the field and the SCADA System Operator in the Control Room are the other vital parts of operations. To this end, such dynamicity allows us to elaborate main characteristics of an STS, which creates a worthwhile background to discuss the potential deadlocks in managing cyber security crises. For instance, the real problem is that these components are inherently interdependent, making cyber security management much more problematic in CIs. The Colonial Pipeline ransomware attack is one of the most remarkable cases demonstrating the severe consequences of interdependence. In the Colonial case, the company had to shut down its entire network due to a ransomware cyberattack. Hackers broke into the company's networks via a leaked and compromised password of an employee from another account [20]. In other words, a flaw in the human component affected different facets of the entire system and its functionality. In this chapter, we identify the three characteristics of CIs relevant to elucidate a sociotechnical approach to managing a cyber crisis in CIs: *Complexity, Interdependency/Interconnectedness, and Uncertainty*. While each of these characteristics has merits, it is critical to elaborate on their relevance in developing a sociotechnical cyber security crisis management perspective.

2.1 Complexity

Although "complexity" is a widely referred to term in identifying characteristics of CIs, as Van Der Lei, Bekebrede, and Nikolic discuss, "lack of shared view on the object of the study or the meaning of the word" is problematic, and there is no shared perspective in understanding the meaning of complexity [21], p. 380. Complexity is a trans-disciplinary concern. Renn sees complexity as the "difficulty of identifying and quantifying the causal links between the multitude of potential causal agents and specific observed effects" [22], p. 75. Schöttl and Lindemann elaborate that while system theory sees complexity as a system property, psychology handles this term as a perceived phenomenon [23]. Page, who assumes a "meaningful part of sociology overlaps with a substantial subset of complexity sciences," sees complexity as a "phenomenological property" [24], p. 24.

To identify these types of systems' features, Page's measurement metrics can be helpful: "Complex systems include diverse entities, these entities interact within an interaction structure, individual behaviors are interdependent and influence each other (and the whole network), the entities (should) adapt and learn [24], pp. 23–25." Besides, *emergent behavior* (the behavior of the system that the behavior of individual agents cannot explain) and *self-organization* (a different overall output different than the total of the internal process) are the other relevant metrics [21]. When we add "*openness*" and "*dynamicity*" to this calculation, we might arrive at seven distinguishing features of complex systems proposed by Gilpin and Murphy [25], p. 25

 (i) Complex systems are composed of individual elements/agents,
 (ii) Agents' interactions alter the system over time,
 (iii) Complex systems are self-organizing,
 (iv) Complex systems are unstable,
 (v) Complex systems are dynamic, with their history an essential feature,
 (vi) Complex systems have permeable and ill-defined boundaries,
 (vii) Complex systems are irreducible.

Oughton et al. investigate how complex systems' characteristics might apply to the (critical) infrastructure systems [26]. While nonprecise and nonisolated boundaries of national infrastructures create a high degree of openness, distributed control is significant peculiarity as there might be a nonlinear hierarchy in the overall management. If CIs are appraised as complex systems, each of these features can potentially create significant reflections in the cyber security crisis management scene, as complexity is one of the system-level properties of CIs. Furthermore, treating these deadlocks stemming from complexity as exclusively technical problems would not facilitate robust cyber security crisis management. For instance, at the internal level (agent level), since a cyber security crisis management team may be composed of various people from different backgrounds and branches, ensuring continuous information flow and a coherent flow of action among varying levels of operation are problematic. At the external level (network level), when a private natural gas company's operations are interrupted due to a cyber security crisis, the crisis management team will face difficulties in simultaneously responding to a "high level of requests from customers, business partners, customers, business partners, vendors, regulators, law enforcement, and the board of directors" [27]. Aside from these complicated interactions among various agents, the cyber security crisis in a CI may not be contained at a single state's level due to the interconnectedness of CIs. As complex systems have ill-defined boundaries, a single vulnerability might have an escalatory potential [28]. Distributed control and management complexity might necessitate the orchestration of different actors' involvement and operationalization of various legal and regulatory frameworks. Moreover, openness makes any cyber security crisis's impact unpredictable.

2.2 Interdependency/interconnectedness

As Rinaldi, Peerenboom, and Kelly elaborate, while dependency shows a "linkage between two infrastructures," interdependency represents the "connections among agents in different infrastructures in a general system of systems." Moreover, interdependency CIs and complex adaptive systems are not monolithic and can present themselves in physical, cyber, geographic, and logical terms [29]. From these aspects, interdependence in CIs can be treated as an essential component in forming the complexity of systems and "risk multipliers" [30]. Nevertheless, it is worth re-investigating when it comes to cyber security crisis management and decision-making since it reduces the predictability, measurability, and ability to control. For instance, Gomez and Whyte highlight that "interconnectedness increases overall complexity that limits our ability to predict points of failure and its corresponding consequences" [31], p. 100.

As Clemente claims, the interdependencies are unlikely to curtail soon if there are sufficient political, economic, and social benefits [32]. Any failure or flow in a single node of a CI can cause second or third-order effects and complicate interdependency. For instance, even though it is unrelated to cyber security, the energy crisis that hit Turkey at the beginning of January 2022 can demonstrate how these effects might operationalize since natural gas flow from Iran to Turkey was halted due to a technical problem. This interruption ended in a government intervention to cut the power on heavy industrial users for 3 days and an inherent economic loss [33]. In addition to the geographic and physical interdependencies, the increasing role of ICT expands cyber interdependencies between different CI systems. Such interdependency is problematic since one of the high-priority tasks of a cyber security crisis management team is to contain the incident. Computer networks create inherent cyber interdependencies. To this end, any malicious activity can spread beyond a single system. Such contingency is an essential obstacle to controlling crisis management tasks. Finally, different types of interdependence (physical, cyber, or geographical) complicate the crisis' peak periods and shall be considered in pre-crisis times since the parties must calculate and map interdependencies between various infrastructures to prevent the domino effect [34].

2.3 Uncertainty

Although the two concepts are sometimes used interchangeably, complexity and uncertainty are two different phenomena. As Renn emphasizes, "uncertainty is different from complexity, but most often results from an incomplete or inadequate reduction of complexity in modeling cause-effect chains" [22], p. 75. Haase sees complexity as a "characteristic of a system (or subsystems)," while he associates uncertainty with a specific state of knowledge [35], p. 6027. Uncertainty is present in every aspect of daily life, complex technical systems, and science. Like complexity, it is a trans-disciplinary term to be investigated. Uncertainty is not a monolithic concept and can be elaborated into different types. For instance, according to Li et al. [36], based on their origins, uncertainty can be classified into two main categories: Aleatory and epistemic uncertainty. For this categorization, aleatory uncertainty is less human-centric and more related to "natural variabilities of the physical world," which reflects "inherent randomness of future" like tossing a coin or earthquakes. In contrast, epistemic uncertainty originates "from human's lack of knowledge of the physical world and lack of ability to measure and model the physical world" [36], p. 2463. Regarding CIs and cyber security crisis management, uncertainty can be associated with decision makers' or crisis responders' imperfect/insufficient knowledge.

While the literature on uncertainty is a widely discussed and interdisciplinary concept to be covered under a single subchapter, within this study, two types of uncertainties shall be elaborated on that complicate cyber security crisis management: *Technical* and *Human-Centric Uncertainty*. Technical uncertainties cannot be eliminated in an STS or CI because it is based on a constantly evolving complex-socio-technical system. Any new hardware, software, or user involved in the system might introduce new vulnerabilities. On the other hand, human-centric uncertainties relate to cognition, decision-making, and flaws in interpreting/processing the massive amount of information the decision-maker receives. Another disadvantage of uncertainty in crisis management is the "termination of crisis," which signifies the point at which the crisis has ended and operations can be normalized. Because as highlighted, cyber security crises have "hazy time boundaries" [37], p. 7, and "just because a system has been restored to working order does not mean that vulnerabilities are gone, and the threats are neutralized" [38], p. 43. For instance, a research report demonstrating the vivid memories of the WannaCry incident through first responders' interviews clearly states that "incident response continues" as the unpatched vulnerabilities persist [39]. Another problem regarding STS (particularly Critical Information Infrastructures using extensive ICT) is that multiple uncertainties are present simultaneously. However, as Ansell and Boin discuss, leaders generally do not have either cognitive not organizational means to tame deep uncertainty in crisis [40], which is often the case for cyber security crises.

3 Grasping the nature of crisis

3.1 On the nature of crisis and crisis management

Crises indicate difficult and rare times in the flow of events. Although Dayton et al. claims crises encompass "a variety of 'un-ness' events" [41], p. 167 which are *unwanted, unexpected, unmanageable,* and *unprecedented,* like many concepts in social sciences, there is no agreement on "what a crisis is." Another problem is "subjectivity" of the crises since "each crisis speaks for itself" [42]. Understanding this subjectivity is necessary for managing CIs, as each sector might have different definitions and thresholds for defining crises for their functionality. For instance, when the CIA (Confidentiality-Integrity-Availability) principle of cyber security is considered, a mass violation of "availability" (of digital services) can be a source of a crisis for the communication industry. In contrast, loss of confidentiality might be more vital for the health sector.

Despite the variations, several commonalities form the characteristics of crises in practice. As Kovoor-Misra suggests, crises might share some common attributes with other forms of change, such as threats, opportunities, ambiguities, urgency, stress, and emotions. However, what makes a crisis problematic is "the degree to which these attributes are present" [43], p. 52. From this aspect, Boin et al. offer peculiarities of a crisis as an event that (a) creates a high degree of uncertainty or ambiguity, (b) evokes a sense of great urgency, and (c) puts high and often conflicting values at stake [44]. Gilpin and Murphy append several attributes to these peculiarities since crisis represents contingency, happenstance, unexpected confluence of unrelated events, and rapidly changing circumstances [25], p. 4.

Current and modern crises distinguish themselves from traditional ones concerning globalization and hyperconnectivity. As OECD's report posits [45], governments should deal with how crises impact their populations, national economies, and reputations. Modern CI management systems are depended on each other's services to function. More interconnectivity and interdependency among cities, regions, global supply chains, people, goods, and capital expand the impact of the current crises prone to produce transboundary outcomes. Since the governments have neither enough local sources nor the knowledge/human capacity to manage the current crisis alone, they need to cooperate with other actors, particularly the private sector, to address any crisis. Boin refers to the changing shape and dynamics of the current crisis as a "*transboundary crisis*" and notes that even though "threat agents might look familiar, the consequences (of transboundary crisis) play out very differently" [46], p. 367.

Notwithstanding, the most remarkable feature of the transboundary crisis is its *contingency* since it can easily transcend geographical and functional borders and exceed time boundaries. In addition to these properties, the increasing digitalization also sustained the transboundary crisis's persistence, increased interdependencies' penetration, and deepened complexity, particularly for the management of CIs. As a result, unexpected and sudden events that affect the design flow will quickly petrify the CI operators. Particularly for CIs, it shall be noted that these systems are primarily designed with efficiency concerns. Such prioritization of "efficiency and continuity" makes critical assets "vulnerable to the stress of extreme events" [47], p. 1.

Cyber security crises can be appraised under modern crisis typologies concerning contemporary crises' transboundary, contingency, and cross-functional features.

Backman concluded that the challenges present in transboundary crisis management could be identified throughout the management of Estonian (2007) and UK cyber security crisis (2017) cases [37]. For instance, while WannaCry represents a transboundary crisis, the most recent Log4j vulnerability transcended national borders and affected multiple sectors [37]. According to Checkpoint, while the spread of this vulnerability affected more than 90 countries and various sectors, beyond 15 countries were impacted, from education to transportation [48]. The cyber security crisis's characteristics, scope, and impact are shifting as interconnectivity and interdependence increase.

In simplest terms, crisis management can be judged as the sum of all efforts that aim to back to normal when an organization, a system, or a community breaks down because of an unexpected, nonroutine, cacophonic, and unstable event. For a more comprehensive definition, it is convenient to refer to Pearson and Clair since they define organizational crisis management as "a systematic attempt by organizational members with external stakeholders to avert a crisis or to effectively manage those [crises] that do occur" [49], p. 61. While there is extensive interdisciplinary literature on crisis management, the challenges for this delicate action can be described with 4Cs, defined by Comfort, as Cognition, Communication, Coordination, and Control [50]. In line with our study, these four elements are essential to recognize traditional and cyber security crisis management. In that sense, while cognition can be imagined as the first step to establishing an appropriate crisis response and providing recovery, it can be specified as "the capacity to recognize the degree of emerging risk to which a community is exposed and to act on that information" [50], p. 189. As the second chief challenge, communication becomes a mere barrier where information sharing is flawed, and a common terminology does not exist among the crisis responders. As the third task, coordination also emerges as a prerequisite for success (or the cause of failure) in crisis times. A decisive and unexpected event might force different teams and people to work harmoniously that may never have collaborated before. Finally, the challenge of controlling is connected to "keeping ongoing action focused on a shared goal" and "getting units to work in the same direction" [51], p. 23.

Based on this framing, it seems reasonable to ask how these 4Cs can affect cyber security crisis management efforts. In truth, the cognition phase might be one of the most challenging tasks for cyber security crisis managers as modern society is already passing through a "cognition crisis" [52]. This problem does not take its roots from the lack of knowledge but the excessive use of technology, changing communication paths of the human brain and external environment, and exposure to the large volume of data that hinders an individual's cognitive limits. For cyber security professionals working in a Security Operation Center (SoC), the cognition task is one of the trickiest since SoC employees might receive dozens of alarms in an ordinary working day. Thus, true alarm validation can be tiresome when many false positives exist. In other words, security monitoring is genuinely "human-centered" and requires an extensive Human-Machine collaboration where "excessive numbers of alarms contribute to alarm desensitization and lack of human responsiveness" [53], p. 2784. However, apart from cognition, communication is a Sisyphean task. Even in ordinary times, Chief Information Security Officers (CISOs) "fall in their communication and are not successful in influencing board decisions" [54], despite being increasingly placed in Board Rooms.

3.2 Walking on the thin ice: Properties and management of cyber security crises

The transboundary and unpredictable impacts of the cyber incidents affecting interconnected systems introduced the "cyber crisis" term into the literature. Several articles and books emphasize the term to emphasize crises originating from the interruption of Critical Information Infrastructures (CIIs). Cyber-attacks have a significant role in the appearance of such terms. In most cases, due to the high level of digitalization, some cyber-attacks substantially affect the system's functionality. Today, although it is possible to come across at least one major cyber security incident in media headlines daily, the cyber security crisis is a poorly investigated concept [28]. Another significant tendency within the field is the ambiguity of concepts and the interchangeable usage of "crisis," "events," and "incidents" terms, even though they represent separate meanings. In that sense, while an incident is "a situation that is on a small scale initially and can be managed with quick actions," a crisis can "pose higher uncertainty and disturb critical activities" [55]. Yet, such a sharp distinction can be delusional since an incident can quickly escalate into a crisis due to many factors, including cognition problems and human errors.

When can a decision maker define an event as a cyber security crisis? Previous crises show that the event's intensity might vary due to the pressure and time frame. This fluctuation might end up with vague definitions and labels for attribution, whether an event is a small-scale incident or a major crisis [38], p. 4. This also brings us to sense-making in crises which is an inherently complex process since the leaders "have to think, and problem-solve in a novel ambiguous situation" [56], p. 307. This turning moment is also essential because how a crisis manager defines an event is the first step in managing it. If every cyber incident is described as a crisis, this requires a proportionate budget/human resources allocation.

The definition of a cyber security crisis varies. For instance, Kaschner posits that we deal with the cyber crisis "whenever a breach of protection goals can result in real danger to the life and limb of people or strategic goals, reputation or survivability of our organization" [57], p. 6. Israel's cyber crisis management guide defines the cyber crisis as "a situation posing a real threat of damage, or actual damage, to a vital cyber asset, which is liable to cause critical damage to routine operations, reputational damage, economic damage and endanger human lives. A cyber crisis has varying degrees of gravity, and, in an extreme situation, substantial damage is caused to core processes and to the functional continuity of an organization/the economy, which is liable to escalate to the point of a national state of emergency [58]."

Our framework concentrated on the CIs as sociotechnical systems with various standards and interdependencies up to their functionality. So, the advancement of crisis could show differences in the type of CI. A Nuclear Power Plant (NPP) crises are remarkably dissimilar to an airport crisis. The dependency and connectedness would increase the vitality of the CIs. Thus, in each sector, the crisis would come in different forms and shapes up to the type of CI. Besides their differentiated impacts on different types of CIs, cyber-originated crises would also appear in a different temporal setting. Crises could stem from natural disasters, sabotage, accidents, and unexpected outages. In some cases, invisible design problems could originate the potential crisis. The crisis might swiftly develop and disrupt the services or emerges slower and affects the organization. Both types of crises have time compression on the social agents (Fig. 2).

As the National Institute of Standards and Technology (NIST) proposes, any computer security incident handling effort can follow the subsequent phases, including preparation (i), detection and analysis (ii), containment, eradication, and recovery (iii), and postincident activity (iv). Thus, detecting "what we are dealing with" is a prerequisite for managing a cyber incident. Yet, visibility makes a crisis in cyber space different from a crisis in the kinetic world. This phenomenon is mainly related to how we perceive and identify cyber threats since crisis managers often confront and try to contain the threats that cannot be seen, heard, or touched. For instance, an experienced fire responder can physically see how big the spread is. Also, the visible scope of the fire gives the opportunity the responder that they might calculate the severity of its impact. However, cyber

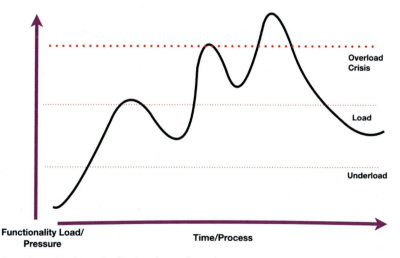

Fig. 2 Crisis and overload graph. *(Authors' own figure.)*

threats are often invisible to the human eye. Once the interconnectivity among different CIs is added to this invisibility dilemma, imagining the potential cascading effects can be an elusive task. For any organization, the visibility problem comes to the forefront as a significant challenge since "deflecting the incident that has not yet developed into a fully-blown crisis can be critical" [59], p. 8. To understand the severity of the invisibility problem, stating the SolarWinds incident can be meaningful since the attackers deployed different tactics to stay invisible in the systems to gain sufficient time for reconnaissance and profile of their targets [60].

Despite promising technological solutions and automatization toward better managing cyber security crises, empowering the human component cannot be overlooked. As Boin, and thus, discusses the influential response aftermath of a significant breakdown will "depend on the adaptive behavior of citizens, frontline workers, and middle managers" [61], p. 50. Besides, decision-making and leadership stress are major problems in both potential and crises. In addition to time compression and uncertainty, the complexity of the CIs would deepen the problem. The majority of CI organizations are socio-technical systems, which are the synergistic combination of humans, machines, environments, work activities, and organizational structures and processes that make up a given enterprise [62]. These facts emphasize the need to improve decision-making and human performance in managing crises.

4 Decision-making and organizational culture in managing cyber security crisis

4.1 Short time, high stakes: General framework on crisis decision-making

Crises are the low rare but high-impact events that represent a significant rupture from the normal. Crisis decision-making is "making hard calls, which involve tough value trade-offs and major political risks" [63]. Decision-making in crises is painful for all organizations since the conventional decision-making procedures and routines might not be sufficient to address an organizational crisis [64]. Besides, in crises, the routine decision management structures might also change. An organization experiencing a crisis might adopt a more centralized approach and rigid command and control understanding. One of the most vital tasks for crisis management in CIs is setting up specific crisis task forces, including a centralized decision-making authority to react to the situation without delay, contain the impact, and ensure recovery [65].

When it comes to crisis management of CIs, stakes are much higher since these systems are so vital that the flaws in their functioning could spread across multiple sectors, organizations, and even individuals. Thus, any decision while managing the crisis in a CI would have significant and unexpected outcomes. For instance, as was witnessed during the Colonial Pipeline incident recently, Company's CEO noted that although it was the most challenging decision in his career, shutting down the gas and paying the ransom was the right decision to make for the country [66]. This decision came with high costs since it led to panic buying that caused fuel shortages and local price spikes along the Colonial Pipeline route.

In its most basic form, decision-making can be defined as selecting one of several alternatives. Decision-making contains multiple subsequential steps, including problem identification and diagnosis, collecting information, formulating alternatives, choosing an option among the alternatives, and executing the decision. Yet, crises can obscure different decision-making steps, as some obstacles that might shape the crisis decision-making environment are identified in Fig. 3. For instance, information-gathering is vital for the decision-maker to evaluate options and finally arrive at the best option. Nevertheless, the information-gathering process might be flawed due to the rapid and interrupted information flow and limited time in crisis since "decisions have to be made quickly and based on incomplete information" [65].

As various interdependencies among CIs were discussed in the previous chapters, management of multiple stakeholders' demands and their expectations might be another barrier in crisis decision-making since "stakeholders tend to demand action very quickly and by doing so create an obligation to act by the organization" [67], p. 12. In any major CI dealing with a crisis, stakeholder involvement in different levels affecting the decision-

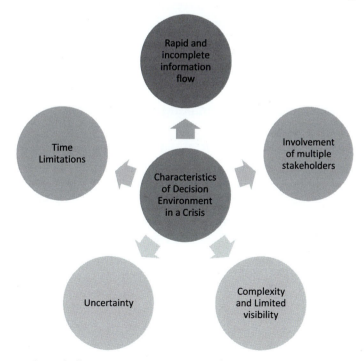

Fig. 3 Decision-making challenges in crisis environment for CIs. *(Authors' own figure.)*

making process might be more visible. For instance, when a country's major natural gas pipeline's operations are broken down because of a cyber incident, the public sector (including different agencies like law enforcement and intelligence services), private sector (that are dependent on the continuous functioning of this pipeline), individual users/citizens can be influential in decision-making processes.

Complexity (and consequently lack of visibility) is a chief characteristic of CIs and a problem in crisis decision-making. Complex systems have multiple interaction nodes with other systems and possess dynamicity. Thus, the interactions are not linear, and any decision made throughout a crisis could create unforeseen and unexpected impacts [68]. As Canyon discusses, "complex situations render decision-makers incapable of determining what will lead an organization to victory" [69]. In other words, when complexity occurs, the capability of forecasting all potential social and economic impacts of a crisis decision becomes a real challenge. Group thinking and other decision biases can be the other deadlocks of crisis decision-making, particularly where uncertainty prevails. Group thinking is a common pathology in crisis decision-making, mainly where a dominant leadership exists and when "there is a strive for unanimous agreement" [67], p. 50. Such a syndrome might result in "failing to weigh up decisions from all angles" [70], p. 214. Group thinking syndrome might delay the optimal decision, and appropriate leadership asks questions and collects all facts. Finally, at the individual level, uncertainty is one of the enemies of rationality since it triggers further cognitive limitations and leads to simplified thinking models or decision biases [71]. Even though sociotechnical systems increasingly rely on automation and other technologies, humans are still at the center of crisis decision-making. As a result, human errors concluded with disastrous events are primarily the outcomes of bad decisions.

4.2 Sense-making and decision-making in cyber security crisis management

With the increasing embeddedness of ICT and automation technologies into business processes, companies, and CIs, organizations are experiencing increasing cyber incidents. According to the statistics provided by Norton, in 2020, the FBI received more than 2.000 internet crime complaints per day [72]. Temple University's Critical Infrastructure Ransomware Attacks (CIRA) recorded 1305 ransomware attacks that hit CI sectors between November 2013 and April 2022 [73]. According to a survey that was performed at the RSA conference in 2018, it is reported that "most enterprises see more than 10,000 (cyber security) alerts per day" [74]. However, only a tiny percentage of all these attacks are genuinely critical or true positives. Since security alerts are on the rise and cyber security has been gaining popularity in media coverage, it is increasingly challenging to understand what factors distinguish an incident from a crisis. It shall also be highlighted that the "crisis threshold" might vary concerning the type of organization. For instance, while a Distributed Denial of Service (DDoS) attack that paralyzes online sales might represent a cyber security crisis for an e-commerce company, a water management utility breach of ICS and loss of control of the valves that control the level of chemicals in water truly represent a crisis. Even in some cases, the impact of an incident can be so major and widespread that it can lead to a declaration of national emergency, as recorded in Costa Rica in May 2022 [75].

In crisis management efforts, *sense-making* is one of the primary tasks and vital input for decisions since the responders need to "make sense of crisis to understand its origins, nature, and implications" [76]. The real problem that decision-makers face in cyber security crises is ambiguity and uncertainty, so "sense-making" remains a contest [77]. In other words, identifying the threshold that represents a rupture from an incident to a crisis is challenging in cyber-induced events since understanding the scope of the infected systems, the potential impact of the attack, the ultimate intent, and the location of the attackers are challenging. As claimed by Van den Berg and Kuipers, the complexity of cyber-enabled systems like CIs hardens the visibility, so for a decision maker, it is challenging "to monitor adequately whether anything is amiss, and if so, where and what exactly" [77]. As a component of crisis decision-making, sense-making also creates another cognitive burden on decision-makers since they need to solve the problem in a new and ambiguous environment under time pressure and stress [56].

In most companies, including CIs, the incidents which represent a lower level of criticality are handled by cyber security teams that are led by CISOs (Chief Information Security Officers) and/or CIOs (Chief Information Officers), who are senior-level officers and executives responsible for "establishing security strategy and ensuring the data assets are protected" [78]. While companies are increasingly creating cyber security management units, many boards are currently actively involved with people with cyber security expertise, making CISOs a critical component of an organization's strategic decision-making processes [79]. At the organizational level, different than the routine times when CISOs make the majority of the cyber security decisions, crises times require the creation of special structures in CIs like the crisis task forces, which are generally headed by the head of the organization (that is, CEO or general manager), including a core team, an extended team, and the advisors. While the core team prepares decisions for the task force head, the advisors might be added where detailed information is needed, such as environmental or legal concerns [65]. It should be noted that, in times of stability, CISOs should develop communication skills in addition to their heavily technical tasks. Internal communication between CISO and board members and external communication between stakeholders and other authorities—including the public—could become more demanding during cyber security crises.

As demonstrated in Fig. 4, in the case of a cyber security crisis, decision-makers simultaneously need to deal with multiple tasks, including coordination of the different teams, performing technical and strategic decisions, containing the impact of the crisis, and recovering the systems' functioning. Furthermore, according to Van den Berg and

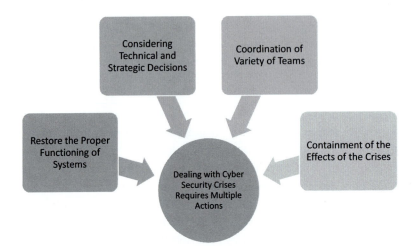

Fig. 4 Handling cyber security crises [80].

Kuipers, the geographical location in a cyber security crisis may not be bounded only by a single area or country, and decision-makers must overcome the challenges of decentralized authorities, barriers to transparent communication and engage in international information sharing and cooperation [77], p. 19.

Following the landscape at the organizational level, it is also worthwhile to discuss cyber security decision-making challenges at the individual level since uncertainty is prone to create various decision-making traps, including decision biases and heuristics. As the human brain makes decisions by relying on past experiences, uncertainty becomes problematic when individuals do not have proper and adequate experience to address and fit with new and unknown conditions [81]. When this hypothesis is re-considered for the cyber security field, as McDermott claims, because of the multiple uncertainties regarding the source of the attack and potential impacts, the chief problems of decision-makers in case of a cyber-attack are "fundamentally and intrinsically psychological in nature" [82], p. 227.

When a situation necessitates a rapid response, like cyber security crises, responders generally apply *System 1* thinking, typically characterized by a "fast, automatic, intuitive approach informed by experience" [83]. Yet, "automatic responses are based on cognitive biases is always a risk" [84]. Davenport argues that "cognitive biases are likely to appear in highly changeable, high-stress environments," which describe the case for the crises' peak moments [85]. Since cyber security crises have high uncertainty and time constraints, decision biases (such as anchoring, confirmation, overconfidence, and representativeness) are most likely to occur throughout the process.

Anchoring bias is likely to occur when the decision maker "locks onto a specific feature among many features" [86]. For example, such a case can emerge throughout a cyber security crisis if "the CISO places a particular cyber threat on higher priority and lower-level employees find themselves anchored to a specific threat instead of assessing the entire threat landscape" [87]. Representativeness bias occurs when "we erroneously group people (or other things) based on qualities considered normal/typical for that group." The "overconfidence effect" is also undesirable as it can mislead decision-makers and create a false sense of security. Finally, emotions might have significant ramifications on cyber security decisions and uncertainty. While Renaud and et al.'s research article proposes that individual experiences related to uncertainty lie at the roots of the negative feelings, McDermott assumes that "specific emotion that characterizes a given cyber conflict can influence the outcome in decisive ways" [88], p. 26. For instance, the emotions triggered by uncertainty might impact the risk-taking behaviors of individuals. If uncertainty triggers anxiety, individuals are likely to demonstrate lower-risk attitudes. Similarly, anger might lead to unnecessary escalation and risk-prone decisions. More importantly, McDermott highlights that "emotions are contagious" and, thus, likely to dominate and affect all teams in case of a cyber security crisis.

4.3 Organizational culture and crisis management efforts

Starting a new position in a company would probably take time to grasp organizational culture since such knowledge is tacit, and the rules are generally unwritten and distinct for every workplace. While culture is a broad term, it comprises and reflects the shared values, beliefs, and norms of members of an organization. Organizational culture also is a leading factor shaping any organization's internal and external communication and leadership attitudes. Schein defines organizational culture as "the pattern of basic assumptions that a group has invented, discovered, or developed in learning to cope with its problems of external adaptation and internal integration, and that has worked well enough to be considered valid, and, therefore, to be taught to new members as the correct way to perceive, think and feel in relation with those problems" [89].

In this vein, organizational culture is a significant phenomenon in understanding the management of CIs since cohesive work culture is found in transportation, power, and communication industries, eventually affecting worker decisions [90]. Cultural parameters also influence how organizations cope with acute changes, such as crisis moments, and how they deal with unprecedented conditions and adapt to the new environment. Some useful typologies exist to understand how different organizational cultures embrace crisis management strategies. For instance, Deverell and Olsson present three typologies (*Fully Adapting Organizations*, *Semi Adapting Organizations*, and *Non-Adapting Organizations*) to understand different organizations' adaptabilities in crises. Deverell and Olsson's conclusion shows that top management and centralization of leadership have a pivotal role in managing sudden changes and the capacity to bounce back [91].

Interestingly, Deverell and Olsson choose a "semi-public power company" (which is also a CI) as a "*Non-Adapting Institution*" to illustrate the challenges of crisis management. They suggest that the decentralized structure of a CI might hinder "fit[ting] public's expectations" in crisis moments, and the pervasive "technical culture" in a CI cannot easily align with the expectations of the management of stakeholder relations. In addition to institutional typologies, it should be highlighted that organizational culture might affect sense-making in crisis management efforts. Sherman and Roberto explore organizational cultures as a sense-making mechanism in crisis. They underline that "crises are filled with ambiguity," which might halt meaningful actions. Organizations need "plausible messages" and vivid sense-making to cope with ambiguity. From this perspective, according to Sherman and Roberto, culture affects not only the messages sent throughout a crisis but also how employees receive these messages [92].

Organizational culture has a substantial role in reaching better outputs within the cyber crisis management process. Organizations should have to design their cyber crisis management responsibility structure as similar to the day-to-day operations. The people in charge of daily operations should lead the team in case of crises. Their leadership would quicken the crisis responses in the facility. Another vital point closer to this approach is the mirror principle. The cyber crisis management team should reflect the organizational structure of the plant or the cooperation as closely as possible. The mirror principle could help the crisis management team to cultivate a multiperspective approach to figure out the problems on the table. One of the major problems in the crisis process is dealing with the crises with outsourcing or an in-house team in the headquarters trying to put out the fire on the plant side. One of the major issues is dealing with the crisis through outsourcing or an in-house team in the headquarters attempting to put out the fire on the plant side. In such a crisis, the team in charge of infrastructure (networking, servers, etc.) should oversee the crisis management process. If a technical team is brought in for specific expertise, that team should work with the teams closest to the crisis. This principle implicitly includes transparent communication among stakeholders. Cooperation of all responsible parties during the cyber crisis is critical to end the process. Thus, many cyber crisis management scenes assume that cyber crisis teams are critical and indispensable to all other developments. On the other hand, the importance of collaboration with all stakeholders within the system would determine the flow of the crises. Cooperation among various stakeholders is also essential for maintaining trust within the organization.

In this context, leadership in cyber crisis management plays a specific role in resolving the institution's crisis. As Bhaduri puts along with motivated employees and teams, "effective leadership has a great effect on averting and controlling the crisis" [93], p. 2. To this end, Bhaduri proposes a conceptual framework combining cultural elements and leadership types and highlights the need for developing leadership competencies and "crisis-prone cultures" for effective crisis management. Bhaduri's research draws on various leadership styles (directive, cognitive, transactional, and transformational). It creates an organizational culture/leadership competency matrix for various stages of crisis (precrisis, crisis, and postcrisis). For example, transformational leadership may be required during crisis damage containment and recovery periods [93], p. 11. This type of leadership can inspire positive changes and re-establish confidence in the group. Similarly, adaptive leadership may be helpful during the cyber incident response period, when leaders must "assist individuals and organizations in adapting and thriving in challenging environments" [94].

Another less touched but critical point regarding the organizational culture is transition and workload. As German philosopher Byung-Chul Han precisely discusses, our society transitions from a disciplinary society to an achievement society [95]. To this end, the workload over the human workforce in the achievement society has created burnout (*müdigkeit*/fatigue) among employees. Understanding his point about changing organizational culture provokes us to consider what might happen if the exhausted human workforce shifts in complex technological production sites. Such a scenario would promise us more cyber security crises in CIs. Using the same exhausted human workforce during the cyber crisis would also deepen our problems.

From this perspective, our approach purports that organizational culture might impact employees' and top management's attitudes, decision-making, and communication efforts when a cyber security crisis occurs. Although studies investigating the link between organizational culture and cyber security crisis management are nascent, existing literature described in this chapter proposes useful findings to contemplate building an effective cyber security crisis management from the organizational culture and leadership perspectives. The rigid and hierarchical organizational cultures may undermine efforts to mitigate cyber security crises where quick decisions and effective solutions are required. To this end, CIs should be able to quickly establish ad hoc cyber security crisis management bodies and ease the decision-making processes. The decentralized management and stakeholders' conflicting interests could be another crisis communication problem. In that sense, CIs should strengthen their preparedness and plan proper crisis communication. A crisis-prone culture, along with the right leadership style, becomes an inherent necessity since cyber-attacks are increasingly targeting CIs.

5 Conclusion: Lessons learned from crises and future prospects

Managing crises in CIs has always been an uphill battle as these systems evolve and change continuously for their technological dynamicity and emerging socio-economic requirements. Besides, expanding the integration of ICT in CIs did not ease but complicated the crisis management efforts. The human ability to predict to potential impacts of cyber replated disruptions remains limited despite the sophistication of visualization, prediction, and modeling techniques. This chapter purports that CIs are beyond being just the mechanical/engineering giants but complex STSs. In this vein, it is necessary to explore how the chief characteristics of STSs (complexity, uncertainty, and interdependence) are aligned with CIs and why these are important to manage cyber security crises.

As suggested throughout this chapter, the nature of the crises is changing concerning the increased global connectedness. The contemporary crises are more contagious and more transboundary. Moreover, the cybered world duplicates existing complex networks and increases the impact and frequency of disruptions and the occurrence of "unknown unknowns." In the face of a cyber security crisis, the distinct characteristics of cyberspace (that is, threats are not physically apparent) predict the full-scale impact, and understanding the reason for the breach and figuring out the attackers' motivation becomes much more demanding.

These multiple uncertainties inherently affect the sense-making and decision-making processes in managing a cyber security crisis. As long as the frequency of cyber security alerts increases, it continues to hinder "signal detection" efforts which form the first phase of managing any crisis. In the later stages of the crisis management phase, decision-making and termination might become more problematic when managing a cyber security crisis. Because crisis decisions might be more biased when uncertainties increase, and coordination of the different teams become more challenging when CIs are subject to decentralized management structures. In other words, when cyber security problems are added to the traditional crisis management agendas, decision-making, mean-making, sense-making, and termination phases within crisis management efforts become much more complicated for the management of CIs. Moreover, the uncontrolled drift toward Byung-Chul Han's burnout society inherently transforms the organizational cultures. For instance, many studies demonstrate how CISOs' burnout and stress create additional cyber security risks [96,97]. Such large-scale changes in socialtechnical systems are not negligible and generate additional burdens on cyber security crisis responders.

Any crisis system should consider five essential elements: Technology, organizational culture, organizational structure, human factors, and top management psychology [93], p. 2. To this end, our chapter suggests that only adopting a multidisciplinary attitude and approaching CIs as STSs could better address the efforts to manage cyber security crises. The cyber crisis management playbook should be prioritized and embedded in the CI's cyber routines and cyber security cultures. The board should prioritize role-playing, cyber drills, table-top-exercises, and interactive training for enhanced cyber crisis preparedness scenarios. The existing (cyber) crisis scenarios should be tested continuously. CIs should adopt customized "cyber crisis management toolboxes" for their specific functionalities and services. Strengthening internal communication is vital because, in crisis times, teams that had not previously cooperated need to work together. Managing external communication, particularly with the public, is also vital since most of the cyber threats targeting CIs also aim to corrode public trust, and crises are the perfect moments for the spread of disinformation.

References

[1] D. Clark, Characterizing Cyberspace: Past, Present, and, Future, 2010, Available from: https://ecir.mit.edu/sites/default/files/documents/%5BClark%5D%20Characterizing%20Cyberspace-%20Past%2C%20Present%20and%20Future.pdf. (Accessed 21 March 2022).

[2] Homeland Security & Governmental Affairs, Full Committee Hearing: Threats to Critical Infrastructure Examining the Colonial Pipeline Cyber Attack, 2021, Available from: https://www.hsgac.senate.gov/hearings/threats-to-critical-infrastructure-examining-the-colonial-pipeline-cyber-attack/. (Accessed 10 March 2023).

[3] National Cyber Security Center of UK, A Sociotechnical Approach to Cyber Security, 2022, Available from: https://www.ncsc.gov.uk/blog-post/a-sociotechnical-approach-to-cyber-security.

[4] M. Malatji, S. Von Solms, A. Marnewick, Socio-technical systems cybersecurity framework, Inf. Comput. Secur. 27 (2) (2019) 233–272, https://doi.org/10.1108/ICS-03-2018-0031.

[5] C.M. Newbill, Defining critical infrastructure for a global application, Indiana J. Glob. Leg. Stud. 26 (2) (2019) 761–780, https://doi.org/10.2979/indjglolegstu.26.2.0761.

[6] M. Sony, S. Naik, Industry 4.0 integration with socio-technical systems theory: a systematic review and proposed theoretical model, Technol. Soc. 61 (2020), https://doi.org/10.1016/j.techsoc.2020.101248.

[7] P. Assogna, et al., Critical infrastructures as complex systems: a multi-level protection architecture, in: R. Setola, S. Geretshuber (Eds.), Critical Information Infrastructure Security, 5508, Springer, Berlin, 2009, pp. 368–375.

[8] R. Setola, E. Luiijf, M. Theocharidou, Critical infrastructures, protection and resilience, in: R. Setola, E. Kyriakides, V. Rosato, E. Rome (Eds.), Managing the Complexity of Critical Infrastructures: A Modelling and Simulation Approach, Springer, Cham, 2016, pp. 1–19.

[9] C. Aradau, Security that matters: critical infrastructure and objects of protection, Secur. Dialogue 41 (2010) 491–514, https://doi.org/10.1177/0967010610382687.

[10] M. Ottens, M. Franssen, P. Kroes, I. Van De Poel, Modelling infrastructures as socio-technical systems, Int. J. Crit. Infrastruct. 2 (2006) 133–145.

[11] G. Ropohl, Philosophy of socio-technical systems, Techné: Res. Philos. Technol. 4 (1999) 186–194, https://doi.org/10.5840/techne19994311.

[12] W. Pasmore, S. Winby, S.A. Mohrman, R. Vanasse, Reflections: sociotechnical systems design and organization change, J. Chang. Manag. 19 (2019) 67–85, https://doi.org/10.1080/14697017.2018.1553761.

[13] IGI Global, What Is Socio-Technical Systems, 2022, Available from: https://www.igi-global.com/dictionary/socio-technical-systems/27578.

[14] T.G. Topcu, Management of Complex Sociotechnical Systems, Virginia Polytechnic Institute and State University, Falls Church, 2020.

[15] A. Kelly, Strategic Maintenance Planning, Butterworth-Heinemann, Oxford, 2006.

[16] G. Baxter, I. Sommerville, Socio-technical systems: from design methods to systems engineering, Interact. Comput. 23 (2011) 4–17, https://doi.org/10.1016/j.intcom.2010.07.003.

[17] T. Abreu Saurin, S. Sosa Gonzalez, Assessing the compatibility of the management of standardized procedures with the complexity of a sociotechnical system: a case study of a control room in an oil refinery, Appl. Ergon. 44 (5) (2013) 811–823, https://doi.org/10.1016/j.apergo.2013.02.003.

[18] I. Gondal, Hydrogen transportation by pipelines, in: R.B. Gupta, A. Basile, T.N. Veziroğlu (Eds.), Compendium of Hydrogen Energy, Woodhead Publishing, 2016, pp. 301–322 (s.l.).

[19] A. Fekete, Common criteria for the assessment of critical infrastructures, Int. J. Disaster Risk Sci. 2 (2011) 15–24, https://doi.org/10.1007/s13753-011-0002-y.

[20] Bloomberg, Hackers Breached Colonial Pipeline Using Compromised Password, 2021, Available from: https://www.bloomberg.com/news/articles/2021-06-04/hackers-breached-colonial-pipeline-using-compromised-password.

[21] T.E. Van Der Lei, G. Bekebrede, I. Nikolic, Critical infrastructures: a review from a complex adaptive systems perspective, Int. J. Crit. Infrastruct. 6 (2010) 380–401.

[22] O. Renn, Risk Governance: Coping With Uncertainty in a Complex World, Earthscan, London, 2008.

[23] F. Schöttl, U. Lindemann, Quantifying the complexity of socio-technical systems—a generic, interdisciplinary approach, Procedia Comput. Sci. 44 (2015) 1–10, https://doi.org/10.1016/j.procs.2015.03.019.

[24] S.E. Page, What sociologists should know about complexity, Annu. Rev. Sociol. 41 (2015) 21–41, https://doi.org/10.1146/annurev-soc-073014-112230.

[25] D.R. Gilpin, P.J. Murphy, Crisis Management in a Complex World, Oxford University Press, New York, 2008.

[26] E.J. Oughton, W. Usher, P. Tyler, J.W. Hall, Infrastructure as a complex adaptive system, Complexity (2018), https://doi.org/10.1155/2018/3427826.

[27] Deloitte, Cyber Crisis Management: Readiness, Response, and Recovery, 2016, Available from: https://www2.deloitte.com/content/dam/Deloitte/no/Documents/risk/cyber-crisis-management.pdf.

[28] M.F. Prevezianou, Beyond ones and zeros: conceptualizing cyber crises, Risk, Hazards Crisis Public Policy (2021) 51–72, https://doi.org/10.1002/rhc3.12204.

[29] S.M. Rinaldi, J.P. Peerenboom, T.K. Kelly, Identifying, understanding, and analyzing critical infrastructure interdependencies, IEEE Control. Syst. Mag. 21 (6) (2001) 11–25.

[30] F. Petit, D. Verner, Critical Infrastructure Interdependencies Assessment, 2016, Available from: https://www.osti.gov/servlets/purl/1400396. (Accessed 28 April 2022).

[31] M.A. Gomez, C. Whyte, Cyber wargaming: grapling with uncertainty in a complex domain, Def. Strateg. Assess. J. 10 (2020) 94–135.

[32] D. Clemente, Cyber Security and Global Interdependence: What Is Critical?, Chatnam House, London, 2013. https://www.chathamhouse.org/sites/default/files/public/Research/International%20Security/0213pr_cyber.pdf.

[33] Financial Times, Turkish Industry Hit by Power Cuts Amid Gas Supply Troubles, 2022, Available from: https://www.ft.com/content/65fd1e40-c2cc-4098-81fc-2f3cbe3f66cf.

[34] H. Menashri, G. Baram, Critical infrastructures and their interdependence in a cyber attack—the case of the U.S., Mil. Strateg. Affairs 7 (1) (2015) 79–100. https://www.inss.org.il/wp-content/uploads/systemfiles/5_Menashri_Baram.pdf.

[35] T.W. Haase, Uncertainty in crisis management, in: A. Farazmand (Ed.), Global Encyclopedia of Public Administration, Public Policy, and Governance, Springer, Cham, 2018, pp. 6027–6031.

[36] Y. Li, J. Chen, L. Feng, Dealing with uncertainty: a survey of theories and practices, IEEE Trans. Knowl. Data Eng. 25 (11) (2013) 2463–2482, https://doi.org/10.1109/TKDE.2012.179.

[37] S. Backman, Conceptualizing cyber crises, J. Conting. Crisis Manag. 29 (2020) 429–438, https://doi.org/10.1111/1468-5973.12347.

[38] European Union Agency for Cybersecurity (ENISA), Report on Cyber Crisis Cooperation and Management, European Union Agency for Cybersecurity (ENISA), Heraklion, 2014.

[39] T. Wheeler, J.L. Alderdice, Cyber Collateral: WannaCry & the Impact of Cyberattacks on the Mental Health of Critical Infrastructure Defenders, The University of Oxford Changing Character of the War Center, 2022. https://www.ccw.ox.ac.uk/blog/2022/10/5/cyber-collateral-wannacry-amp-the-impact-of-cyberattacks-on-the-mental-health-of-critical-infrastructure-defenders-by-tarah-wheeler-and-john-alderdice.

[40] C. Ansell, A. Boin, Taming deep uncertainty: the potential of pragmatist principles for understanding and improving strategic crisis management, Adm. Soc. 51 (7) (2019) 1079–1112, https://doi.org/10.1177/0095399717747655.

[41] B.W. Dayton, et al., Managing crises in the twenty-first century, Int. Stud. Rev. 6 (1) (2004). In press.

[42] Britannica, Crisis Management, 2022, Available from: https://www.britannica.com/topic/crisis-management-government.

[43] S. Kovoor-Misra, Crisis Management: Resilience and Change, Sage, Thousand Oaks, 2019.

[44] A. Boin, P. Hart, E. Stern, B. Sundelius, The Politics of Crisis Management: Public Leadership Under Pressure, Cambridge University Press, Cambridge, 2005.

[45] OECD, The Changing Face of Strategic Crisis Management, OECD Reviews of Risk Management Policies, Paris, 2015.

[46] A. Boin, The new world of crises and crisis management: implications for policymaking and research, Rev. Policy Res. 26 (2009) 367–377, https://doi.org/10.1111/j.1541-1338.2009.00389.x.

[47] L.K. Comfort, Managing Critical Infrastructures in Crisis, Oxford Research Encyclopedia of Politics, 2020.

[48] Checkpoint, Protect Yourself Against the Apache Log4j Vulnerability, 2021, Available from: https://blog.checkpoint.com/2021/12/11/protecting-against-cve-2021-44228-apache-log4j2-versions-2-14-1/. (Accessed 28 April 2022).

[49] C.M. Pearson, J.A. Clair, Reframing crisis management, Acad. Manag. Rev. 23 (1) (1998) 59–76.

[50] L.K. Comfort, Crisis management in hindsight: cognition, communication, coordination, and control, Public Adm. Rev. 67 (2007) 189–197, https://doi.org/10.1111/j.1540-6210.2007.00827.x.

[51] J. Wolbers, K. Boersma, Key challenges in crisis management, in: R.P. Gephart, C.C. Miller, K. Svedberg (Eds.), The Routledge Companion to Risk, Crisis and Emergency Management, Routledge, New York, 2018, pp. 17–34.

[52] L.D. Rosen, A. Gazzaley, The Distracted Mind: Ancient Brains in a High-Tech World, MIT Press, Cambridge, 2016.

[53] B. Alahmadi, L. Axon, I. Martinovic, 99% False Positives: A Qualitative Study of SOC Analysts' Perspectives on Security Alarms, USENIX, 2022.

[54] CISO Mag, Are CISOs failing in their communication to the Board? 2019. https://cisomag.com/ciso-communication/. (Accessed 12 October 2022).

[55] Scalefusion Blog, What Is the Difference Between Incident and Crisis Management? 2021, Available from: https://blog.scalefusion.com/incident-vs-crisis-management/.

[56] I.A. Combe, D.J. Carrington, Leaders' sensemaking under crises: emerging cognitive consensus over time within management teams, Leadersh. Q. 26 (3) (2015) 307–322, https://doi.org/10.1016/j.leaqua.2015.02.002.

[57] H. Kaschner, Cyber Crisis Management: The Practical Handbook on Crisis Management and Crisis Communication, Springer, Berlin, 2021.

[58] Cyber Israel Prime Minister's Office National Cyber Directorate, National Cyber Concept for Crisis Preparedness and Management, Cyber Israel Prime Minister's Office National Cyber Directorate, 2022.

[59] L. Ezioni, G. Siboni, Cyber crisis management and regulation, in: L. Ezioni, G. Siboni (Eds.), Cyber-security and Legal-Regulatory Aspects, World Scientific, 2021, pp. 1–21.

[60] R. Lakshmanan, Here's How SolarWinds Hackers Stayed Undetected for Long Enough, 2021, Available from: https://thehackernews.com/2021/01/heres-how-solarwinds-hackers-stayed.html. (Accessed 16 April 2022).

[61] A. Boin, A. McConnell, Preparing for critical infrastructure breakdowns: the limits of crisis management and the need for resilience, J. Conting. Crisis Manag. 15 (2007) 50–59, https://doi.org/10.1111/j.1468-5973.2007.00504.x.

[62] G. Østby, et al., A socio-technical framework to improve cyber security training: a work in progress, CEUR Workshop Proceedings (2019) 81–96. https://ntnuopen.ntnu.no/ntnu-xmlui/handle/11250/2624957.

[63] Britannica, Making Critical Decisions, 2022, Available from: https://www.britannica.com/topic/crisis-management-government/Making-critical-decisions.

[64] A. Sommer, C.M. Pearson, Antecedents of creative decision making in organizational crisis: a team-based simulation, Technol. Forecast. Soc. Chang. 74 (8) (2007) 1234–1251, https://doi.org/10.1016/j.techfore.2006.10.006.

[65] Federal Ministry of the Interior, Protecting Critical Infrastructures-Risk and Crisis Management, Federal Ministry of the Interior, Berlin, 2017.

[66] NPR, The Colonial Pipeline CEO Explains the Decision to Pay Hackers a $4.4 Million Ransom, 2021, (Online). Available from: https://www.npr.org/2021/06/03/1003020300/colonial-pipeline-ceo-explains-the-decision-to-pay-hackers-4-4-million-ransom.

[67] The Royal Commission Into National Natural Disaster Arrangements, Decision Making During a Crisis: A Practical Guide, The Royal Commission into National Natural Disaster Arrangements, Canberra, 2020.

[68] D.J. Snowden, M.E. Boone, A Leader's Framework for Decision Making, 2007, Available from: https://hbr.org/2007/11/a-leaders-framework-for-decision-making. (Accessed 10 May 2022).

[69] D. Canyon, Strategic Approaches to Simplifying Complex Adaptive Crises, 2018, Available from: https://apcss.org/wp-content/uploads/2021/01/Nexus-Canyon-simplifying-complexity.pdf. (Accessed 12 May 2022).

[70] D.C. Grube, A. Killick, Groupthink, oolythink and the challenges of decision-making In cabinet government, Parliam. Aff. 76 (1) (2023) 211–231, https://doi.org/10.1093/pa/gsab047.

[71] C. Strauss Einhorn, How to Make Rational Decisions in the Face of Uncertainty, 2020, Available from: https://hbr.org/2020/08/how-to-make-rational-decisions-in-the-face-of-uncertainty.   (Accessed 10 March 2023).

[72] Norton, Emerging Threats, 2021, Available from: https://us.norton.com/internetsecurity-emerging-threats-cyberthreat-trends-cybersecurity-threat-review.html. (Accessed 13 April 2022).

[73] A. Rage, Critical Infrastructure Ransomware Attacks (CIRA) Dataset. Version 12, 2022, Available from: https://sites.temple.edu/care/cira/. (Accessed 15 May 2022).

[74] Bricata, How Many Daily Cybersecurity Alerts Does the SOC Really Receive? 2022, Available from: https://bricata.com/blog/how-many-daily-cybersecurity-alerts. (Accessed 4 May 2022).

[75] Bleeping Computer, Costa Rica Declares National Emergency After Conti Ransomware Attacks, 2022, Available from: https://www.bleepingcomputer.com/news/security/costa-rica-declares-national-emergency-after-conti-ransomware-attacks/. (Accessed 9 May 2022).

[76] J.P. Kalkman, Sensemaking questions in crisis response teams, Disaster Prev. Manag. 28 (5) (2019) 649–660, https://doi.org/10.1108/DPM-08-2018-0282.

[77] B. Van den Berg, S. Kuipers, Vulnerabilities and Cyberspace: A New Kind of Crises, Oxford Research Encyclopedia of Politics, 2022.

[78] ZDNET, What Is a CISO? Everything You Need to Know About the Chief Information Security Officer Role, 2020, Available from: https://www.zdnet.com/article/what-is-a-ciso-everything-you-need-to-know-about-the-chief-information-security-officer/. (Accessed 14 May 2022).

[79] Fortinet, Landing a Seat on a Corporate Board as a CISO: A Conversation With Joyce Brocaglia, 2020, Available from: https://www.fortinet.com/blog/ciso-collective/how-a-ciso-can-land-a-seat-on-a-corporate-board-a-conversation-with-joyce-brocaglia-from-boardsuited. (Accessed 11 May 2022).

[80] Agence Nationale de la Sécurité des Systèmes d'information, Organizing a Cyber Crisis Management Expertise, 2021, Available from: https://www.ssi.gouv.fr/uploads/2021/09/anssi-guide-organising_a_cyber_crisis_management_exercise-v1.0.pdf. (Accessed 10 May 2022).

[81] K. Burch, How to Deal With Life's Uncertainty and the Stress It Causes, 2020, Available from: https://www.insider.com/how-to-deal-with-uncertainty.

[82] R. McDermott, in: National Research Council (Ed.), Decision Making Under Uncertainty, The National Academies Press, Washington, DC, 2010, pp. 227–242.

[83] C. Barlow, When a Cyber Crisis Hits, Know Your OODA Loops, 2017, Available from: https://securityintelligence.com/when-a-cyber-crisis-hits-know-your-ooda-loops/. (Accessed 2 May 2022).

[84] P. Carpenter, Five Cognitive Biases That Can Threaten Your Cybersecurity Efforts, 2021, Available from: https://www.forbes.com/sites/forbesbusinesscouncil/2021/12/30/five-cognitive-biases-that-can-threaten-your-cybersecurity-efforts/?sh=324a42559e31. (Accessed 3 May 2022).

[85] T.H. Davenport, How to Make Better Decisions About Coronavirus, 2020, Available from: https://sloanreview.mit.edu/article/how-to-make-better-decisions-about-coronavirus/. (Accessed 12 May 2022).

[86] Z. Zorz, How Human Bias Impacts Cybersecurity Decision Making, 2019, Available from: https://www.helpnetsecurity.com/2019/06/10/cybersecurity-decision-making/. (Accessed 10 May 2022).

[87] S. Durbin, 10 Cognitive Biases That Can Derail Cybersecurity Programs, 2022, Available from: https://www.securitymagazine.com/articles/96918-10-cognitive-biases-that-can-derail-cybersecurity-programs. (Accessed 12 May 2022).

[88] R. McDermott, Emotional Dynamics of Cyber Conflict, 2019, Available from: https://ndisc.nd.edu/assets/290142/emotional_cyber.pdf. (Accessed 8 May 2021).

[89] E.H. Schein, Coming to a New Awareness of Organizational Culture, 1985, Available from: https://sloanreview.mit.edu/article/coming-to-a-new-awareness-of-organizational-culture/.

[90] K.C. Muzyczka, K. Chapman, C. McCarty, Culture in interdependent critical infrastructure, J. Org. Cult. Commun. Confl. 23 (1) (2019) 1–14.

[91] E. Deverell, E. Olsson, Organizational culture effects on strategy and adaptability in crisis management, Risk Manag. 12 (2010) 116–134.

[92] W.S. Sherman, K.J. Roberto, Are you talkin' to me? The role of culture in crisis management sense-making, Manag. Decis. 58 (10) (2020) 2195–2211, https://doi.org/10.1108/MD-08-2020-1017.

[93] R.M. Bhaduri, Leveraging culture and leadership in crisis management, Eur. J. Train. Dev. 43 (5/6) (2019) 554–569, https://doi.org/10.1108/EJTD-10-2018-0109.

[94] Cambridge Leadership Associates, Adaptive leadership. https://cambridge-leadership.com/adaptive-leadership/. (Accessed 14 October 2022).

[95] B.C. Han, The Burnout Society, Stanford University Press, Stanford, 2015.

[96] CNBC, Chief Information Security Officers Say Stress and Burnout, Not Job Loss as a Result of a Breach, Are Their Top Personal Risks, 2022, Available from: https://www.cnbc.com/2022/09/08/cisos-say-stress-and-burnout-are-their-top-personal-risks.html. (Accessed 8 September 2022).

[97] The Cyberwire, Overworked CISOs May Be a Security Risk, 2022, Available from: https://thecyberwire.com/stories/f43cd7160ec34a49b92cf549b265a3f3/overworked-cisos-may-be-a-security-risk. (Accessed 15 October 2022).

PART III

Engineering of critical infrastructures

CHAPTER 7

Event-based digital-twin model for emergency management

Rabia Arslan[a] and Mehmet Akşit[b]
[a]Department of Engineering, TOBB University of Economics and Technology, Ankara, Turkey
[b]Department of Computer Science, University of Twente, Enschede, The Netherlands

1 Introduction

Destruction of critical infrastructure by a natural or man-made cause may have a major negative impact on the citizens especially in metropolitan areas [1]. Emergency management procedures [2] are defined as healing actions to minimize the impact of damages. We consider emergency management systems to be critical infrastructures as well. With the advancements in computer technology, more and more control systems are adopted for this purpose [3]. Naturally, due to distribution of critical infrastructures, such control architectures are generally associated with geographical topology. Various GIS systems have been introduced to provide the necessary facilities for systems that need topology support.

Unfortunately, current GIS systems are not expressive enough to represent the data that are required in effective emergency management. For example, GIS models for hospitals, schools, government buildings, factories, roads, bridges, and water, electricity, gas, and communication networks are lacking.

In general, control systems [3] incorporate models of the actual parameters of the processes that they control. These parameters are compared with the references and control actions are undertaken if there are deviations. Digital twins [4] are suitable in online modeling of the controlled processes since they are kept consistent with the physical realities. As a general architectural style, event-driven approaches [5] are common in realizing such systems.

Current GIS systems are basically dedicated database systems with topology support. They lack the necessary primitives to support event-driven architectural styles. To overcome this limitation, in general, GIS databases are mapped onto "Object Runtime Systems," in which the necessary events are implemented. Unfortunately, due to the distributed and heterogeneous nature of emergency management systems and lack of standardization, multiple and incompatible object runtime systems are likely to be adopted by the different subsystems of the architecture. This may cause inconsistencies in referring to the shared events as a consequence of simultaneous read-write and write-write operations [6].

Management and Engineering of Critical Infrastructures
https://doi.org/10.1016/B978-0-323-99330-2.00008-8

This chapter introduces three novel contributions. Firstly, as an underlying GIS model, CityGML 3.0 [7] is extended with a set of new abstractions to express the necessary data for effective emergency management. Secondly, a model-checking-based analysis is carried out with the model of a distributed and heterogeneous emergency management system to find out the possible conditions that cause inconsistencies. Thirdly, an architectural style is proposed to extend CityGML with an event-driven layer supporting atomic publisher-subscriber protocol [8]. To the best of our knowledge, there has not been any work in the literature which extends GIS systems with new data models and event-driven layers for the purpose of emergency management of critical systems in case of earthquakes.

2 Example case, background work, and problem statement

2.1 An example case: Earthquake management system

2.1.1 Representing the earthquake related terms

Assume that a large-scale earthquake has occurred in a city of reasonable size, where different kinds of structures have been damaged. Also, it is considered important to represent the effect of the earthquake within the GIS database. For example, the following terms are relevant for this purpose:

- Physical objects: Base Station, Bridge, Communication Network, Control Center, Electricity Network, Electricity Station (Transformers), Factory, Financial Center, Firefighter Station, Gas Network, Gas Pump Station, Government Building, Historical Buildings, Hospital, Logistic Center, Museum, Office Building, Police Station, Railway, Religious Building, Repair Center, Rescue Center, Residence, Restaurant, Road, School, Shop, Shopping Center, Square, Track, Tunnel, Vegetation, Water Network, Water Pump Station, Waterways.
- Damage: Collapse, Fire, Flood, Landslide, Tsunami, Unhealth.
- Intensity of the Damage: Low, Medium Low, Medium, Medium High, High
- Task Forces: Security, Rescue, Firefighter, Health, Repair.

Each of these terms must be represented in the database as a first-class abstraction, meaning that they must be directly supported by the facilities of the database. Each term may have its own specific attributes. In this chapter, we assume that the database size is large enough to store all damages that have occurred in the city.

2.1.2 Digital-twin-based emergency management

In Fig. 1, the architecture of a management system consists of n control centers, m task forces, object runtime (OR) systems, and a GIS database as shown in white, light gray, and dark gray colors, respectively. The architecture is layered and divided into a distributed management system and a distributed digital-twin system. This is an event-based distributed architecture in which all events are generated and registered in the digital-twin system. The events that are generated in the twin system are published using a

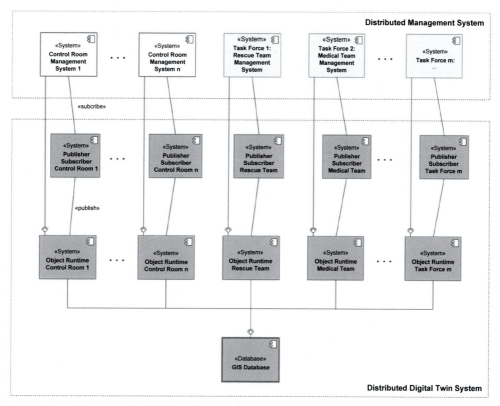

Fig. 1 The architecture of a distributed emergency management system.

publisher-subscriber system. Distributed management system subscribes to the events that are needed. To assure consistency of data, the structure and the attributes of the relevant geography are kept in the same GIS database which is shown at the bottom of the figure. Due to the heterogeneous architecture and lack of standardization, it is not considered realistic to have a common OR system for all management systems.

The distributed digital-twin system should be consistent with the external and internal reality. The term external reality refers to the relevant events and states which are not generated by the system shown in Fig. 1 but occur in the real world. For example, a collapse of a building after an earthquake in the real world is considered as a part of the external reality. The other relevant events and states belong to the internal reality. They can be generated in the management systems or in the digital twin. To assure consistency, each relevant event in the external and internal reality must be represented in the digital twin. Similarly, any reference which is made to the events and states of the digital twin must be consistent with the actual values in the digital twin.

Fig. 2 shows the interactions between the modules of the control center in the UML sequence diagram. The OR, publisher subscriber, and control center processes are shown

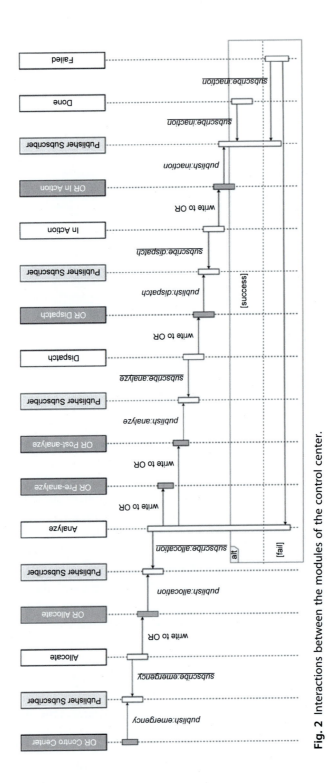

Fig. 2 Interactions between the modules of the control center.

in dark gray, light gray, and white colors, respectively. The OR processes, which represent the event-based twin system, publish the events to which control center processes subscribe. Subsequently, when a process is completed, the control center writes the status on the OR system. Within this configuration, the control center executes in order, the processes Allocate, Analyze, Dispatch, In Action, Done, and Failed. Here, the processes Done and Failed are the two alternatives which are shown at the bottom of the figure within a separate rectangle named "alt."

Fig. 3 depicts the interactions between the modules of the rescue task force. Dark gray, light gray, and white colors represent the OR, publisher subscriber, and rescue task force operations, respectively. The OR processes, which represent the event-based digital-twin system, publish the events that are subscribed to by the rescue task force processes. When a process is finished, the rescue task force updates the OR system with the actual status. In this architecture, the rescue task force conducts the processes Call Register, Check Availability, In Operation, and End subsequently. The processes OR Failed and OR End are the two options shown at the bottom of the figure within a separate rectangle labeled "alt."

Sequence diagrams 2 and 3 represent distributed executions on a heterogenous architecture where possibly simultaneous read/write actions are carried out on shared data. Moreover, OR systems also read and write on the GIS database which is not shown in Figs. 2 and 3. It is known that uncontrolled simultaneous read-write and write-write actions cause data inconsistencies [9]. For this reason, the architecture shown in Fig. 1 must be analyzed if there can indeed be data inconsistencies as anticipated.

2.2 Background work

Design, maintenance, and protection of critical infrastructures have attracted many research in different disciplines. Emergency management systems are also critical infrastructures which aim at realizing rescue operations to minimize the negative effect of incidents, for example on critical infrastructures. Software intensive emergency management is supported by a system architecture which incorporates software/hardware technology to increase efficiency and effectiveness. We consider GIS systems and digital-twin-based control systems as the essential elements of software intensive emergency management systems.

A Geographic Information System (GIS) [10,11] is a tool for storing, querying, modifying, and analyzing data concerning geographic coordinates and attributes. During the last decades, several GIS systems have been proposed for a large category of applications. Examples are logistics, city planning, disaster management, etc. Our focus on GIS systems is from the perspective of emergency management.

CityGML, an application-independent information model and interchange format, is a commonly used international standard for defining the content of 3D city and

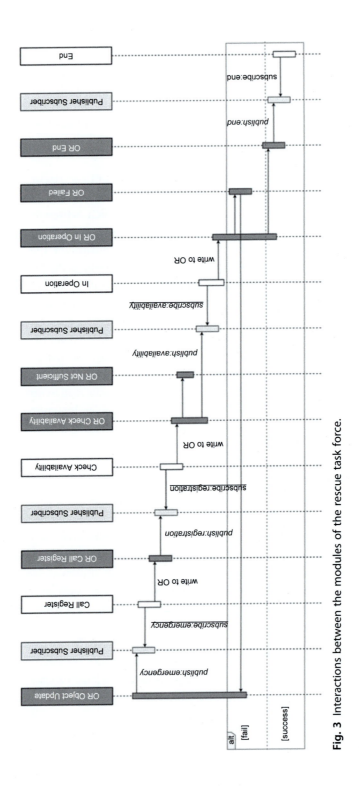

Fig. 3 Interactions between the modules of the rescue task force.

landscape models [12]. It provides facilities to extend data models by using UML-like extensions. ADE is a framework for adding new classes and attributes to the data model of CityGML [13]. To the best of our knowledge, CityGML and other GIS systems have not been extended to support digital-twin system-based emergency management before.

Intersection of GIS and emergency management are handled in a few textbooks. For example, in Ref. [14], various GIS concepts such as coordinate systems, map projections and cartography are studied within the context of emergency management. The focus of this book is not on software intensive systems. In Refs. [15,16], GIS-based studies are given to manage floods in certain areas. Several research works [17,18] have been in combining GIS systems and emergency management with the help of remote sensing technologies. A comprehensive evaluation of GIS within the context of emergency management is presented in Ref. [19]. This publication states that new research activities are needed in GIS metrics, educational approaches and data integration techniques. Several researchers [20] have proposed to include additional information to GIS databases, such as above ground and subsurface information of buildings, utility lines, and transportation networks to increase efficiency in rescue operations.

Digital-twin systems [4] are online digital representations of "some reality" which are generally used for control purposes. The digital representations can be used to monitor, alter, predict failures, and optimize real-world processes [21]. In recent years, digital twins have been applied in a variety of industrial sectors [22]. There are a few research works [23,24] on the application of digital twins for disaster management. These are mainly preliminary ideas. In Ref. [25], a digital-twin system is proposed for a virtual-reality-based training in crisis management.

The event-based approach is a natural architectural style for control systems. The components of a control system interact with each other by creating and receiving event notifications, where an event is any occurrence of a state change. The publisher-subscriber pattern [26] and the associated protocols are commonly used to transfer the occurrence of events among the system components [27].

To the best of our knowledge, there are no publications that address the expressivity and inconsistency problems which one may experience in GIS and digital-twin-based emergency management systems.

A considerable amount of research work [6,28,29] has been carried out to address the data inconsistency problem in simultaneous accesses to share data. In general, semaphores [30] are used to address this problem in centralized computing. In distributed processing, atomic transactions are commonly utilized [31,32,33]. To cope with various communication modalities, dedicated mechanisms are introduced such as atomic broadcasts and publisher subscriber protocols. To overcome the inconsistency problems, these techniques can be adopted in the realization of distributed digital twins. This requires, however, common and/or compatible protocols among the possible heterogeneous nodes of the system.

2.3 Research questions and the research method

Based on the examples shown in Sections 2.1.1 and 2.1.2, the following research questions are defined:

RQ1. How can GIS models be extended to represent vulnerable (critical) objects, damage, intensity of damage, and task forces as defined in Section 2.1.1?

RQ2. What are the data inconsistency challenges in implementing digital twin-system based emergency management architecture as defined in Section 2.1.2?

RQ3. How to represent events and actions for the purpose of emergency management?

RQ4. Within the context of the digital-twin approach, how to design an emergency management architecture without violating the data inconsistency constraints?

To seek answers to these research questions, we adopt two approaches:

1. UML based: to represent the static concepts of emergency management, UML extension mechanisms are used.
2. Formal notation based: to represent the dynamic concepts a state-based model is adopted. Further, model-checking techniques are applied to validate the specified correctness criteria of the designed architecture.

3 The approach

3.1 Extending GIS with earthquake management concepts

To support earthquake management, we consider it necessary to extend CityGML with the following abstractions:

- The required physical objects
- The disaster types which can be caused by earthquakes
- The intensity of disaster types
- The task forces
- The necessary attributes of these abstractions

3.1.1 Extending with physical objects

The physical objects that are required to be represented are introduced in Section 2.2. Some of these objects such as control centers and electricity stations are considered critical objects. We need to relate these objects to the concepts of CityGML and accordingly use the appropriate extension mechanism. Table 1 shows how the physical objects are related to the CityGML concepts.

In this table, we consider six kinds of physical objects to be modeled. The first row lists 23 building classes that are related to AbstractBuilding of CityGML. In CityGML, these can be grouped under the Building module. From the second to the fourth row, the physical objects correspond to the equivalent CityGML abstractions. They all have their own specific modules. Physical objects of the fifth row relate to AbstractTransportationSpace of

Table 1 Relation among physical objects and CityGML concepts.

No.	The physical objects to be modeled	Abstractions of CityGML	CityGML module
1	Base Station, Control Center, Electricity Station, Factory, Financial Center, Firefighter Station, Gas Pump Station, Government Building, Historical Buildings, Hospital, Logistic Center, Museum, Office Building, Police Station, Religious Building, Repair Center, Rescue Center, Residence, Restaurant, School, Shop, Shopping Center, Water Pump Station	AbstractBuilding	Building module
2	Bridge	AbstractBridge	Bridge module
3	Tunnel	AbstractTunnel	Tunnel module
4	Vegetation	AbstractVegetationObject	Vegetation module
5	Road, Track, Railway, Square, Waterways	AbstractTransportationSpace	Transportation module
6	Water Network, Gas Network, Electricity Network, Communication Network	GenericUnoccupiedSpace	Generic module

CityGML. The networks shown in row six are related to GenericUnoccupiedSpace. The physical objects from row two to five have correspondence in CityGML, therefore they can be directly modeled with the existing abstractions of CityGML. However, to represent the building classes in row 1 and the network classes in row 6, CityGML must be extended.

3.1.1.1 Extending CityGML with building classes

In Table 1, row 1 shows the physical objects which are defined as the extensions of AbstractBuilding. In Fig. 4, this extension is depicted using a UML diagram:

AbstractBuilding class is shown on the top-left of the figure. The extension is defined as an association from AbstractBuilding to ADEOfAbstractBuilding. This is termed Application Domain Extension (ADE) in CityGML. To this aim, the superclasses ADEOfAbstractBuilding and BuildingTypeExtension are used. ADEOfAbstractBuilding is a predefined class in CityGML. In Fig. 4, all predefined classes are shown in gray color. Class BuildingTypeExtension is used to represent the common attributes of all Building classes. The Building classes which are listed in row 1 of Table 1 extend class BuildingTypeExtension. The attributes of these classes can be categorized as (a) the inherited attributes from the superclasses such as AbstractBuilding, (b) the attributes

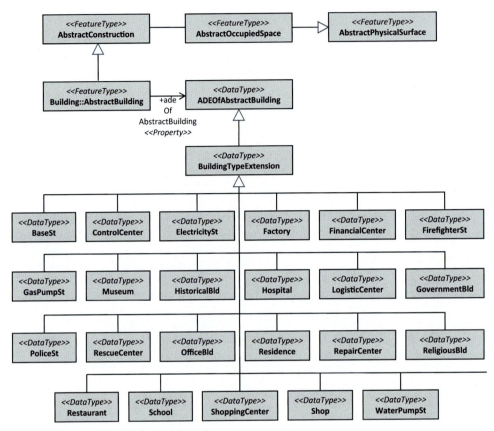

Fig. 4 UML class diagrams for building classes.

specific to building type, and (c) the attributes related to emergency management, for example, positioning of specific sensors such as seismic sensors for earthquake detection. We consider the specification of these attributes out of the scope of this chapter.

AbstractConstruction, AbstractOccupiedSpace, and AbstractPhysicalSurface form a predefined chain of superclasses which extend AbstractBuilding. For simplicity, we do not show the superclass of AbstractPhysicalSurface here.

3.1.1.2 Extending physical objects with a disaster model

Disaster modules are considered extensions of AbstractPhysicalSurface which are shown in middle-left of Fig. 5. The extension is defined as an association from AbstractPhysicalSurface to ADEOfAbstractPhysicalSurface. Class DisasterProperties extends ADEOfAbstractPhysicalSurface and declares zero or more objects of DisasterType.

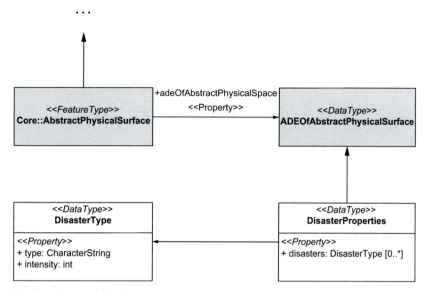

Fig. 5 UML class diagram for disaster types.

DisasterType defines attributes that characterize disasters such as Collapse, Fire, Flood, Landslide, Tsunami, Unhealth. DisasterType also specifies the intensity of the corresponding disaster.

Since AbstractPhysicalSurface is the superclass of AbstractBuilding, AbstractTunnel, AbstractBridge, AbstractVegetationObject, and AbstractTransportationSpace, the disaster model is inherited by all physical objects.

3.1.1.3 Extending physical objects with a disaster model

There have been several publications [34,35] on extending CityGML with UtilityNetworks. Their proposal is to extend class _CityObject for this purpose. However, class _CityObject is not supported in the new version of CityGML. Instead, as shown in Fig. 6, we propose to use an association relation between the class GenericUnoccupiedSpace and ADEOfGenericUnoccupiedSpace. Specification of the attributes of classes are considered out of the scope of this chapter.

3.2 Model checking of the digital-twin based emergency management system

To understand the consistency issues related to the architecture given in Section 2.1.2, we analyze formally the processing steps that are carried out. Fig. 7 shows the state model of one of the control centers in the notation of the UPPAAL [36] system. The start and end states are shown on the left and right sides of the figure, respectively. The states which are tagged with OR_<state name> belong to the system Object Runtime. The result of every action is registered first in the Object Runtime system and consequently, the

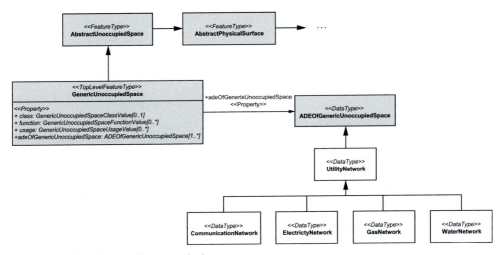

Fig. 6 UML class diagram for network classes.

database is updated where necessary. The states are triggered by the events that are generated in the Object Runtime system.

The state *allocate* is activated by *OR_control_room* when an emergency incident is registered in the database. The state *allocate* determines the tasks and resources that are necessary to cope with the emergency condition and this is registered in Object Runtime. This results in a database update operation and subsequently, the state *analyze* is triggered. The result is registered into the Object Runtime system and the state *optimize_dispatch* is activated. The result is registered first in the Object Runtime system and later in the database and the *in_action* state is activated. The state of task forces is continuously queried from the Object Runtime system until the condition (a) the mission is ended without any problem, (b) the mission is inconsistent, or (c) the mission is failed. The inconsistent state indicates that there is a deviation in value between the estimated task duration as derived from the digital twin model and the actual duration measured within the physical reality. This requires an analysis of the emergency handling process again. The mission can fail for many reasons such as defects in the equipment. In this case, the emergency handling process is carried out all over again.

Fig. 8 shows the state model of the management process of the rescue task force. Like in the previous example, the Object Runtime system is indicated with the tag OR_ <state name> and the start and end states are shown on the left and right of the figure, respectively. A request for a rescue team is detected by the state *OR_task_detect* after querying the database. This results in triggering the state *call_register* which initiates an update operation on the Object Runtime system. Consequently, the state *check_availability* is activated to determine if there are enough resources as it was estimated initially. In case resources are not sufficient due to unexpected changes, the system polls the database

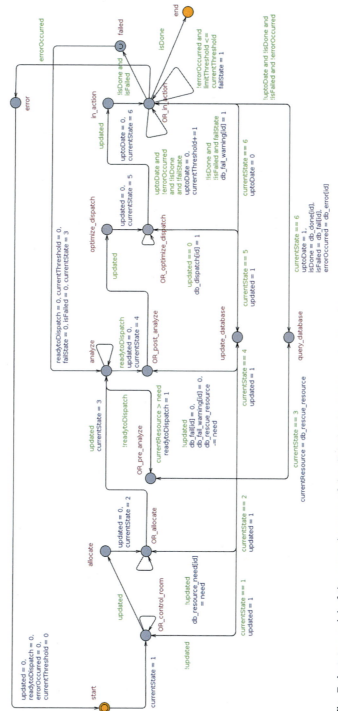

Fig. 7 A state model of the control center of the digital twin system in Section 2.1.2.

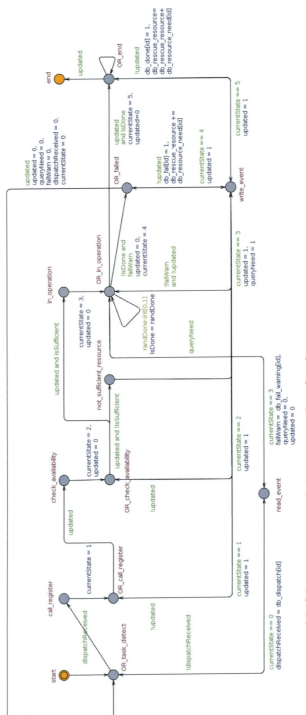

Fig. 8 A state model of the rescue team with the corresponding digital-twin system.

through the states *OR_not_sufficient_resource* and *OR_check_availability*. If there are enough resources, the state *in_operation* is triggered which activates the state *OR_in_operation*. This state polls the database until the mission is completed successfully or the deadline is reached. In the latter case, the state *time_exceed* is triggered and the whole process restarts again.

A critical issue in designing such a system is to assure that the following constraints are not violated:

1. From the start state the end state is reachable, which is expressed in the following UPPAAL formula:

$$E <> controlCenter.end \tag{1}$$

Here, *controlCenter* is variable and the end is the state defined in the UPPAAL model. This expression checks if the end state of the control center is reachable.

2. The mutual exclusion of read/write and write/write operations on the critical regions of the digital-twin system must be assured.

To test this constraint, the following UPPAAL expression checks if *rescueTeam1* and *rescueTeam2* can be in the operation state simultaneously:

$$E <> controlCenter1.in\_action\ and\ rescueTeam1.in\_operation\ and$$
$$controlCenter2.in\_action\ and\ rescueTeam2.in\_operation \tag{2}$$

Here, *rescueTeam1* and *rescueTeam2* are the variables and *in_operation* is the state in both rescue team models.

The following parameters are used in verifying this expression:

- The resource need of rescueTeam1: 3
- The resource need of rescueTeam2: 3
- Available resources: 5

Given these parameters, *rescueTeam1* and *rescueTeam2* cannot be *in_operation* phase due to a lack of resources. However, the UPPAAL system indicates that even with these parameter values, both rescue teams can be in *in_operation* state. This indicates that the mutual exclusion of operations in the critical region cannot be assured.

One may claim that by using locks for each access, inconsistencies can be avoided. Unfortunately, the heterogeneous and distributed nature of such systems and the lack of standards make it not practical to adopt a common locking scheme for the overall architecture.

3.3 Extension GIS with event-based digital twins

To overcome the inconsistency problem, the architecture depicted in Fig. 9 is proposed. At the top of the figure, the distributed management system is shown as a dashed rectangle

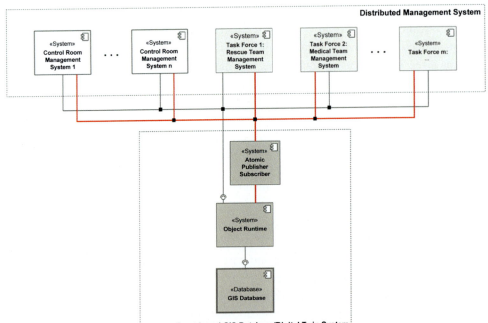

Fig. 9 The proposed architecture is based on event based GIS database.

where its subsystems, control rooms, and task forces are shown symbolically in white and light gray colors, respectively. The event-based GIS database/digital twin system is shown at the bottom of the figure. In contrast to the architecture that is given in Section 2.1.2, here, a common OR system is adopted as a layer at the top of the GIS database. The event-based communication between the subsystems of the distributed management system and OR system is shown by a thick line in red color which is realized through an atomic publisher-subscriber subsystem. The processes in the management system subscribe to the required events and they are notified by the publisher-subscriber system when the events are available. The notification process is realized by an atomic broadcast protocol [31,32,33] where no simultaneous read/write and write/write accesses can occur in between the updates. The subsystems of the distributed management system can also directly call on the OR system to register their state of execution. This is shown as a thin line.

The state model of the architecture of the control center is shown in Fig. 10. The differences between the state models of the new architecture and the architecture shown in Fig. 1 are due to the introduction of atomic actions. To this aim, the model shown in Fig. 11 is defined. Here, there are two states: *unlock_state*, and *lock_state*. When a critical operation has to be carried out in the control center for example, such as in the state *OR_optmize_dispatch* in Fig. 10, the signal *critical_region_lock* is executed. This causes

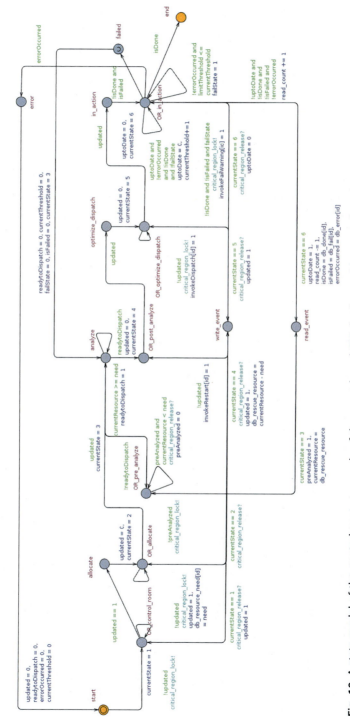

Fig. 10 A state model of the control center example of the event-based GIS database with atomic broadcast.

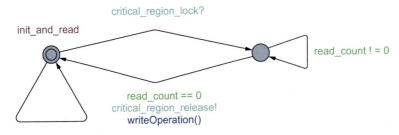

Fig. 11 A state model of the lock mechanism with read priority.

the transition to *lock_state* shown in Fig. 11, if the locking mechanism is in *unlock_state*. In *lock_state*, the critical operation is delayed until there are no active read operations. This strategy implements read priority. When the critical region is available for the write operation, the transition to *unlock_state* is realized. In Fig. 11, *writeOperation()* is an abstraction of a series of operations. The parameters to be updated in the database are denoted for example, by the expression *invokeDispatch*. For brevity, we do not show all the implementation details here.

Similar lock mechanisms are also defined for the state model of the rescue team as shown in Fig. 12. For example, to enter their critical regions, the states *OR_in_operation*, *OR_failed*, and *OR_end* execute first the signal *critical_region_lock*, wait for the acknowledgment to write in the critical state, then they write and release the critical region. Otherwise, the models shown in Figs. 8 and 12 are basically the same.

To verify the consistency of the data in the event-based GIS database with atomic broadcast, we execute the UPPAAL statement F2. In the previous model of the architecture shown in Fig. 1, this expression evaluates True which indicates data inconsistency. Nevertheless, the combined evaluation of the models shown in Figs. 10–12 with two control centers and rescue teams show that a path cannot be found for the formula (2). This indicates data consistency.

4 Discussion and related work

This chapter is based on the assumption that the emergency management system is implemented by a heterogeneous architecture where no standard locking protocols exist. Furthermore, the emergency management system is based on an event-based digital-twin system in which the management system is activated by a set of events that are generated from the digital twin. If these assumptions are not valid, the research questions RQ2 to RQ4 may not be relevant.

Naturally, the UPPAAL models are defined based on certain assumptions over reality. The details of executions that may happen in actual implementations are abstracted away. Furthermore, the models do not reflect all the details of atomic broadcast protocols in the

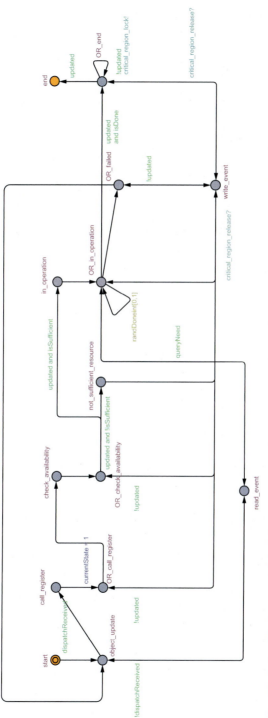

Fig. 12 A simplified state model of the rescue team of the event-based GIS database with atomic broadcast.

publisher subscriber system. If required, the reader may refer to the publication [37], which proposes a UPPAAL model for publisher subscriber systems. We assume that the details of the UPPAAL models shown in this chapter are adequate for the research questions presented in Section 2.3.

We observe two possible challenges in the realization of the proposed architecture that is shown in Fig. 8. A central implementation of event-based GIS database/digital-twin system may suffer from poor time performance in the case of a large management system with high data access. To overcome this problem, a distributed implementation of the system with better performance values can be investigated. Secondly, the parties involved in emergency management must agree in using the proposed architecture as a common platform.

There are many works that address the subproblems related to this chapter. For example, several researchers have proposed extending CityGML for certain domains, such as utility networks [34,35], and energy [38]. Recently, a substantial amount of research work has been published on digital twins [4,21,22,23,24]. The concurrency control issues for simultaneous read/write and write/write accesses have been studied in computer science for decades [39,40,41]. There are also many publications on system design including atomic actions and event-based communication mechanisms [26,27,28]. Model checking techniques have been successfully applied to system verification [37,42]. In this chapter, we benefited from all these research contributions. To the best of our knowledge, there has not been any publication which addresses the same research questions. There is no related work that exactly matches our proposal.

5 Conclusion

This chapter is formulated around four research questions. The first research question is about extending GIS models to represent the missing information related to earthquakes. To address this question, Section 2.1.1 first identifies the entities to be represented. In Section 3.1.1, the necessary extensions to CityGML are presented. By using the UPPAAL system, Section 3.2 demonstrates the possible data inconsistency conditions if no standard locking protocols are defined in heterogeneous twin-system-based architectures. Section 3.3 proposes an architecture in which the extended GIS database is combined with the event-based layer for earthquake management. It is formally proven that this approach can solve the data inconsistency problem.

Acknowledgment

This work has been generously supported by the TÜBİTAK BİDEB program 2232 International Fellowship for Outstanding Researchers.

References

[1] EU Commission, Communication from the commission on a European Programme for critical infrastructure protection, in: COM (2006) 786, 2006.

[2] B.L. Golden, A.A. Kovacs, E.A. Wasil, Vehicle routing applications in disaster relief, in: Vehicle Routing: Problems, Methods, and Applications, second ed., Society for Industrial and Applied Mathematics, 2014, pp. 409–436 (Chapter 14).

[3] D.S. Naidu, Optimal Control Systems, CRC Press, 2002.

[4] I. Errandonea, S. Beltrán, S. Arrizabalaga, Digital twin for maintenance: a literature review, Comput. Ind. 123 (2020) 103316.

[5] S. Malakuti, M. Akşit, Event-based modularization of reactive systems, in: Concurrent Objects and Beyond, Springer, Berlin, Heidelberg, 2014, pp. 367–407.

[6] P.-J. Courtois, F. Heymans, D.L. Parnas, Concurrent control with "readers" and "writers", Commun. ACM 14 (10) (1971) 667–668.

[7] T. Kutzner, K. Chaturvedi, T.H. Kolbe, CityGML 3.0: new functions open up new applications, PFG J. Photogramm. Remote Sens. Geoinf. Sci. 88 (1) (2020) 43–61.

[8] R. Rajkumar, M. Gagliardi, L. Sha, The real-time publisher/subscriber inter-process communication model for distributed real-time systems: design and implementation, in: Proceedings Real-Time Technology and Applications Symposium, IEEE, 1995.

[9] R.H. Katz, et al., Implementing a cache consistency protocol, ACM SIGARCH Comput. Arch. News 13 (3) (1985) 276–283.

[10] G.I. Cooperative, F. Collins, The unique qualities of a geographic information system: a commentary, Photogramm. Eng. Remote. Sens. 54 (11) (1988) 1547–1549.

[11] K.-.T. Chang, Geographic information system, in: International Encyclopedia of Geography: People, the Earth, Environment and Technology: People, the Earth, Environment and Technology, John Wiley & Sons, 2016, pp. 1–9.

[12] K.A. Ohori, et al., Modeling cities and landscapes in 3D with CityGML, in: Building Information Modeling, Springer, Cham, 2018, pp. 199–215.

[13] F. Biljecki, K. Kumar, C. Nagel, CityGML application domain extension (ADE): overview of developments, Open Geospat. Data Softw. Stand. 3 (2018) 1–17.

[14] B. Tomaszewski, Geographic Information Systems (GIS) for Disaster Management, Routledge, 2020.

[15] S.H. Abbas, et al., GIS-based disaster management: a case study for Allahabad Sadar Sub-District (India), Manag. Environ. Qual. 20 (2009) 33–51.

[16] P. Tran, et al., GIS and local knowledge in disaster management: a case study of flood risk mapping in Viet Nam, Disasters 33 (1) (2009) 152–169.

[17] A. Frantzova, Remote sensing, GIS and disaster management, in: Third International Conference on Cartography and GIS, Nessebar, Bulgaria, 2010.

[18] L. Montoya, Geo-data acquisition through mobile GIS and digital video: an urban disaster management perspective, Environ. Model. Softw. 18 (10) (2003) 869–876.

[19] B. Tomaszewski, et al., Geographic information systems for disaster response: a review, J. Homel. Secur. Emerg. Manag. 12 (3) (2015) 571–602.

[20] D.F. Laefer, A. Koss, A. Pradhan, The need for baseline data characteristics for GIS-based disaster management systems, J. Urban Plann. Dev. 132 (3) (2006) 115–119.

[21] S. Weyer, et al., Future modeling and simulation of CPS-based factories: an example from the automotive industry, IFAC-PapersOnLine 49 (31) (2016) 97–102.

[22] F. Tao, et al., Digital twin in industry: state-of-the-art, IEEE Trans. Industr. Inform. 15 (4) (2018) 2405–2415.

[23] C. Fan, et al., Disaster city digital twin: a vision for integrating artificial and human intelligence for disaster management, Int. J. Inf. Manag. 56 (2021) 102049.

[24] Z. Xianyong, T. Zhihong, P. Shirui, Research on the construction of smart city emergency management system under digital twin technology: taking the practice of new coronary pneumonia joint prevention and control as an example, in: Proceedings of the 2020 4th International Seminar on Education, Management and Social Sciences (ISEMSS 2020), 2020.

[25] P.K. Kwok, et al., User acceptance of virtual reality technology for practicing digital twin-based crisis management, Int. J. Comput. Integr. Manuf. 34 (7–8) (2021) 874–887.

[26] K. Palanivel, V. Amouda, S. Kuppuswami, Publisher-subscriber: an agent system for notification of versions in OODBs, in: 2009 International Conference on Intelligent Agent & Multi-Agent Systems, IEEE, 2009.

[27] G. Mühl, L. Fiege, P. Pietzuch, Distributed Event-Based Systems, Springer Science & Business Media, 2006.

[28] D.S. Parker, et al., Detection of mutual inconsistency in distributed systems, IEEE Trans. Softw. Eng. 3 (1983) 240–247.

[29] C. Danilowicz, N.T. Nguyen, Consensus methods for solving inconsistency of replicated data in distributed systems, Distrib. Parallel Databases 14 (1) (2003) 53–69.

[30] M. Yabandeh, et al., CrystalBall: predicting and preventing inconsistencies in deployed distributed systems, in: The 6th USENIX Symposium on Networked Systems Design and Implementation (NSDI'09), 2009.

[31] P.J. Marandi, et al., Ring Paxos: a high-throughput atomic broadcast protocol, in: 2010 IEEE/IFIP International Conference on Dependable Systems & Networks (DSN), IEEE, 2010.

[32] M. Poke, T. Hoefler, C.W. Glass, Allconcur: Leaderless concurrent atomic broadcast, in: Proceedings of the 26th International Symposium on High-Performance Parallel and Distributed Computing, 2017.

[33] A. Gągol, et al., Aleph: Efficient atomic broadcast in asynchronous networks with byzantine nodes, in: Proceedings of the 1st ACM Conference on Advances in Financial Technologies, 2019.

[34] T. Becker, C. Nagel, T.H. Kolbe, Semantic 3D modeling of multi-utility networks in cities for analysis and 3D visualization, in: Progress and New Trends in 3D Geoinformation Sciences, Springer, Berlin, Heidelberg, 2013, pp. 41–62.

[35] I. Hijazi, et al., Initial investigations for modeling interior utilities within 3D geo context: transforming IFC-interior utility to CityGML/UtilityNetworkADE, in: Advances in 3D Geo-Information Sciences, Springer, Berlin, Heidelberg, 2011, pp. 95–113.

[36] G. Behrmann, et al., Developing UPPAAL over 15 years, Softw. Pract. Exp. 41 (2) (2011) 133–142.

[37] Q.-Q. Lin, et al., Modelling and verification of real-time publish and subscribe protocol using Uppaal and Simulink/Stateflow, J. Comput. Sci. Technol. 35 (6) (2020) 1324–1342.

[38] G. Agugiaro, et al., The energy application domain extension for CityGML: enhancing interoperability for urban energy simulations, Open Geospat. Data Softw. Stand. 3 (1) (2018) 1–30.

[39] E.W. Dijkstra, Solution of a problem in concurrent programming control, in: Pioneers and Their Contributions to Software Engineering, Springer, Berlin, Heidelberg, 2001, pp. 289–294.

[40] D.E. Knuth, Additional comments on a problem in concurrent programming control, Commun. ACM 9 (5) (1966) 321–322.

[41] N.G. de Bruijn, Additional comments on a problem in concurrent programming control, Commun. ACM 10 (3) (1967) 137–138.

[42] Y. Fei, H. Zhu, X. Li, Modeling and verification of NLSR protocol using UPPAAL, in: 2018 International Symposium on Theoretical Aspects of Software Engineering (TASE), IEEE, 2018.

CHAPTER 8

Analyzing systems product line engineering process alternatives for physical protection systems

Bedir Tekinerdogan[a] **and İskender Yakın**[b]

[a]Information Technology Group, Wageningen University & Research, Wageningen, The Netherlands
[b]ASELSAN A.Ş., Ankara, Turkey

1 Introduction

Critical infrastructures play a pivotal role in maintaining the safety, security, and overall functioning of modern societies. They encompass a wide range of sectors, including energy, transportation, telecommunications, water supply, and more. These infrastructures are often interdependent and vulnerable to various threats, both natural and human-induced. As a result, ensuring the protection of these vital systems is paramount to national security and public welfare.

Physical protection systems (PPS) serve as an essential component in the preservation and security of critical infrastructures. These systems are designed to prevent unauthorized access, detect intrusion attempts, and respond to security breaches effectively.

A PPS is a combination of people, procedures, and equipment designed to protect assets and facilities from malicious attacks [1–3]. These systems are used in a variety of settings, including airports, rail hubs, highways, medical facilities, bridges, electrical grids, dams, oil refineries, and water systems. The design of an effective PPS requires careful evaluation of the resources needed to provide the necessary protection. Without accurate assessment and design, significant resources may be wasted on unnecessary protection or, worse, fail to offer adequate security in critical areas of the system that need protection. To design and analyze PPSs effectively, a variety of PPS design methodologies have been proposed in the literature.

A PPS offers the key functionalities of deterrence, detection, delay, and response to prevent malicious acts by adversaries. Therefore, a PPS design methodology explicitly considers these concerns and specifies how to effectively address them. While effective PPSs have been designed using the methods currently in use, there is a potential for large-scale reuse in the development of various PPSs. This paper focuses on the context of an industrial large-scale systems engineering company that is developing a wide range of PPSs. While each system that needs protection is unique, recurring development

Management and Engineering of Critical Infrastructures
https://doi.org/10.1016/B978-0-323-99330-2.00002-7

activities can be observed over the entire systems engineering lifecycle [4,5]. Thus, there is a significant potential for reuse to support the development process.

In many industrial practices, reuse has been an important objective and has been extensively examined in the literature. Initially, reuse was mainly focused on a small scale, and the approach was ad hoc. However, the most significant and extensive benefits are realized through a systematic, large-scale approach to reuse. This concept has led to the development of the Systems Product Line Engineering (SPLE) approach, which aims to exploit reuse throughout the entire lifecycle process [6,7]. A product line refers to a group of systems that share a managed set of features, which satisfy the particular needs of a specific market segment or mission. These systems are developed from a common set of core assets in a defined manner.

Numerous studies have analyzed and discussed the advantages of adopting a product line approach, which includes large-scale productivity gains, improved product quality, lowered product risk, reduced time to market, enhanced market agility, increased customer satisfaction, efficient use of human resources, ability to achieve mass customization, maintain a market presence, and sustain unprecedented growth. The PLE process can be applied to develop a product line for any domain, and it is independent of the domain. In the literature, different PLE approaches have been introduced, including single PLE, multiple PLE, and feature-based PLE. Selecting one of these three PLE process alternatives for PPSs is not trivial. However, guidance on selecting PLE process alternatives for PPSs has not been addressed in the literature. This paper presents the background and relevant concepts of the PPS domain, describes the three types of PLE approaches, and presents a method for selecting the proper PLE approach. Finally, the paper illustrates the application of the method to a real industrial case study of a large-scale systems company.

The paper is organized as follows: Section 2 provides background information on PPS, Section 3 describes the three different PLE process alternatives, Section 4 presents the method for selecting these approaches for PPS, Section 5 presents related work, and Section 6 concludes the paper.

2 Physical protection systems

A PPS typically consists of various physical security measures such as intrusion detection sensors, cameras, access control devices, barriers, and response protocols. Automated subsystems within the PPS transmit information and video footage to a central alarm station (CAS) where operators can take appropriate action based on specific information. The metamodel shown in Fig. 1 outlines the fundamental components of a PPS. Although PPS may vary in their specific details, there is a consensus that the system should include three critical elements: detection of potential threats, delaying the adversary, and response

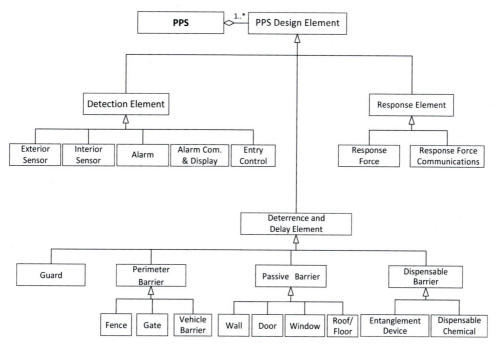

Fig. 1 PPS design elements.

by security personnel. These three elements must be carried out in the order of detection, delay, and response to ensure an effective PPS. It is important that the time required for detection and response is shorter than the time needed for the adversary to complete their task.

Fig. 2 illustrates the primary activities involved in the PPS design process. The process consists of three main stages: identifying PPS objectives, designing the PPS, and evaluating the PPS. The first stage involves characterizing the facility, defining threats, and specifying the target to be protected. Designing the PPS involves three crucial activities: detection, delay, and response while keeping in mind operational, safety, legal, and economic constraints. Detection should be as far from the target as possible, and delay should be near the target. During PPS design, the designer should consider the associations between detection and assessment, as well as between response and response force communications, and use equipment combinations that complement each other to protect against any weaknesses. The final stage is evaluating the design PPS using techniques such as Path Analysis, Scenario Analysis, and System Effectiveness Analysis. The outcome of the evaluation is a system vulnerability assessment that can identify previously unnoticed weaknesses or confirm that the design is feasible and effectively achieves the protection objectives.

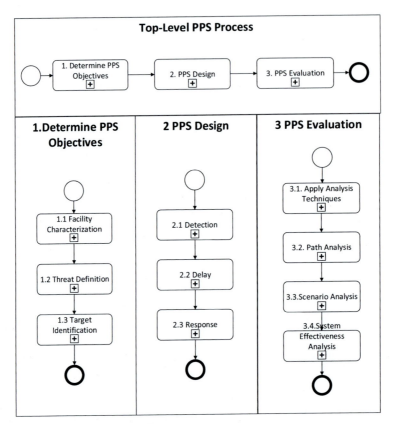

Fig. 2 Physical protections system development process.

3 Product line engineering process alternatives

3.1 Single system development—Non-PLE

There are various methodologies for developing Physical Protection Systems (PPS), one of which is the traditional single system engineering approach (shown in Fig. 3) that does not incorporate Product Line Engineering (PLE) practices. In this approach, there may be a product portfolio, but each system is designed separately without utilizing PLE practices such as explicit commonality variability modeling, a product family architecture, and a shared asset base. The fundamental task in this approach is to locate a previously manufactured or delivered PPS system that most closely matches the formal requirements and needs of a potential new customer. Once identified, the engineering artifacts of the existing PPS are reused and modified to fully satisfy the new PPS's requirements. This approach can lead to a number of challenges, such as the inability to capture and manage the commonalities and variabilities across the product portfolio. Furthermore, the cost and time to develop new PPS may not decrease significantly, as each new system requires

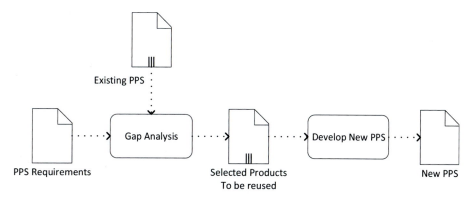

Fig. 3 Ad hoc, non-PLE reuse strategy for developing PPS.

significant effort to modify existing engineering artifacts to fit the specific customer's needs.

On the other hand, Product Line Engineering (PLE) offers a more systematic approach to PPS development by exploiting commonalities and variabilities across the product line, thereby reducing development time and cost. PLE also supports the reuse of engineering artifacts across the product line, enabling efficient development and maintenance of the PPS.

3.2 Single product line engineering

The conventional approach to PPS development overlooks the opportunity for reusing common elements across different systems. Despite their differences, PPSs have certain shared characteristics and structures that can be leveraged for reuse. By adopting a domain-driven design approach [8], a company can develop a product line of PPSs using a common set of core assets in a predetermined manner. This approach offers several advantages over single system development, such as increased efficiency, consistency, and cost-effectiveness.

However, adopting a product line engineering approach requires additional investments compared to the traditional approach. The initial investment may only result in a return on investment (ROI) after developing more than one product, which is commonly known as the break-even point. Although different PLE processes have been proposed, they all involve domain engineering, product line architecture, and application engineering. In the domain engineering phase, a reusable platform and product line architecture are developed, while the application engineering phase utilizes the results of the domain engineering process to develop the products. Technical and organizational management processes are employed to manage the overall development process. Fig. 4 provides an overview of a typical product line engineering process.

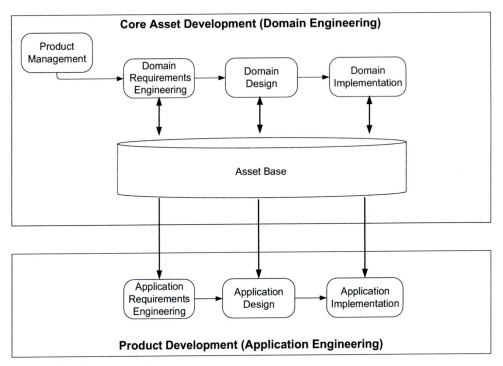

Fig. 4 Product line engineering process.

3.3 Multiple product line engineering

In order to tackle the challenges that arise from the complexity of a large system, a multiple product line engineering (MPLE) approaches is utilized, which permits products to be constructed from various, yet interrelated product lines [9]. This is achieved through the use of composite patterns for creating product lines, which can either be flat software product lines or composite product lines (CPLs) that contain other product lines. The general structure of this approach is depicted in Fig. 5.

It should be noted that in this approach, each product line is characterized by a two-life cycle process, which involves domain engineering with product management and application engineering. As a result, two separate architectures are developed, namely the product line architecture and the application architecture. The former represents the common and variant structures of a set of products within the selected product line, while the latter represents the architecture of a single system. The individuals responsible for defining these architectures are referred to as the product line architect and the application architect.

It is important to recognize that the MPLE approach necessitates additional investments in comparison to traditional single-system development. Nevertheless, it can

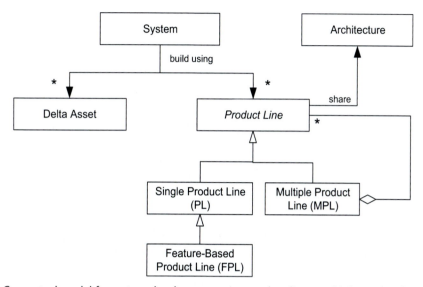

Fig. 5 Conceptual model for system development using product lines, multiple product lines.

potentially result in a higher return on investment due to the possibility of reusability and the ability to efficiently develop multiple products within interrelated product lines.

3.4 Profile-based product line engineering

The third alternative for PLE is to utilize feature profiles within feature modeling to represent the domain by explicitly showing common and variable features. Features refer to relevant system properties used to capture similarities or differences. A feature model is usually represented as a tree structure, with the root node representing the system and its descendant nodes being the features. The selection of features in a feature diagram represents a feature configuration, which often corresponds to a system instance. Not all configurations are feasible, and feature constraints are used to indicate this explicitly.

In the application engineering process, ensuring a thorough and accurate configuration of the SPL that meets the requirements is crucial. However, selecting the correct and complete configuration based solely on a feature diagram can be challenging. To tackle this problem, staged configuration can be utilized (Fig. 6). This approach involves

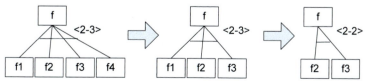

Fig. 6 Example specialization of a feature model.

selecting features in stages from a family feature model until a final feature model is obtained that precisely describes the configuration of a single application.

Reusable intermediate configurations in the staged configuration are termed feature profiles. Feature profiles are semicomplete feature models in which the features for the required (sub)product family have been selected during the domain engineering process. Feature profiles can be defined to support reuse and reduce the effort required in the application engineering process. For a large product line scope in which we can identify product family categories, it is convenient to develop feature profiles for these categories. The overall process that we define for feature-driven PLE is shown in Fig. 7, where the overall domain is decomposed into several subdomains for which profiles are developed, and in the application engineering process, the profiles are reused instead of directly using the family models.

Note that each feature model defines a set of configurations, and the use of feature profiles can reduce the number of configurations that developers must manually select, reducing the effort required in the application engineering process.

When considering a broad scope, there are implications for adopting a PLE approach. One option is to continue with a single PLE approach, utilizing a single domain engineering and application engineering process with one reusable asset base. Alternatively, a multiscope perspective using MPLE or profile-driven PLE may be more appropriate.

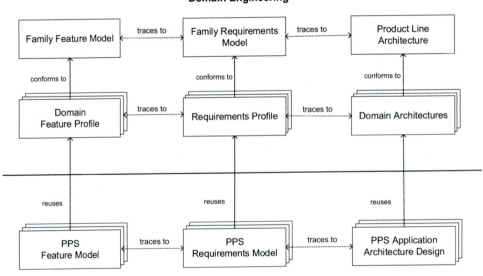

Fig. 7 Profile-based PLE.

It is important to carefully select and justify the chosen method as it will have both technical and organizational impacts.

To address this decision-making process, we have developed an overall research method, illustrated in Fig. 8. This includes parallel analysis of the PPS method, PLE methods, and a case study. These activities can be performed in any order. Following completion of the analysis phase, the three different PLE methods are evaluated, and a decision is made on the most feasible approach. Our research also aims to answer the following questions:

RQ1. What are the current PPS methods?

RQ2. What are the current PLE methods?

RQ3. How can we analyze the PLE methods and determine the best method for PPS?

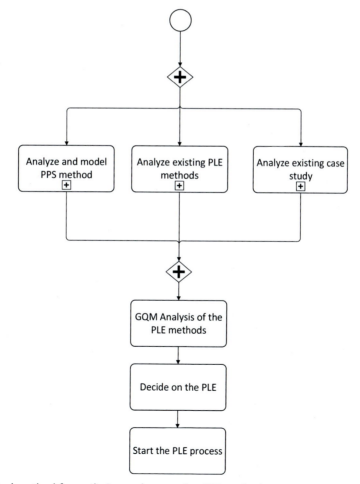

Fig. 8 Adopted method for analyzing and comparing PLE method.

To select the most suitable PLE method, we adopt the Goal-Question-Metric approach, a measurement model promoted by Basili and others [10]. This approach assumes that to measure in a purposeful way, an organization must first specify project goals, define a set of questions for each goal, and finally define metrics associated with each question to measure progress in a meaningful way.

To effectively implement the Goal Question Metric (GQM) approach, it is recommended to follow a six-step process that involves identifying business goals, generating quantifiable questions based on models, specifying relevant measures, developing data collection mechanisms, analyzing data in real time for corrective action, and assessing conformance and making recommendations for future improvements [10,11]. Properly defining the goals, questions, and metrics is crucial in this approach, where GQM goals consist of five facets: object, purpose, focus, viewpoint, and environment.

In this study, we examine four different development approaches, including the absence of Product Line Engineering (PLE), single PLE, multiple PLE, and Profile-based PLE.

Table 1 outlines the five goals and their corresponding questions and metrics for measuring the success of the PLE approach. The goals include optimizing asset reuse, increasing productivity, managing complexity, easing organizational management, and easing product line management. The questions focus on aspects such as manpower requirements, commonality and variability, and organizational structure. The metrics include percentages of reused assets per product, depth of feature diagram, and maintenance effort in man-months. Subjective evaluations by project management are also considered for some metrics. By defining these goals, questions, and metrics, the GQM approach can be used to make informed decisions and drive improvements in the PLE approach.

4 Evaluation method

Selecting systems within a product line is a deliberate and strategic process that considers their potential economic benefits and efficient development with core assets. Effective product management is a critical aspect of Product Line Engineering (PLE) and encompasses the control of the development, production, and marketing of the product line and its applications. Input for product management is provided by top management, who define the company's goals. The primary objective of product management is to contribute to the company's entrepreneurial success by integrating the development, production, and marketing of products that meet customer needs. A crucial responsibility of product management is managing the company's product portfolio, which includes the types of products offered by the product line organization. Portfolio management is a continuous decision-making process that evaluates and updates the portfolio based on market and business requirements.

Table 1 GQM results for the targeted PLE approaches.

Goal	Questions	Metric
Enhance asset reuse	• How much are assets reused in the development process? • What is the distribution of reused assets across different products? • Is there any functional overlap among different product lines?	• % of reused assets per product • Asset distribution over products • Common features over assets (in multiple PLs)
Boost efficiency and output	• How many people are needed for domain engineering activities? • How many people are needed for application engineering activities?	• Man month per PL • Man month per application
Streamline complexity management	• What level of complexity is involved in the commonality and variability of the system? • Have the boundaries of the domain been appropriately defined and separated across the product lines? • Is each product line focused on a single domain, ensuring proper separation of domains across product lines?	• Depth of feature diagram • Features • PL per domain • Domain per PL
Simplify organizational management	• What is the organizational structure required for product line development? • Are the teams properly defined and separated for domain-specific product line activities? • Are the product line development activities well-aligned with the organizational teams?	• Organization hierarchy depth
Facilitate product line management	• How much effort is required to add, remove, or update product features within a product line? • How much effort is required to add, remove, or update product features across different product lines? • Can the organizational structure easily handle the addition of new products or domains?	• Man month for total maintenance activities per PL • Man month for total maintenance activities per CPL • Subjective evaluation by project management
Simplify product composition	• What is the effort required to compose products from different product lines?	• Subjective evaluation by project management

Adapted from B. Tekinerdogan, O. Ozkose Erdogan, O. Aktug, Supporting incremental product development using multiple product line architecture, Int. J. Knowl. Syst. Sci. 5 (2014) 1–16.

The product management subprocess involves defining a product roadmap that outlines the estimated product line and identifies the significant common and variable features of all product line applications. This product roadmap is then handed over to domain requirements engineering, which is responsible for defining the product requirements used to design the product line architecture. It is important to note that effective product management requires continuous evaluation of the product line and its applications to ensure that they meet the customer's needs and are aligned with the company's goals.

Scoping techniques are utilized in product management to define the boundaries of the product line and determine which products are included and excluded. The success of the product line heavily depends on an appropriate scope, as a scope that is too large may result in excessive variation among products, making it challenging to achieve commonality and variability. Conversely, a scope that is too small may impede future growth and prevent the realization of economies of scope, resulting in a stagnant product line. Thus, scoping should be carefully considered to mitigate these risks.

The term PPS is broad and encompasses several subdomains, with a many-to-many relationship between domains and systems. Systems may not cover an entire domain and could belong to multiple domains. Similarly, a domain can be utilized in various systems, and multiple domains may be integrated into one system. While Fig. 9 illustrates the

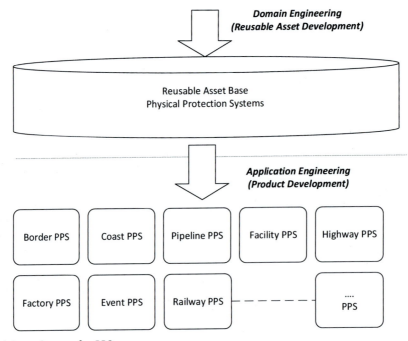

Fig. 9 Adopted scope for PPS.

overall domain with the identified subdomains considered in the project, it is important to note that the global scope is not limited to these domains.

The evaluation of the four alternatives according to the defined goals is presented in Table 2, where all six goals from Table 1 hold equal weight. The evaluation scale includes five values ranging from very negative to very positive ($-$, $- -$, $+$, $++$, $+-$), and the results are based on these criteria.

From Table 2 we can observe that the first alternative of not using a PLE approach has been evaluated negatively. It fails to optimize reuse, and as such does not increase productivity, while complexity is high. On the other hand, organizational management is considered neutral, product line management is doable because of the focus of a single system. The composition of products is evaluated positively because no subdomains are specified, and products can be configured without constraints. The 3 PLE approaches have been evaluated positively implying the need for systematic reuse. A single PLE provides the benefits of reuse and higher productivity but for a rather broad PPS scope, the complexity management will become challenging. An MPLE approach is not favored due to the difficulty of managing separate PLEs. In the end, the profile-based PLE was found most feasible offering both reuse and avoiding the added complexity of MPLE.

After the selection of a profile-based PLE, we had to decide on the product domain categories. Initially, it was stated that a broad PPS domain would be selected. But the question remained of which PPS subdomains the profiles would be developed. Based on company objectives the following product domains have been selected for which profiles would be developed.

- Border Security

  Involves the protection of the national borders from the illegal movement of weapons, drugs, illegal imports, and people, while promoting lawful trade and travel.

Table 2 Evaluation matrix for PLE alternatives.

Goals	No PLE	Single PLE	MPLE	One PLE profiles
Optimize reuse (reduce overlapping functionality)	$- -$	$++$	$+$	$++$
Increase productivity	$- -$	$+$	$- -$	$+$
Manage complexity	$- -$	$+-$	$-$	$+$
Ease organizational management	0	0	0	0
Ease product line management	$+-$	$+-$	$+-$	$+$
Ease composition of products across business units	$++$	$+$	$+$	$+$

- Pipeline Security
    Refers to the measures taken to safeguard land-based pipelines from acts of sabotage, illegal tapping, and terrorist attacks.
- Critical Settlement Security
    Protection of civil settlements
- Coastal Security
    Protection of the coast and territorial waters
- Mobile Platform Security
    Protection of the adopted mobile platforms

This means that for these five domains the feature profiles, requirements profiles and the design profiles would be made reusable in the asset base. For feature profiles this is illustrated in Fig. 10. Here, each profile is in essence a subset of the general PPS feature model. To develop the profile, it is necessary to select or deselect the features that are needed for the corresponding product category (for example, Border PPS). On the other hand, not all features will be determined and as such variability will be left that can only be decided at the application engineering for the specific product of the product category.

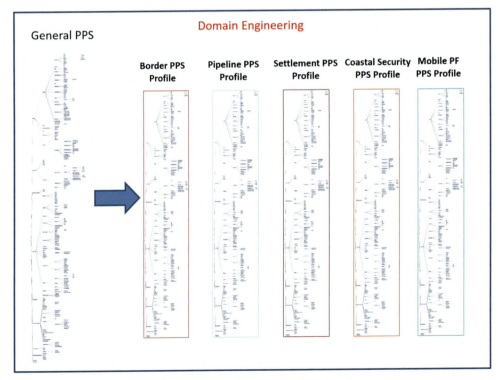

Fig. 10 Illustration of profile development for five different PPS domains.

5 Related work

The field of physical protection systems (PPS) has been extensively studied, with several studies addressing the design and evaluation of PPS [2,3,11,12]. However, many of these studies have not considered the potential for systematic reuse through systems product line engineering (PLE), which can lead to reduced development costs, improved quality, and faster time-to-market. In our earlier study, we provided a family feature modeling approach that explicitly shows the common and variant features for developing PPSs [13] but did not consider the process activities. To address this, we proposed an integrated process for PLE of PPSs that combines model-based development and PLE [14], providing a more comprehensive approach to PPS design.

To ensure a proper PPS design, a well-defined architecture framework with corresponding viewpoints is necessary to address the required concerns [15]. Building on this, we presented an architecture framework for PPS that includes six viewpoints to model the system's architecture from different perspectives. This framework aligns with the vision of model-based systems engineering and provides a comprehensive approach to PPS architecture modeling.

In addition to general PPS design, there have been extensive discussions on intrusion detection systems (IDS) in relation to PPS. An IDS monitors a network or system for malicious activity or policy violations, and its detection functionality primarily uses the techniques proposed by IDS and IDPS.

6 Conclusions

Physical protection systems (PPS) are commonly used to secure areas, facilities, and assets. Due to the similarities among PPS features, there is potential for systematic reuse using systems product line engineering (PLE). PLE can reduce development costs, and time-to-market, and improve quality. This paper introduces three PLE approaches: single PLE, multiple PLE, and profile-based PLE, and presents a systematic approach for selecting the most suitable method. The proposed method is applicable not only to PPS but also to other systems. The paper provides a case study to illustrate the use of the method in an industrial context.

The novelty of the paper lies in its systematic approach to selecting the appropriate PLE method for a given industrial context. By presenting three different PLE approaches and outlining their strengths and limitations, the paper offers decision-makers a framework for making informed choices. The selection of a PLE approach is critical for achieving cost-effectiveness, reducing time-to-market, and improving quality in PPS development. Our systematic approach to the qualitative analysis of PLE methods takes into account various factors, such as the size of the product line, the complexity of the system, and the organizational context, to determine the feasibility of different PLE methods. This approach enables organizations to make informed decisions and select the most suitable PLE approach for their specific needs. It is important to note that our qualitative analysis method can also be applied to other systems besides PPS, making it a valuable contribution to the field of systems product line engineering.

References

[1] L. Fennelly, Effective Physical Security, fifth ed., Butterworth-Heinemann, USA, 2016.
[2] M.L. Garcia, Vulnerability Assessment of Physical Protection Systems, Elsevier Butterworth-Heinemann, Amsterdam, The Netherlands, 2006.
[3] M.L. Garcia, The Design and Evaluation of Physical Protection Systems, second ed., Amsterdam, The Netherlands, Elsevier Butterworth-Heinemann, 2008.
[4] International Atomic Energy Agency (IAEA), Handbook on the Physical Protection of Nuclear Material and Facilities, IAEA-TECDOC-127, March 2000.
[5] INCOSE, Guide to the Systems Engineering Body of Knowledge (SEBoK), October 2016.
[6] P. Clements, L. Northrop, Software Product Lines: Practices and Patterns, Addison-Wesley, Boston, MA, USA, 2002.
[7] E. Tüzün, B. Tekinerdogan, M.E. Kalender, S. Bilgen, Empirical evaluation of a decision support model for adopting software product line engineering, Inf. Softw. Technol. 60 (2015) 77–101.
[8] B. Tekinerdogan, M. Akşit, Classifying and evaluating architecture design methods, in: Software Architectures and Component Technology, Springer, Berlin/Heidelberg, Germany, 2002, pp. 3–27.
[9] B. Tekinerdogan, O. Ozkose Erdogan, O. Aktug, Supporting incremental product development using multiple product line architecture, Int. J. Knowl. Syst. Sci. 5 (2014) 1–16.
[10] V.R. Basili, G. Caldiera, H.D. Rombach, The goal question metric approach, Science 2 (1994) 1–10.
[11] R. Solingen, E. Berghout, Goal/Question/Metric Method, Mcgraw Hill Higher Education, New York, NY, USA, 1999.
[12] J.D. Williams, Physical Protection System Design and Evaluation. IAEA-CN-68/29, 1997.
[13] B. Tekinerdogan, K. Özcan, S. Yagiz, I. Yakin, Feature-driven survey of physical protection systems, in: Proceedings of the 11th Complex Systems Design & Management (CSD&M) Conference, Paris, France, 2–4 December, 2020.
[14] B. Tekinerdogan, S. Yagiz, K. Özcan, I. Yakin, Integrated process model for systems product line engineering of physical protection systems, in: B. Shishkov (Ed.), Business Modeling and Software Design, vol. 391, Springer, Cham, Switzerland, 2020, ISBN: 978-3-030-52305-3.
[15] B. Tekinerdogan, K. Özcan, S. Yagiz, I. Yakin, Systems engineering architecture framework for physical protection systems, in: Proceedings of the International Symposium on Systems Engineering (ISSE), Vienna, Austria, 12–14 October, 2020.

CHAPTER 9

Reference architecture design for machine learning supported cybersecurity systems

Cigdem Avci[a], Bedir Tekinerdogan[a], and Cagatay Catal[b]
[a]Information Technology Group, Wageningen University & Research, Wageningen, The Netherlands
[b]Department of Computer Science and Engineering, Qatar University, Doha, Qatar

1 Introduction

Cybersecurity is an increasingly important concern in the digital age as technology continues to evolve and become more widespread. The increasing reliance on digital systems has led to a rise in security threats, from malware to phishing and other cyberattacks. To address these threats, many organizations have invested in cybersecurity solutions designed to protect their systems and data. However, these solutions are often developed in isolation, with little consideration for integrating machine learning and other advanced technologies.

The need for more understanding among software architects of machine learning and machine learning experts of software architecture design further compounds the cybersecurity problem, which has led to the development of machine learning-based models that need to be better suited to the software-intensive systems commonly encountered in cybersecurity. However, software architects have yet to explicitly consider cybersecurity with machine learning support in designing software-intensive systems, resulting in systems not being equipped to deal with the latest security threats.

This study aims to develop a reference architecture design for machine learning-supported cybersecurity systems. By adopting a holistic approach that considers both the engineering design and machine learning perspectives, we aim to address the current limitations of cybersecurity solutions. Our approach involves first performing a domain analysis, followed by an architecture design stage in which decomposition and deployment views are developed based on the Views and Beyond architecture development framework.

The proposed reference architecture is built based on the synthesized architecture design approaches and machine learning applications used in cybersecurity. As a result, it provides a comprehensive and effective solution for machine learning-supported cybersecurity systems.

Management and Engineering of Critical Infrastructures
https://doi.org/10.1016/B978-0-323-99330-2.00014-3

We demonstrate that the proposed reference architecture design and the holistic approach effective for machine learning-supported cybersecurity systems. By providing a comprehensive solution that considers both the engineering design and machine learning perspectives, we aim to address the current limitations of cybersecurity solutions and help organizations better protect their systems and data from security threats.

The chapter's remaining organization is as follows: Section 2 presents the background. Section 3 discusses various reference architectures for cybersecurity systems. Section 4 describes reference architectures for machine learning systems. Section 5 introduces the reference architecture for machine learning-supported cybersecurity systems. Section 6 includes the related work. Section 7 covers the discussion, and Section 8 concludes the chapter.

2 Background

This section will provide the background related to machine learning-based cybersecurity systems. First, in Section 2.1, the notion of cybersecurity will be described. Then, Section 2.2 will provide the background on reference architecture design.

2.1 Cybersecurity

The ISO/IEC 27032 standard for Information Technology – Security Techniques – Guidelines for Cybersecurity was developed to meet users' needs for protecting the integrity, confidentiality, and availability of information in cyberspace [1]. The standard provides a comprehensive framework for ensuring information security. Cyberspace is a virtual, complex, and multi-dimensional environment comprising individuals, organizations, and activities that occur on interconnected devices and networks [1,2]. The standard outlines the relationships between cybersecurity and other security domains and emphasizes the importance of stakeholders in maintaining and improving the trustworthiness and usefulness of cyberspace (Fig. 1).

The building blocks of cybersecurity include application security, information security, network security, and internet security. These concepts and their relations are defined as follows:

Information security refers to the protection of data and information systems against unauthorized access, use, disclosure, disruption, modification, or destruction. It encompasses a wide range of security domains, including network security, internet security, cybersecurity, application security, and many others.

Application security is specialized for protecting software applications and the data they process, store, and transmit. It involves securing the underlying code, data, and infrastructure that support applications and protecting against threats such as application-level attacks, SQL injection, and cross-site scripting.

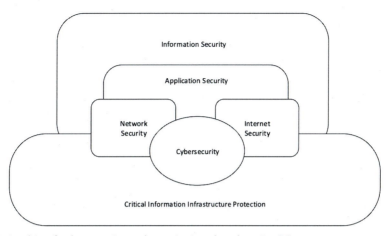

Fig. 1 Relationship of cybersecurity and security in other domains [3].

Network security protects networks and their associated infrastructure, including routers, switches, firewalls, and other networking equipment. It is concerned with ensuring the confidentiality, integrity, and availability of network traffic and data and protecting against hacking, malware, and unauthorized access.

Internet security refers to the security of the Internet and the protection of online activities, including electronic commerce, online banking, and online communication. It is focused on securing the underlying technology and infrastructure of the Internet as well as protecting against online threats such as phishing, spam, and viruses.

Cybersecurity refers to protecting the Internet and computer systems from theft, damage, or unauthorized access. It is a broad field that encompasses a range of security domains, including network security, internet security, and application security. Cybersecurity is concerned with ensuring data and systems' confidentiality, integrity, and availability and protecting against cyber threats such as hacking, malware, and cyber-attacks.

Thus, while all these security domains are related, they each have specific focuses and concerns. For example, information security provides a broad framework for protecting data and information systems. In contrast, network security, internet security, cybersecurity, and application security are specific domains within this broader framework, each with its unique focus and concerns. To be effective, a comprehensive approach must address all these security domains and ensure that they work together to provide a holistic and effective security solution.

In addition to the system's design, human factors must also be considered. The key cybersecurity challenges include building trust, establishing a cybersecurity culture, and overcoming the shortage of cybersecurity professionals [4]. Building trust involves creating awareness of cybersecurity issues, detecting gaps, and improving the system's maturity. The obstacles to secure and usable authentication include storing user credentials and weak passwords. Cybersecurity also involves continuous preparation, detection, response, and recovery to maintain the system's resilience. Finally, the need for talented professionals, certification, and standardization is essential for assessing and maintaining the maturity of the system's cybersecurity.

Regarding current solutions, requirements, challenges, and trends in cybersecurity, traditional management and analysis approaches include risk management, malware detection, intrusion prevention and detection, and using network perimeter gateways. In the area of security intelligence platforms, some of the solutions include IBM's QRadar and Infosphere BigInsights, Teradata Aster App Center for Security, BotCloud Platform, Beehive framework, Dofur Platform, LogRhythm's security intelligence platform, blue coat security platform, and WINE platform.

In cybersecurity, data science and machine learning are becoming increasingly important [5]. Machine learning techniques are being applied to develop intelligent cybersecurity solutions that can be used for autonomous flight operations security, adversarial and uncertainty modeling, airport security monitoring, anomaly detection, and software reliability and security certification. Some of the machine learning and deep learning methods used in these solutions include support vector machine, k-nearest neighbor, decision tree, deep belief network, recurrent neural networks, and convolutional neural network.

The application of intelligent cybersecurity solutions can significantly enhance the capabilities of traditional cybersecurity management and analysis approaches. By leveraging advanced machine learning algorithms, it is possible to identify and respond to security threats more quickly and effectively and improve cyberspace's overall trustworthiness and usefulness. However, it is essential to note that while these solutions can be highly effective, they are not a substitute for a comprehensive approach to cybersecurity that includes not just technical solutions but also the development of a strong cybersecurity culture, the involvement of stakeholders, and the training and certification of professionals.

2.2 Reference architecture design

Software architecture is a critical aspect of software engineering that provides the high-level design of a software system. It is the high-level structure of a software system, which defines its components, relationships, and interactions. Software architecture should satisfy a system's functional and non-functional requirements, provide a blueprint for development, and ensure a software system's scalability, maintainability, and adaptability over its lifetime.

One widely used approach for defining and documenting software architectures is the "Views and Beyond" approach [6]. In this approach, the architecture of a system is described through multiple viewpoints that concentrate on specific quality features.

The number of viewpoints selected depends on the complexity of the system and the stakeholder's areas of use. The viewpoints consolidate architectural patterns, templates, and limitations, making understanding the system's architecture easier.

Before constructing the software architecture, domain analysis, and commonality or variability analysis are usually performed to understand the system's requirements and constraints [7], and a domain-driven architecture design approach is adopted [8–11]. Next, quality attributes, such as availability, scalability, and security, are identified and used to make design decisions. For example, if availability is a critical quality attribute, the architecture should consider tactics such as detecting faults, recovering from faults, and preventing faults. The logic for constructing design decisions based on quality attributes can be broken down into several steps, starting with identifying the quality attributes and then listing the available tactics and methods for achieving those attributes.

Software architecture and reference architecture have a relationship in that the reference architecture often guides the software architecture of a system for the domain it belongs to. The reference architecture provides the high-level guidelines and templates that the software architect uses to create the software architecture. In contrast, the software architecture defines the specific details of the system.

In other words, reference architecture provides a roadmap for software architecture design and helps ensure that the system is aligned with industry standards and best practices. In addition, the reference architecture can be used to evaluate the design of the software architecture, identify areas for improvement, and ensure that the architecture meets the requirements of the specific domain.

The main goal of reference architecture design is to reduce the complexity and uncertainty of developing systems. This is achieved by providing a clear and concise description of the system's components, relationships, and constraints. In addition, reference architectures are intended to be flexible, reusable, and adaptable to changing requirements and new technologies.

2.3 Reference architecture design for cybersecurity systems

A security system reference architecture provides a comprehensive framework for designing and implementing cybersecurity solutions. The architecture offers guidelines for selecting components, technologies, and processes to secure systems, networks, and applications from cyber threats. It helps organizations understand the security requirements and provides a blueprint for meeting them.

Designing a reference architecture for cybersecurity systems is a complex process that requires a comprehensive understanding of cybersecurity, its components, and the technologies that support it. The following steps outline the process of designing a reference architecture for cybersecurity systems:

- *Define the scope and objectives*: The first step in designing a reference architecture is to define the scope and objectives of the architecture. The scope defines the systems, networks, and applications that the architecture will cover. The objectives define the goals and requirements that the architecture must meet.
- *Identify security requirements*: The next step is to identify the security requirements for the systems, networks, and applications covered by the scope. This includes defining the desired quality attributes, such as confidentiality, integrity, and availability, and determining the level of protection required for each.
- *Review existing architectures*: The next step is to review existing reference architectures for cybersecurity systems. This helps to understand best practices and common design patterns in developing cybersecurity systems.
- *Define the components and technologies*: The next step is to define the components and technologies needed to meet the security requirements. This includes defining the components and technologies used to implement security services, such as authentication, authorization, and access control, and determining how the components and technologies will be integrated into the overall architecture.
- *Develop the architecture*: The next step is to develop the architecture. This involves defining the relationships between the components and technologies and documenting the architecture in a format that organizations can easily understand and use.
- *Validate the architecture*: The final step is to validate the architecture by testing it against the security requirements and the scope of the architecture. This helps to ensure that the architecture meets the desired quality attributes and that the systems, networks, and applications can be secured to meet the requirements.

In this chapter, we have focused on existing reference architectures in the literature and designed an enhanced reference architecture. The list of reference architectures that are reviewed is presented in Table 1.

In addition to these general cybersecurity reference architectures, we have also identified the reference architectures for machine learning which are listed in Table 2.

In the following section, we will describe these reference architectures separately.

3 Reference architectures for cybersecurity systems

3.1 National Institute of Standards and Technology (NIST) cloud computing security reference architecture

The NIST Cloud Computing Security Reference Architecture (NCC-SRA) is designed to address the unique security requirements of cloud computing ecosystems (Fig. 2). The architecture is implemented based on three dimensions: service models (IaaS, PaaS, SaaS), cloud deployment models (Public, Private, Hybrid, and Community), and actors. By selecting the appropriate service and deployment models, security components can be implemented to identify, protect against, detect, respond to, and recover from threats.

Table 1 List of cybersecurity reference architecture.

Paper ID	Year	Reference architecture	Details
1	2013	NIST Cloud Computing Security Reference Architecture (DRAFT) [7]	Vendors of various cloud services can readily map their business models to the NIST architecture supported by the flexible formal model provided by the NIST Reference Architecture and the overarching Security Reference Architecture.
2	2015	IBM Security Framework [12]	The Advanced Security and Threat Research infrastructure serves as the basis for the capabilities outlined by the IBM Security Framework.
3	2017	Cisco Security Control Framework (SCF) Model [13]	Maximizing visibility into network devices and events and maintaining control over people, devices, and network traffic are two core tenets of the Cisco Security Control Framework.
4	2019	Microgrid Cyber Security Reference Architecture (V2) [14]	The microgrid cyber security reference architecture offers guidelines and security recommendations for implementing secure microgrid control systems. The design approach supports cyber security for microgrid control systems by implementing functional segmentation to provide defense-in-depth.
5	2013	Trend Micro Cybersecurity Reference Architecture for Operational Technology [15]	The built-in IoT security tool Trend Micro IoT Security (TMIS) keeps an eye out for potential threats like data theft and ransomware assaults and guards against them. In addition to preventing damage to IoT devices, this also decreases device maintenance costs and safeguards reputation. Furthermore, it also assures firmware integrity.
6	2013	DXC Cyber Reference Architecture (CRA) [16]	The Open Group Architecture Framework (TOGAF), Sherwood Applied Business Security Architecture (SABSA), Control Objectives for Information and Related Technology (COBIT), National Institute of Standards and Technology (NIST), and International Organization for Standardization are among the security standards and methodologies that are aligned with DXC's CRA framework, which describes security holistically. A defined taxonomy and terminology are also features of CRA.
7	2015	Oracle Security in Depth Reference Architecture [17]	The important emphasis areas of security from the inside out are gathered in the conceptual view of the architecture.
8		Microsoft Cybersecurity Reference Architecture [17]	The Microsoft Cybersecurity Reference Architecture (MCRA) is a collection of Microsoft's cybersecurity features and how they work together to ensure online safety.

Table 2 List of reference architectures for machine learning.

Paper ID	Year	Reference architecture	Details
1	Not specified (2018–2023)	ML Reference Architecture [18]	The machine-learning reference model represents the architecture-building pieces that might be found in a machine-learning solution. The objective of the following metamodel is to offer a condensed but helpful overview of several areas of information architecture (IT), particularly machine learning.
2	2020	The conception of a reference architecture for machine learning in the process industry [18]	The IT structure and data flow from an asset arriving at the ML and application level makes up the architecture's fundamental structure.

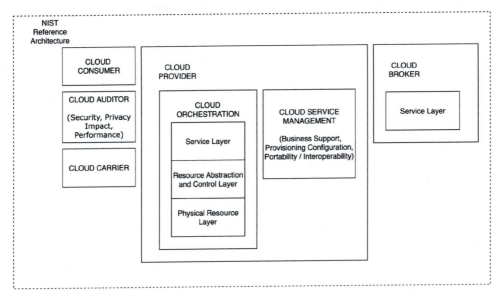

Fig. 2 NIST reference architecture. *(Adapted from M. Akşit, B. Tekinerdogan, F. Marcelloni, L. Bergmans, Deriving object-oriented frameworks from domain knowledge, in: Building Application Frameworks: Object-Oriented Foundations of Framework Design. Wiley & Sons, 1999, pp. 169–198.)*

To ensure the security of each application instance, a risk-based approach is used to investigate the cloud ecosystem with all involved actors [7]. This process is supported by a risk management framework that includes selecting, monitoring, assessing security controls, and categorizing information systems.

When implementing the NCC-SRA, all actors must prioritize risk management, business continuity, physical security, compliance with standards, transparency, and user account control. Among the actors, the cloud consumer receives the service via a cloud provider or broker, with the provider responsible for service deployment, orchestration, management, privacy, and security, and the broker managing service use, performance, delivery, and provider-consumer relationships.

Thanks to the flexible formal model provided by the NIST Reference Architecture and the overarching Security Reference Architecture, vendors of various cloud services can easily align their business models with the NCC-SRA.

The security reference architecture is a comprehensive approach to address the unique security requirements of cloud computing ecosystems. The formation methodology starts with data analysis, which includes data collection, data aggregation and validation, and deriving security responsibilities for both the provider and the broker. Security components are then mapped to security control families, followed by empirical data analysis and generating a heat map.

The formal model of the security reference architecture is composed of multiple layers, including the consumer layer, provider layer, broker layer, carrier layer, and auditor layer. The consumer layer is responsible for cloud consumption management, with support for business, configuration, portability, interoperability, and orchestration within the cloud ecosystem. The provider layer is responsible for secure cloud system orchestration, with deployment, service, resource abstraction, and control, as well as secure cloud service management, with provisioning, configuration, portability, interoperability, and business support. The broker layer is responsible for technical and business mediation, secure service aggregation, cloud service management, intermediation, and arbitrage. The carrier layer ensures the secure transfer of data between clouds, while the auditor layer monitors and evaluates the overall security of the cloud ecosystem.

By using this formal model, organizations can better understand their roles and responsibilities in the cloud ecosystem and implement appropriate security measures to protect against threats and vulnerabilities.

3.2 IBM security framework

IBM's security framework provides a comprehensive and business-driven approach to security, taking into account the organization's risk posture (Fig. 3). The framework is structured around several layers, including advanced security and threat research, people, data, applications, infrastructure, security intelligence and analytics, and a security-maturity model.

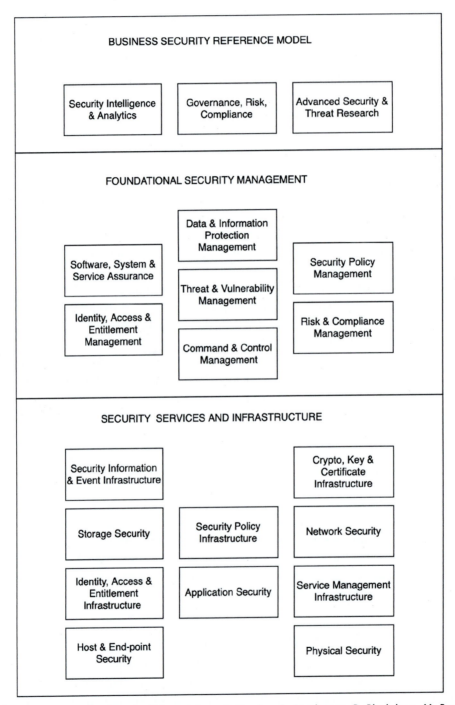

Fig. 3 IBM security framework. *(Adapted from A. Buecker, S. Arunkumar, B. Blackshaw, M. Borrett, P. Brittenham, J. Flegr, et al., Using the IBM Security Framework and IBM Security Blueprint to Realize Business-Driven Security. IBM Redbooks, 2014.)*

At the core of the framework is the Advanced Security and Threat Research infrastructure, which provides the basis for the capabilities offered within the security domains and additional levels [12]. This infrastructure includes software, hardware, and services that can be used to deliver the solutions offered by the framework.

Identity management and access control are critical aspects of protecting assets and services. Authorized roles within the organization should be clearly defined, and unauthorized access should be denied in accordance with business agreements and policies. Record keeping and compliance should also be encouraged to ensure that actions are traceable to real people.

Data and information should be classified according to their value and risk, and policies should be implemented to protect them. The organization should generate and maintain the knowledge required to manage assets, inventory, services, data, and information. Protection should be applied using a plan aligned with the information lifecycle, and the effectiveness of the security plan should be measured using selected metrics.

Security considerations should be incorporated into the design and implementation of applications. The lifecycle of critical applications should be assessed and monitored against security policies, and threats should be eliminated during design, development, testing, and production.

The organization's infrastructure should be monitored and managed to ensure protection from emerging threats. Network, server, and endpoint infrastructure elements should be protected, and virtualization, web-based and non-web-based applications, and failure or loss of physical infrastructure or service should be considered in the implementation of protection policies and processes.

The security intelligence and analytics layer of the framework collects, aggregates, and analyzes information such as security events and application logs. This platform enables users to log, view, analyze, alert, and report across domains, model risks, prioritize vulnerabilities, detect incidents, and carry out impact analysis.

3.3 Cisco security control framework (SCF) model

The Cisco SAFE architecture is a modular framework that aims to address the complex security requirements and interoperations of modern network structures (Fig. 4). Its design ensures the security of day-to-day operations and extends the equipment lifetime, taking into consideration the network elements that can impact the system's lifespan.

One of the key features of the SAFE architecture is its unified strategy for security intelligence using policies, which allows for a common control strategy that can monitor, analyze, and respond to threats. The framework adopts a standardized approach based on principles and actions, selecting the appropriate technologies and best practices for a given situation.

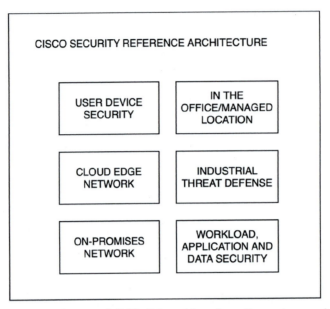

Fig. 4 Cisco security control framework (SCF). *(Adapted from https://www.cisco.com/c/en/us/solutions/ enterprise/design-zone%20security/landing_safe.html#~overview.)*

The Cisco Security Control Framework, which is at the heart of the SAFE architecture, is built on two core tenets: maximizing visibility into network devices and events, and maintaining control over people, devices, and network traffic. The security implementation covers the planning, design, implementation, operation, and maintenance stages.

The SAFE Security Architecture provides detailed modular designs, enabling the aggregation of threat detection information and offering step-by-step architecture/ platform-specific guidance. This design approach allows for the administration of different security technologies in a common manner, deployed under the SAFE architecture, and supporting network availability.

Moreover, the SAFE architecture offers support services for the system's entire lifecycle, including strategy and assessment, deployment and migration, remote management, security intelligence, and security optimization.

3.4 Microgrid cybersecurity reference architecture (V2)

The ongoing research on cyber security in microgrid control system networks is summarized in [15] The microgrid cyber security reference architecture provides comprehensive guidelines and security recommendations for the implementation of secure microgrid control systems (Fig. 5). The approach to the design supports cyber security for microgrid control systems by leveraging functional segmentation to achieve

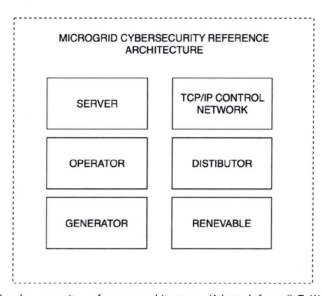

Fig. 5 Microgrid cyber security reference architecture. *(Adapted from IIoT World, Trend Micro Cybersecurity Reference Architecture for Operational Technology—White Paper 2017, http://iiot-world. com/wp-content/uploads/2017/12/Trend-Micro-Cybersecurity-Reference-Architecture-for-Operational-Technology.pdf. Accessed 12 June 2022.)*

defense-in-depth. This approach takes advantage of the limited set of permitted operations for control networks while protecting the vulnerable components they rely on.

A basic microgrid control system comprises several components, such as transformers, generators, switches, and client/server systems, which communicate over a network. The microgrid cyber security reference architecture can be applied to configure a secure network, for example, by integrating a client interface with the permission operation for a connect/disconnect function. Alternatively, a microgrid control system can use the user interface to forward control messages to the power network.

To address network security concerns, network segmentation can be applied to operate certain parts of the microgrid control system. Segmentation can be implemented based on the domain of the data exchanged to achieve reliable operation. Segmentation may require additional layers with switches/routers, and intrusion prevention systems may need to be involved in the new layer.

3.5 Trend micro cybersecurity reference architecture for operational technology

Trend Micro IoT Security (TMIS) [15] is an integrated IoT security solution that proactively identifies and protects against potential threats like data breaches and ransomware attacks (Fig. 6). The tool ensures the integrity of firmware, reduces device maintenance costs, and safeguards the reputation of the organization.

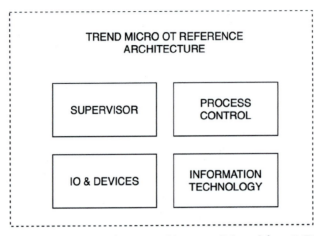

Fig. 6 Trend micro OT cybersecurity reference architecture. *(Adapted from IIoT World, Trend Micro Cybersecurity Reference Architecture for Operational Technology—White Paper 2017, http://iiot-world. com/wp-content/uploads/2017/12/Trend-Micro-Cybersecurity-Reference-Architecture-for-Operational-Technology.pdf. Accessed 12 June 2022.)*

To ensure the security of operational technology (OT) systems, it is recommended that they are separated from the network architecture. The network traffic on these two architectures is different, and OT systems should not have internet access or remote access as these connections pose a security risk. Further partitioning can be achieved by network segmentation and isolation, which involves identifying critical parts through operational risk analysis and minimizing access to these components, systems, and resources to protect against attacks or incidents. Network segmentation can be managed at gateways, and security controls can be implemented at various points, such as firewalls, routers, virtualization, sandboxing, intrusion detection/prevention, and other endpoints or nodes to achieve boundary protection. Cybersecurity controls, such as monitoring and intrusion detection, can also be implemented at the hardware, software, or application levels.

3.6 DXC cyber reference architecture (CRA)

The DCX CRA is a comprehensive security architecture that aims to monitor and respond to threats while also ensuring that security objectives are met through the deployment of the best security implementations (Fig. 7). It provides a framework and blueprint for addressing architectural challenges in a straightforward and focused manner.

The architecture aligns with established security standards and methodologies such as TOGAF, SABSA, COBIT, NIST, and ISO, providing a holistic approach to security that includes a defined taxonomy and terminology.

The architecture consists of three levels: tactical and operational, technical, and strategic, with domains logically grouped under these levels. Each domain is supported by a

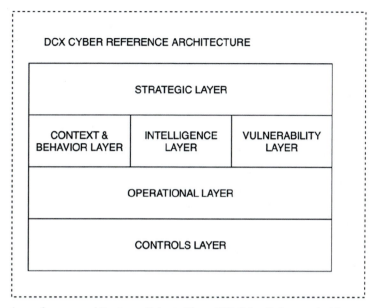

Fig. 7 DXC cyber reference architecture. *(Adapted from DXC Cyber Reference Architecture White Paper, Security in Depth Reference Architecture. O. E. T. S. (2013). https://assets1.dxc.technology/secu-rity/downloads/DXC-Security-Cyber_Reference_Architecture.pdf.)*

set of objectives and is further divided into subdomains and capabilities, which represent security requirements and the organization's capacity to meet them.

Blueprints are sets of reference architectures that include a conceptual view, key functional areas, and mapping layers, while work packages are defined using a conceptual view. The DCX CRA also supports organizations in maintaining their cyber maturity.

By selecting security objectives, defining security requirements, deploying the best security implementations, and supporting the plan, the DCX CRA helps organizations to achieve their security goals in a comprehensive and strategic manner.

3.7 Oracle security in depth reference architecture

The Oracle Security in Depth Reference Architecture provides a comprehensive approach to security, covering compliance enablement, data security, and fraud detection (Fig. 8). The architecture is designed with multiple layers to ensure that security is provided at every level of the system.

At the top level, the architecture includes platforms and infrastructure, security framework, and security interfaces. The security framework is divided into several layers, including management and administration, fraud detection and compliance enablement-related components, security services, and security information. The architecture recommends the

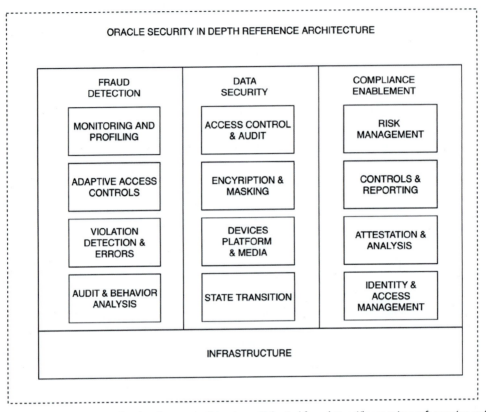

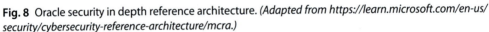

Fig. 8 Oracle security in depth reference architecture. *(Adapted from https://learn.microsoft.com/en-us/ security/cybersecurity-reference-architecture/mcra.)*

use of open standards such as SAML, LDAP, and XACML to ensure interoperability between different systems.

The logical architecture of the security framework includes components for auditing, access control, and encryption. It includes a secure DBMS with row and column-level access control that is administered by privileged users. Common access control services allow the system to be managed holistically. The architecture also provides support for filtering, alerting, reporting, and monitoring capabilities, which can be configured according to the organization's needs.

The fraud detection logical architecture covers database security and platform components to implement detective and preventive controls. The architecture ensures that unprivileged users do not perform actions that are beyond the scope of their role, and administrative user activities are monitored and audited against system manipulation. The architecture also examines requests at gateways and provides support for ad-hoc investigation and investigation with patterns.

The compliance enablement logical architecture covers capabilities related to compliance enablement and other common cybersecurity components, such as access control and auditing, identity and access governance, and reporting. By incorporating these capabilities, the architecture enables organizations to achieve compliance with relevant regulations and standards.

3.8 Microsoft cybersecurity reference architecture

Microsoft 365 cybersecurity services provide comprehensive coverage of the NIST Cybersecurity Framework (Fig. 9), including its core functions of identify, protect, detect, and respond [19]. Microsoft's cybersecurity solutions for Microsoft 365 are divided into four dimensions: identity and access management, threat protection, information protection, and security management.

The identify function is supported by asset management, governance, risk management strategy, risk assessment, business environment, and supply chain risk management. The protect function is achieved through data security, access control, information protection processes and procedures, awareness and training, protective technology, and

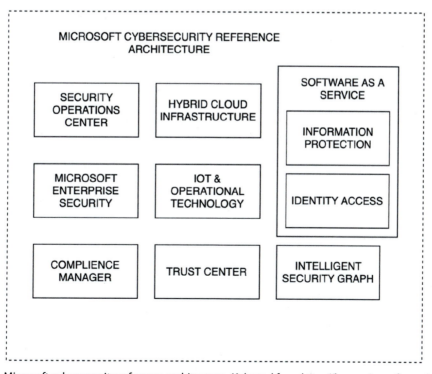

Fig. 9 Microsoft cybersecurity reference architecture. *(Adapted from https://learn.microsoft.com/en-us/security/cybersecurity-reference-architecture/mcra.)*

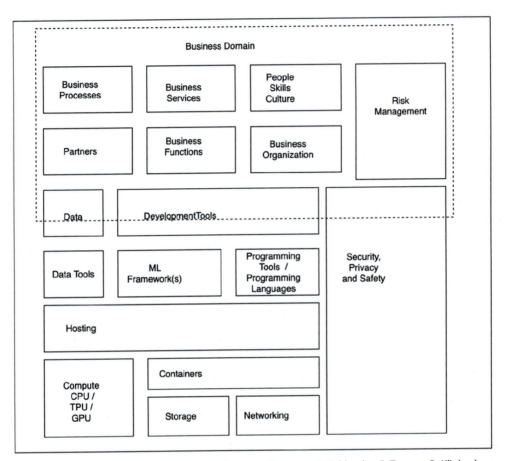

Fig. 10 ML reference architecture. *(Adapted from R. Wöstmann, P. Schlunder, F. Temme, R. Klinkenberg, J. Kimberger, A. Spichtinger, J. Deuse, Conception of a reference architecture for machine learning in the process industry, in: 2020 IEEE International Conference on Big Data (Big Data), December. IEEE, 2020, pp. 1726–1735.)*

maintenance. The detect function is covered by continuous security monitoring, anomalies and events, and detection processes. The response function includes communications, response planning, analysis, improvements, and mitigation. The recover function involves improvements, recovery planning, and communications.

The Microsoft Cybersecurity Reference Architecture (MCRA) [13] is a collection of Microsoft's cybersecurity features and how they work together to ensure online safety. The reference architecture is presented in Fig. 10 and consists of several critical elements such as Microsoft Defender and XDR, IoT Operations and Security, People Security, Security Operations, Microsoft Azure Defender, Security Score, Compliance Portal, and Microsoft Cloud Security Broker. Each of these components plays a unique role in threat identification, protection, and cross–platform visibility.

The MCRA architecture also supports the protection of the hybrid cloud infrastructure, utilizing Microsoft Threat Intelligence to handle data aligned with customer requirements. Using AI-based implementations, the data is analyzed and published. The Microsoft Security Development Lifecycle further supports the implementation of secure applications.

4 Reference architectures for machine learning systems

4.1 Machine learning (ML) reference architecture

To successfully apply machine learning, a comprehensive full-stack strategy is necessary. This approach requires experts to have a thorough understanding of the entire technical stack, including both vertical and horizontal machine learning. The former refers to the transition from hardware to machine learning applications, while the latter encompasses the entire toolchain for each process step.

To guide the development of machine learning applications, a machine-learning reference model has been developed [20]. This model offers a concise yet informative overview of various areas of information architecture, with a particular focus on the complex field of machine learning. By providing a structured approach, the reference model can help ensure that machine learning applications are developed in a thoughtful and efficient manner.

4.2 The conception of a reference architecture for machine learning in the process industry

The architecture's fundamental structure is based on a basic level that addresses the IT structure and data flows from an asset arriving at the ML and application level. This architecture comprises different tiers that can be analyzed in more detail, depending on the specific needs and requirements of the ML deployment and the integration of edge devices.

The foundation layer for data collection, analysis, control, and production processes includes tools such as sensors and actors (assets). Additionally, the reference architecture is versatile and can be used with any other data sources that define assets accurately.

The next layer of the architecture deals with data preparation and storage, where data is cleansed, integrated, and stored in preparation for the ML process. The ML process layer is the core of the architecture, where algorithms are developed and trained. This layer also includes model evaluation and selection, as well as model deployment.

The application layer provides the ML output for business use, including insights and predictions. Finally, the top layer of the architecture addresses the governance and management of the overall ML system, including issues such as security, privacy, compliance, and ethical considerations.

In summary, the machine-learning reference model provides a comprehensive overview of the different areas of information architecture required for machine-learning applications.

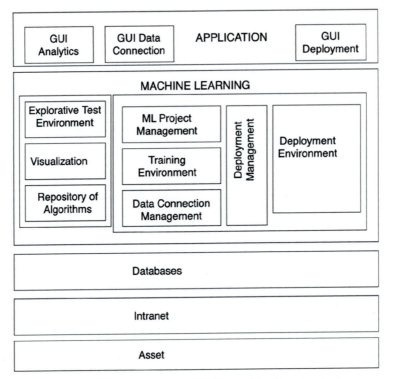

Fig. 11 The conception of a reference architecture for machine learning in the process industry. *(Adapted from R. Wöstmann, P. Schlunder, F. Temme, R. Klinkenberg, J. Kimberger, A. Spichtinger, J. Deuse, Conception of a reference architecture for machine learning in the process industry, in: 2020 IEEE International Conference on Big Data (Big Data), December. IEEE, 2020, pp. 1726–1735.)*

The architecture's fundamental structure includes tiers for data collection, data preparation and storage, ML process, application, and governance and management (Fig. 11).

5 Designing a reference architecture for machine learning-supported cybersecurity systems

The reference architecture developed in this study is a comprehensive solution that covers various components, users, and workforce. It is designed to provide a generic architecture that encompasses the key elements of the reference architectures discussed in previous sections.

At the core of the architecture is a high-level component architecture, which consists of several sub-components, including people and clients, information, actions, and security. These components are selected based on their commonality among the different reference architectures and their potential to function in all application architectures. The architecture core is depicted in both the decomposition and deployment diagrams, and it is proposed to be deployed on a private cloud infrastructure, which will manage load-balancing activities for intensive tasks and network traffic. The network is represented as connections among the nodes in the deployment diagram.

Security is a critical aspect of the architecture, and several security-related components are listed in the architecture core. These components include endpoint security, physical security, operations security, converged security, data security, and development lifecycle security. Each of these components is defined in detail to ensure that the architecture is robust and secure.

In addition to the architecture core, the reference architecture includes a resilient architecture, a resilient workforce, high-level infrastructure components, and the cloud. The cloud infrastructure will have its orchestrator to manage cloud servers and local devices. Servers can be configured as separate entities, replicated, or distributed. Local devices can communicate with personal devices or the cloud infrastructure.

Overall, the reference architecture provides a comprehensive and scalable solution that can be adapted to meet the specific requirements of different organizations and industries. By implementing this architecture, organizations can improve their data management, enhance their security posture, and ensure that their workforce is prepared to handle the challenges of modern-day computing.

The cybersecurity reference architecture is configured to manage access control, including firewalls to prevent circumvention, barriers, physical security to prevent tampering, security zones controlling physical access, proxies, reverse proxies by hiding IP addresses, virtual private network servers, network access control to prevent bypass, VLAN, switching, domain network services, and others (Figs. 12 and 13).

A cybersecurity architecture's security policy may rely on several frameworks that comply with regulations, such as ISO 27001, ISO/IEC Technical Standard 19249:2017, The Protection of Information in Computer Systems, ITIL, GDP, and others. Furthermore, it must comply with statutory and regulatory requirements.

Having a resilient workforce is essential for maintaining a functioning system. The workforce must be available in the event of system failure, and their knowledge must be easily transferable among one another as needed.

The implementation of the security architecture should start with a risk assessment to provide an overview and understanding of the system's risks. This allows the architect to design a cybersecurity architecture that can prevent, mitigate, or minimize those risks. Risks that are costly to mitigate are assessed, and if the cost of the proposed solution is high, these risks are accepted. When selecting technologies to map to the reference architecture and use in the application architecture, the priority should be compliance with requirements, rather than the technology's novelty.

When designing the architecture, the ten design principles of Saltzer and Schroeder [5] should be considered, including economy of mechanism (simplifying security functions into smaller components), fail-safe defaults (denying access or action if not explicitly

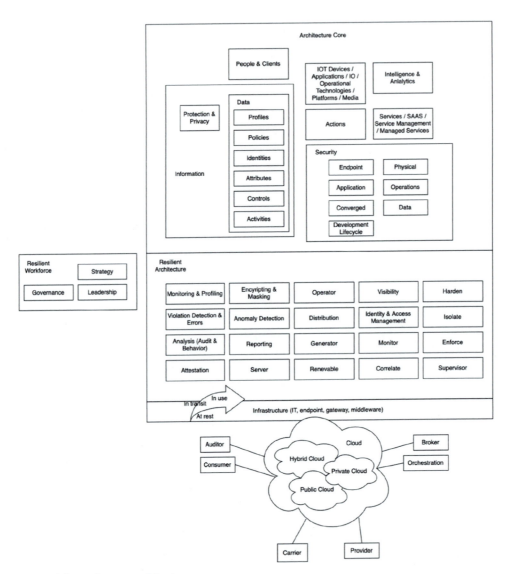

Fig. 12 Cyber reference architecture.

approved), complete mediation (re-authentication while moving to a higher trust zone), open design (publicly available design methods/architecture), separation of privilege (implementing the least privilege principle), least privilege (minimizing privilege to avoid hazards), least common mechanism (not automating trust mechanisms but explicitly defining them), psychological acceptability (ensuring user acceptance so they won't try to bypass security mechanisms), work factor (keeping security implementation costs reasonable), and compromise recording (recording intrusions into the system) (Fig. 14).

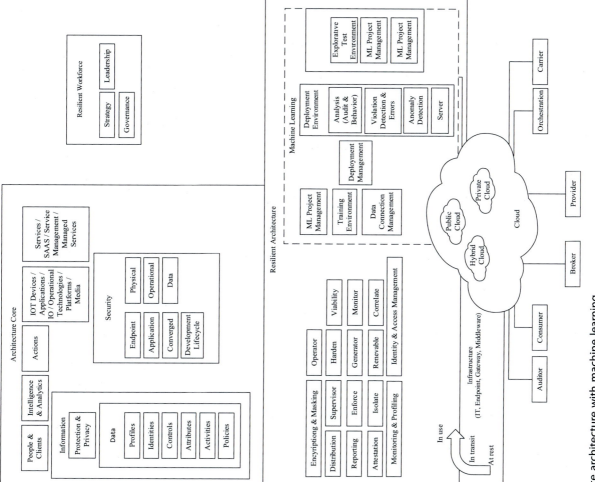

Fig. 13 Cyber reference architecture with machine learning.

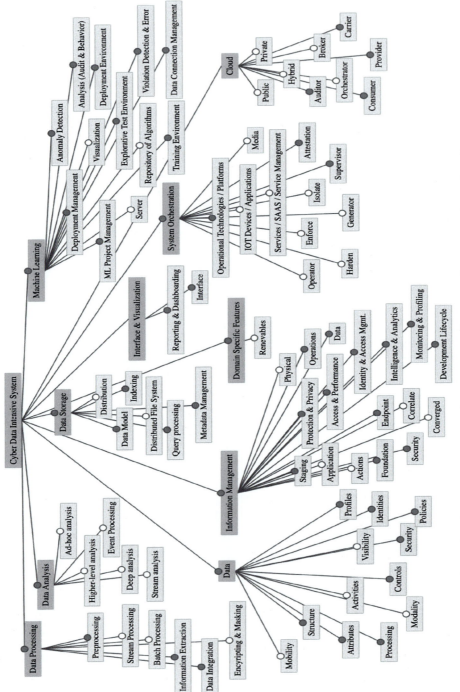

Fig. 14 Feature model of the cyber data-intensive system of the cyber reference architecture with machine learning.

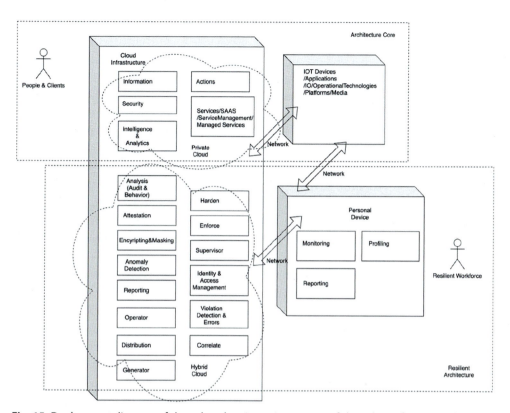

Fig. 15 Deployment diagram of the cyber data-intensive system of the cyber reference architecture with machine learning.

ISO/IEC 19249:2017 architectural principles [6] could also form a basis for reference architecture design that are virtualization (to lower the costs of resource use), domain separation (for access control among domains), redundancy (to prevent single-point failure), encapsulation (for access control among entities), layering (intruder needs authentication at each layer and due to the segmentation of the layers by network access, the trail to be followed becomes more complicated. and controlled), least privilege (minimize privilege to avoid an intentional or unintentional hazard), attack surface minimization (the area of interaction with other systems is to be controlled by disabling, authenticating, and monitoring), centralized parameter validation (validate inputs), centralized general security services and preparing for errors and exceptions (Fig. 15).

6 Related work

In the literature, reference architectures are designed as aligned with the domain that they are applicable. For example, in [21] a reference architecture is designed for smart homes focusing on the cybersecurity solution of the attack surface analysis problem. The architectural mapping of the devices within the reference architecture enables the security analysis of the smart home system. The architecture supports the formation of a pool of vulnerabilities such that crucial vulnerabilities could be recorded to be provided as a basis to develop the cyber defense strategies of the smart home systems. Autonomic architectures are proposed within the cybersecurity landscape for the systems that are expected to be self-protecting [19]. The proposed architecture in [19] maintains three threat-resilient system features: resource virtualization, autonomic managers hierarchy, and alarm protection. The study [20] proposed a novel IoT architecture for online monitoring of the gas-insulated switchgear status against cyber-attacks. An open architecture is described in [22] as a solution to the cybersecurity challenges, which is utilizing machine learning algorithms in a hybrid way to improve the classification performance, combining the supervised methods and their causality with unsupervised methods, providing the evaluation of the results, and the method by a human expert such that the causality could be integrated to the models.

The Machine Learning Operations (MLOps) pipeline is recommended as a good set of practices throughout the deployment, operation, and maintenance of a machine learning-based system. MLOps contains practices both from Machine Learning and DevOps processes. Where MLOps is also applicable to this study's proposed cybersecurity system reference architecture, an application of privacy-aware MLOps for learning analytics is discussed in [23]. A simplified version of MLOps is presented in [24], and its pipeline consists of data science (build, experiment, and evaluate the model), ML model, and production parts (deploy the model). The idea of MLOps is having a high level of communication among parts of the system, reproducible results, and reusable tools.

7 Discussion

This study highlights the need to integrate machine learning into cybersecurity reference architectures. Our research found that while significant work has been done on cybersecurity reference architectures, the integration of machine learning has yet to be thoroughly explored. This lack of integration presents a potential gap in overall security. This study presents a reference architecture design for machine learning-supported cybersecurity systems to address this issue.

The results of the domain analysis and architecture design stage demonstrate the effectiveness of the proposed reference architecture. The architecture integrates engineering and machine learning perspectives to provide a holistic approach that enhances the overall security of software-intensive systems. However, our findings also highlight the need for more discussion in the literature on DevOps or MLOps practices, which could impact the maturity and improvement of the cybersecurity architectures.

The architecture core of the cybersecurity system reference architecture consists of a common set of entities, including people, clients, and a resilient workforce. However, the resilience of the architecture is variable, with the set of resilient entities varying depending on the design approach. Our study presents a machine learning integrated cybersecurity reference architecture, including architecture components for project management, deployment management, and the test environment. A sample deployment configuration is also presented, though it is important to note that depending on the configuration, the complexity of the architecture may increase, leading to potential vulnerabilities.

Threats to validity in this study include selection bias, data collection and analysis limitations, and generalizability limitations. Selection bias may have occurred in the selection of studies reviewed in the literature, leading to a potential oversimplification or omission of essential aspects of reference architecture designs for cybersecurity systems. Limitations in data collection and analysis may have resulted in an incomplete or inaccurate understanding of the reference architecture designs and their applicability to machine learning-supported cybersecurity systems. Finally, the results of this study may only be generalizable to some cybersecurity systems or all organizations, as the specific context, needs, and constraints of each system and organization may vary. Further studies may be needed to validate the proposed reference architecture design in various contexts and with different stakeholders.

8 Conclusion

In conclusion, the proposed reference architecture design for machine learning-supported cybersecurity systems has demonstrated its effectiveness in enhancing the overall security of software-intensive systems. Furthermore, by integrating engineering and machine learning perspectives, the design provides a holistic approach that addresses the issue of gaps in overall security caused by the isolated design of cybersecurity systems and the use of machine learning.

The domain analysis and architecture design stage results and the synthesis of architecture design approaches and machine learning applications used in the cybersecurity domain have provided valuable lessons. Firstly, the importance of considering engineering and machine learning perspectives in designing cybersecurity systems has been emphasized. Secondly, the results have shown the need for a flexible and adaptable reference architecture that can accommodate the changing needs of different organizations.

The proposed reference architecture design could be further developed and evaluated for future work through case studies and implementation in real-world settings. Additionally, the architecture could be extended to cover other aspects of cybersecurity, such as privacy and data protection. Finally, integrating different machine learning algorithms and techniques could also be explored to improve software-intensive systems' overall security. Overall, the proposed reference architecture design represents a valuable contribution to machine learning-supported cybersecurity systems and provides a foundation for future research and development.

Availability of data and material

Data sharing does not apply to this article because no datasets were generated.

Conflict of interest

The authors declare that they have no competing interests.

Authors' contribution

All authors have contributed equally to finish this paper. All authors read and approved the final manuscript.

References

[1] A. Kudrati, C. Peiris, B. Pillai, Microsoft Cybersecurity Reference Architecture and Capability Map, 2022.

[2] S. Fischer-Hübner, C. Alcaraz, A. Ferreira, C. Fernandez-Gago, J. Lopez, E. Markatos, et al., Stakeholder perspectives and requirements on cybersecurity in Europe, J. Inf. Secur. Appl. 61 (2021) 102916.

[3] NIST Cloud Computing Security Working Group, NIST Cloud Computing Security Reference Architecture (No. NIST Special Publication (SP) 500-299), National Institute of Standards and Technology, 2013.

[4] A.B. Garcia, R.F. Babiceanu, R. Seker, Artificial intelligence and machine learning approaches for aviation cybersecurity: an overview, in: 2021 Integrated Communications Navigation and Surveillance Conference (ICNS), April, IEEE, 2021, pp. 1–8.

[5] J.H. Saltzer, M.D. Schroeder, The protection of information in computer systems, Proc. IEEE 63 (9) (1975) 1278–1308.

[6] ISO/IEC 27032, Information Technology–Security Techniques–Guidelines for Cybersecurity, 2012.

[7] M. Akşit, B. Tekinerdogan, F. Marcelloni, L. Bergmans, Deriving object-oriented frameworks from domain knowledge, in: Building Application Frameworks: Object-Oriented Foundations of Framework Design, Wiley & Sons, 1999, pp. 169–198.

[8] B. Tekinerdogan, M. Akşit, Classifying and evaluating architecture design methods, in: M. Akşit (Ed.), Software Architectures and Component Technology. The Springer International Series in Engineering and Computer Science, vol. 648, Springer, Boston, MA, 2002. https://doi-org.ezproxy.library.wur.nl/10.1007/978-1-4615-0883-0_1.

[9] R. Dinter van, B. Tekinerdogan, C. Catal, Reference architecture for digital twin-based predictive maintenance systems, Comput. Ind. Eng. 177 (2023) 109099, https://doi.org/10.1016/j.cie.2023.109099.

[10] A. Kassahun, R.J.M. Hartog, B. Tekinerdogan, Realizing chain-wide transparency in meat supply chains based on global standards and a reference architecture, Comput. Electron. Agric. 123 (2016) 275–291, https://doi.org/10.1016/j.compag.2016.03.004.

[11] S. Nazir, S. Patel, D. Patel, Autonomic computing architecture for SCADA cyber security, Int. J. Cogn. Inform. Nat. Intell. 11 (4) (2017) 66–79, https://doi.org/10.4018/IJCINI.2017100105.

[12] A. Buecker, S. Arunkumar, B. Blackshaw, M. Borrett, P. Brittenham, J. Flegr, et al., Using the IBM Security Framework and IBM Security Blueprint to Realize Business-Driven Security, IBM Redbooks, 2014.

[13] https://www.cisco.com/c/en/us/solutions/enterprise/design-zone security/landing_safe.html#~ overview.

[14] C.K. Veitch, J.M. Henry, B.T. Richardson, D.H. Hart, Microgrid Cyber Security Reference Architecture (No. SAND2013-5472), Sandia National Lab. (SNL-NM), Albuquerque, NM, United States, 2013.

[15] IIoT World, Trend Micro Cybersecurity Reference Architecture for Operational Technology—White Paper, 2017. http://iiot-world.com/wp-content/uploads/2017/12/Trend-Micro-Cybersecurity-Reference-Architecture-for-Operational-Technology.pdf. (Accessed 12 June 2022).

[16] DXC Cyber Reference Architecture, White Paper, Security in Depth Reference Architecture. O. E. T. S, 2013. https://assets1.dxc.technology/secu-rity/downloads/DXC-Security-Cyber_Reference_Architecture.pdf.

[17] https://learn.microsoft.com/en-us/security/cybersecurity-reference-architecture/mcra.

[18] R. Wöstmann, P. Schlunder, F. Temme, R. Klinkenberg, J. Kimberger, A. Spichtinger, J. Deuse, Conception of a reference architecture for machine learning in the process industry, in: 2020 IEEE International Conference on Big Data (Big Data), December, IEEE, 2020, pp. 1726–1735.

[19] M. Elsisi, M.-Q. Tran, K. Mahmoud, D.-E.A. Mansour, M. Lehtonen, M.M.F. Darwish, Towards secured online monitoring for digitalized GIS against cyber-attacks based on IoT and machine learning, IEEE Access 9 (2021) 78415–78427.

[20] https://nocomplexity.com/documents/fossml/architecture.html.

[21] K. Ghirardello, C. Maple, D. Ng, P. Kearney, Cyber security of smart homes: development of a reference architecture for attack surface analysis, in: Living in the Internet of Things: Cybersecurity of the IoT—2018, London, 2018, pp. 1–10, https://doi.org/10.1049/cp.2018.0045.

[22] J. Navarro, V. Legrand, A. Deruyver, et al., OMMA: open architecture for operator-guided monitoring of multi-step attacks, EURASIP J. Inf. Secur. 2018 (2018) 6, https://doi.org/10.1186/s13635-018-0075-x.

[23] B.Y. Ozkan, M. Spruit, Cybersecurity standardisation for SMEs: the Stakeholders' perspectives and a research agenda, in: Research Anthology on Artificial Intelligence Applications in Security, IGI Global, 2021, pp. 1252–1278.

[24] P. Niemelä, B. Silverajan, M. Nurminen, J. Hukkanen, H. Järvinen, LAOps: learning analytics with privacy-aware MLOps, in: Proceedings of the 14th International Conference on Computer Supported Education—Volume 2: CSEDU, 2022, pp. 213–220. ISBN 978-989-758-562-3, ISS 2184-5026 https://doi.org/10.5220/0011113300003182.

CHAPTER 10

An architectural framework for the allocation resources in emergency management systems

Umur Togay Yazar[a] and Mehmet Akşit[b]
[a]Department of Engineering, TOBB University of Economics and Technology, Ankara, Turkey
[b]Department of Computer Science, University of Twente, Enschede, The Netherlands

1 Introduction

Critical infrastructures provide assets that are essential for the functioning of a society and economy. Examples of critical assets are gas, electricity, and water supply facilities, road networks, railways, telecommunication infrastructure, hospitals, fire stations, and emergency operations centers. The resilience of critical infrastructure is considered crucial.

In the literature, critical infrastructures are studied from various perspectives such as system design, preparation, prevention, mitigation, resource management, emergency management, etc. In this chapter, as an example case, we focus on critical aid activities to manage emergency conditions caused by natural events such as earthquakes. As such, we consider systems for dealing with emergency management also as critical infrastructures [1].

Aid activities can be modeled as tasks that are assigned to resources. For example, if a building collapses due to an earthquake, several aid activities can be assigned to task forces: *Security forces* are called to protect the area, *rescue teams* to remove the people from under the rubble, *ambulances* to transfer the wounded ones to hospitals, etc. The required tasks and resources depend obviously on the type of emergency conditions, which are generally specified in a set of predefined procedures. As such, the execution of aid activities can be planned by scheduling the relevant activities.

In large-scale emergency conditions, there may not be enough resources to satisfy the needs of all tasks. In such cases, it may be necessary to prioritize and/or trade-off the assignment of resources to the tasks. In the case of prioritization, resources are first assigned to the tasks with higher priorities. In case of a trade-off, a compromise is made in assigning resources, for example, by assigning a reduced number of resources than the required ones. A mixed strategy can also be adopted by grouping tasks. The first group of tasks, for example, is assigned to the required tasks as desired, the second group can only receive say 80% of the desired resources, and so on.

This chapter introduces a novel emergency control system architecture for earthquake management. The emergency instances which are obtained for example from a

Management and Engineering of Critical Infrastructures
https://doi.org/10.1016/B978-0-323-99330-2.00003-9

dedicated IoT network are first processed and converted to sets of tasks for aid activities. Second, the required resources are determined. If the scheduling of tasks over the available resources is not possible without exceeding the deadlines, various prioritization and/ or trade-off strategies are applied. To facilitate flexibility, the architecture is designed as an open system where new strategies can be introduced if necessary. As a case study, emergencies caused by earthquakes are studied. To the best of our knowledge, there has not been a significant publication on an architectural approach for the automated scheduling of tasks and resolving conflicts in earthquake management.

This chapter is organized as follows. The following section introduces an example case, gives an overview of the background work, and defines the research questions that are addressed in this chapter. Section 3 explains the approach undertaken. To demonstrate the utility of the proposed architecture, Section 4 presents two case studies: Management of earthquakes with minor and major impacts. Section 5 includes a discussion on the validity of the approach. Finally, Section 6 gives conclusions.

2 Example case, background work, and research questions

2.1 Example case: Scheduling in earthquake management

Consider a midsize town that consists of the following structures: 1000 residences, 5 schools, 2 hospitals, 2 government buildings, 1 shopping center, 20 shops, 5 finance centers, 1 sports center, 2 buildings for religious practices, 1 museum, 1 historical building, 5 bridges, and 20 office buildings. Naturally, the town incorporates communication, electricity, gas, water networks, and facilities for firefighters and security forces. The latter ones are considered high-priority critical infrastructures. There is also one emergency operations center that receives reports on possible emergency conditions.

To evaluate the problem addressed in this chapter and to determine the applicability of the proposed approach we assume two disaster scenarios. In the first scenario, after an earthquake, we assume that there is a 2% probability of damage per structure. In the second scenario, this probability is assumed to be 20%. We assume that after an earthquake one or more of the following task forces are composed: rescue team, medical team, firefighter, repair team, and security forces.

2.2 Background work

An emergency condition is defined as a situation that may cause risk to health, life, property, or the environment [2]. Emergency operations centers are established as command and control facilities responsible to carry out the necessary aid operations to minimize the negative effect of emergency conditions [3]. Critical infrastructures are essential assets in which damage to them may have a negative impact on the well-being of the corresponding society. They are a considerable number of publications related to the different aspects of establishing and operating emergency operations centers [4–6]. The focus of most of these publications is on designing, implementing, and running emergency management processes, which are in general not fully automated. Resource allocation and resolution of resource conflicts are generally realized by human operators.

There are research approaches that consider aid operations as an optimization problem of routing [7]. They consider maximizing the efficiency of transporting the goods to the required locations. In general, these approaches are studied academically, and as such, they are not incorporated as an architectural framework in IoT-driven emergency management systems.

Several approaches have been proposed for conflict resolution in resource allocation. Zeng et al. [8] adapt cross-organizational workflow models with the help of dedicated Petri-net models. Fiedrich et al. [9] propose a simulated annealing algorithm for minimizing the rescue time so that number of casualties is kept low. Altay [10] apply integer programming algorithms to implement capability-based resource allocation. Sherali et al. [11] suggest branch-and-bound relaxation to linear programming for managing risks in emergency situations. Basu et al. [12] propose a rather different approach to resource allocation. They adopt a utility-driven model based on information extracted from microblogs. In Yin et al. [13], a graph-based conflict resolution algorithm is defined for option prioritization among conflicts. In Gao et al. [14], a Petri-net-based approach is proposed to model emergency management processes for the purpose of optimizing resource allocation. Wang and Wu [15] extend emergency workflow models with resource models to verify workflows under resource constraints using state-based analysis techniques. In Tecle et al. [16], conflicts in resource allocation are studied among multiple decision-makers. There are also several proposals to allocate flight routes to demanding aircraft [17,18]. To train the operators for resource allocation after earthquakes, Fiedrich [19] describes an agent-based simulation environment.

The optimization algorithms adopted in these publications are defined and implemented for a specific disaster definition. Most of them offer dedicated models and fixed solutions to the resource allocation problem. However, the impact of a severe earthquake cannot be anticipated precisely. As such, the resource allocation problem must be derived online from actual emergency events. It is, therefore, necessary to adopt an IoT-based open architectural approach that can be programmable for different categories of emergencies.

2.3 The research questions and method

Within the context of the resource allocation problem in emergency management and to address the limitation of the current approaches as described in Sections 2.1 and 2.2, we formulate the following research questions:

Q1. What kind of an open architectural framework must be defined to fulfill the task allocation requirements and address the conflict resolution problems for each different kind of emergency condition?

Q2. What are the conditions that indicate the resource allocation conflicts?

Q3. Which automated resource allocation and conflict resolution technique is suitable to deal with the emergency management problem?

As for the research method, we take three approaches:

1. Architecture-based:

   To seek open and generic solutions to different emergency management conditions, suitable architectures are investigated which provide the desired modularity.

2. Mathematical modeling:

   The resource allocation problem is investigated mathematically to offer precise solutions where possible.

3. Heuristic modeling:

   There are several situations where exact mathematical models cannot be defined to allocate resources and resolve conflicts due to human-made procedures and the complexity of the problems. In such cases, heuristic rule-based modeling is applied.

3 The approach

This section describes the approach undertaken. The following subsection introduces a model for emergency instances. Section 3.2 presents the control architecture for emergency management. Section 3.3 determines the aid operations based on emergency conditions. Section 3.4 formulates aid operations as a scheduling problem. Detecting conflict situations are presented in Section 3.5. Finally, Section 3.6 defined the control strategies.

3.1 Definition of an emergency instance

In this section, we define the terminology that is used in this chapter. When an earthquake has occurred, we assume that each emergency instance that was caused by the earthquake should be reported to the emergency operations center. An emergency instance is a quintuple:

$$e_i = (L_i, K_i, I_i, BC_i, NP_i) \tag{1}$$

where,

- e_i is the ith emergency instance,
- L_i is the location attributes of the ith emergency instance,
- K_i is the kind of ith emergency instance,
- I_i is the intensity of the ith emergency instance,
- BC_i is the building class of the ith emergency instance,

- NP_i is the number of persons affected by the ith emergency instance.

Zero or more emergency instances may arrive at the emergency operations center. The location attribute is used by the task forces. The kind, intensity, number of persons, and building classes are defined as a set of symbols, where for each instance a single symbol must be selected:

Kind (K_i) = Collapse, Fire, Flood, Landslide, Unhealthy.

Intensity (I_i) = Low, Medium Low, Medium, Medium High, High.

Number of Persons (NP_i) = Low, Medium Low, Medium, Medium High, High.

Building Class (BC_i) = Residence, School, Hospital, Government Building, Shopping Center, Shop, Finance Center, Sports Center, Religious Building, Museum, Historical Building, Bridge, Office Building and Communication, Electricity, Gas, Water networks and Facilities for Firefighters and Security Forces.

Of course, multiple kinds of emergency conditions may appear in the same location, where each of them is represented as a separate emergency instance.

3.2 The control architecture

For resource allocation and conflict resolution, we propose a feedback control architecture. Typically, a single loop feedback control architecture consists of the following modules [20]:

Sensor: Measures the necessary parameters of the system to be controlled.

Feedback: Transforms the feedback data for analysis.

Analysis: Evaluates if the error (conflict) conditions have occurred.

Control strategies: Implements actions to minimize the effect of errors.

Adaptor: Communicates with the system to be controlled so that errors can be minimized.

We adopted the control architecture which is shown in Fig. 1. In the mid-left of the figure, the component *Emergency Sensor* is shown with two inner components: *Data Buffer* and *Data Retriever*. The reported emergency conditions after an occurrence of a disaster are registered in *Data Buffer*. The emergency conditions can for example be gathered through an IoT network. Data gathering is out of the scope of this chapter. Upon demand by the component *Feedback*, the emergencies are retrieved through the *Data Retriever*.

The component *Feedback* consists of three inner components: *Task Generator*, *Task to Job Transformer*, and *Jobs Buffer*. The component *Task Generator* accepts *Emergency Handling Procedures* and the received emergency instances as input and creates a set of the corresponding tasks. *Task to Job Transformer* groups tasks that belong to the same location. The heuristic rules for the transformation are defined by the actor *Definer of Procedures* which are stored in the component *Emergency Handling Procedures*. The generated jobs are stored in the component *Jobs Buffer*.

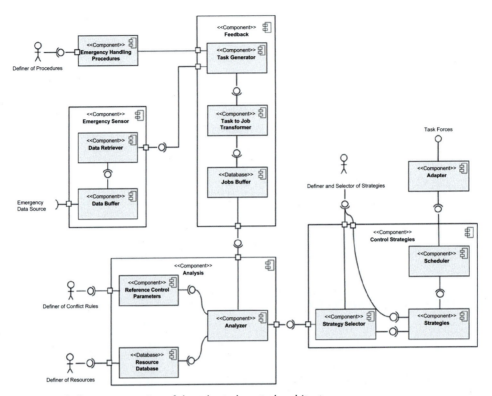

Fig. 1 A symbolic representation of the adopted control architecture.

The component *Analysis* consists of three components *Reference Control Parameters*, *Resource Database*, and *Analyzer*. The rules that refer to the conflict conditions are defined by the actor *Definer of Conflict Rules*. The actor *Definer of Resources* enters the available resources for disaster management in *Resource Database*. The component *Analyzer* accepts the resource control parameters and available resource information and decides if there is a conflict condition in allocating resources.

The component *Control Strategies* consist of three components: *Strategy Selector*, *Strategies*, and *Scheduler*. In case of a conflict condition, the *Strategy Selector* decides on one of the control strategies which are stored in the component *Strategies*. The actor *Definer and Selector of Strategies* program these two components according to the desired control actions. The component *Scheduler* orders the tasks that are delivered by the component *Strategies*. If *Scheduler* cannot find a solution in ordering the tasks for example, due to the complexity of the problem, it warns the component *Strategy Selector* to restart the process again. The component *Adaptor* communicates the list of scheduled tasks to the task forces.

3.3 Determining the aid operations based on emergency conditions

We define emergency aid operations as a series of tasks carried out by the task forces. Tasks must be allocated to resources for; (i) dispatching them to the necessary task forces so that the negative impact of emergencies can be minimized, and (ii) maximizing the efficiency of aid operations. For case i, emergency handling procedures are utilized. For case ii, scheduling techniques are adopted.

The emergency handling procedures are defined as a set of heuristic rules of symbols that are stored in the component *Emergency Handling Procedures* shown in Fig. 1. We adopted the following set of symbols *low, medium-low, medium, medium-high, and high*, in which their effect can be increased or decreased by using additional symbols like *very*. A symbolic representation of rules provides flexibility in their interpretation. For example, depending on the needs, crisp sets, fuzzy sets, probabilistic distribution functions, or multi-valued logical interpretation can be used. In the following, we show a selection from a rule table to illustrate how emergencies can be converted into tasks.

In Section 3.1, an *emergency instance* is defined by the formula (1), which contains a set of attributes including the *kind of emergency instance*, the *number of persons*, and the *intensity* of the *emergency instance*. The *kind* of *emergency instance* can be for example *collapse*.

Consider Table 1 which presents a set of rules to determine the size of the *rescue team* depending on the attributes *Kind, Number of Persons* affected, and the *Intensity* of the collapse. These rules are defined and maintained by experts and are stored in the rule database. Consider the following rule given in the table: "If the structure under consideration is *collapsed*, the intensity of the collapse is measured *high* and the *number of persons* in the structure is *low*, then, the size of the *rescue team* must be in *medium* size."

Of course, if necessary, experts can modify these rules according to their judgments. In the database, rules are defined also for the task forces of *firefighters, medical teams, security forces*, and *repair teams*.

Each inferred task is further prioritized by applying a set of rules defined in Emergency Handling Procedures. Consider, for example, the following prioritization rules:

- Option 1 considers only the attribute *Building Class* of an emergency instance and assigns a priority value according to its importance.
- Option 2 considers the attributes of *Building Class* and *Number of Persons* of emergency instances and assigns a priority value proportional to the combination of the importance of buildings and the number of persons. The higher the importance of a building and the number of persons involved, the higher the priority of the corresponding task.
- Option 3 considers only the attribute *Number of Persons* of an emergency instance and assigns a priority value according to its importance.

Table 1 Rules for defining the size of a *rescue team* using the attributes *kind, number of persons,* and *intensity*, in which the last column indicates the size of the *rescue team* to be allocated.

Kind	Number of persons	Intensity	Rescue team size
Collapse	Low	Low	Low
Collapse	Medium low	Low	Medium low
Collapse	Medium	Low	Medium
Collapse	Medium high	Low	Medium high
Collapse	High	Low	High
Collapse	Low	Medium low	Low
Collapse	Medium low	Medium low	Medium low
Collapse	Medium	Medium low	Medium
Collapse	Medium high	Medium low	Medium high
Collapse	High	Medium low	High
Collapse	Low	Medium	Low
Collapse	Medium low	Medium	Medium low
Collapse	Medium	Medium	Medium
Collapse	Medium high	Medium	Medium high
Collapse	High	Medium	High
Collapse	Low	Medium high	Medium low
Collapse	Medium low	Medium high	Medium
Collapse	Medium	Medium high	Medium high
Collapse	Medium high	Medium high	High
Collapse	High	Medium high	High
Collapse	Low	High	Medium
Collapse	Medium low	High	Medium high
Collapse	Medium	High	Medium high
Collapse	Medium high	High	High
Collapse	High	High	High

3.4 Formulating aid operations as a scheduling problem

We will now elaborate on the components in the *Feedback* module shown in Fig. 1. The component *Task Generator* accepts *Emergency Handling Procedures* and *Emergency Instances* as input and generates the corresponding tasks. A task is defined as,

$$t_k = (TW_k, TD_k, CR_k, PT_k) \tag{2}$$

where,

- t_k is an instance of a task,
- TW_k is the worst-case execution time of the task k,
- TD_k is the deadline of the task k,
- CR_k is the demanded resource class for the task k,
- PT_k is the priority of the task.

An emergency condition may require a set of aid operations composed of various task forces. For example, a high-intensity collapse of a large building with many persons inside may require a substantial number of *rescue teams, firefighters, medical teams, security forces*, and *repair teams*. The aid operations of these task forces must be ordered to minimize the overall duration of rescue operations. Aid operations are therefore formulated as a scheduling problem.

The component *Task to Job Transformer* groups the related tasks that belong to the "same" geographical location into *jobs*. A *job, therefore*, consists of a task or sequence of tasks:

$$j = (t1, t2, ..., tn) \tag{3}$$

All the generated *jobs* are stored in the component *Job Buffer* to be utilized by the *Analysis* module.

3.5 Analysis of jobs to detect conflicts

The component *Analyzer* which is placed in the *Analysis* module shown in Fig. 1, evaluates the scheduling problem based on the generated jobs, the rules stored in the component *Reference Control Parameters*, and the available resources in *Resource Database*. The number of rules is not fixed and can be changed by the actor *Definer of Conflict Rules*. Assume for example that the following rules are defined:

1. The required resource types by the *jobs* stored in the *Job Buffer* do not match the available resources stored in *Resource Database*. If the rule evaluates FALSE, the next rule is executed. Otherwise, the definition of the corresponding job is denoted as ERROR and the operator is notified.
2. The number of required resources for the jobs is less than the available resources. If the rule evaluates FALSE, the component *Strategy Selector* in *the Control Strategies* module is notified that there is a conflict. Otherwise, the next rule is executed.
3. The generated jobs are evaluated as schedulable with respect to the properties of the adopted scheduling algorithm to complete the process within a specified deadline. If the rule evaluates FALSE, the component *Strategy Selector* is notified that there is a conflict. Otherwise, the next rule is executed.

If the last rule evaluates TRUE, the component *Strategy Selector* is notified that there is no conflict.

Each resource in *Resource Database* is defined as a pair:

$$rl = (URl, CRl) \tag{4}$$

- *rl* is the *l*th resource instance.
- *URl* is the unique identifier denoting the *l*th resource instance.
- *CRl* is the resource class of the *l*th resource instance.

For resources, the unique identifier is necessary to select, allocate, track, and release the resources. A resource class is a set of names of the task forces (Police, Rescue, Road, Repair, and Medical). Each of these resource classes refers to a team of dedicated persons and the necessary equipment.

3.6 Control strategies for scheduling

If the component *Strategy Selector* is notified that there is a conflict, one of the predefined strategies is selected. Consider, for example, the following strategies:
1. The tasks are ordered according to their priorities and the resources are allocated starting from the task with the highest priority.
2. The resources required by the jobs are allocated in reduced size determined by the compromise factor.
3. The jobs are divided into groups in which different strategies are applied for each group. For example, one group may be prioritized, the other may be compromised by a lower value, and the third one is compromised by a higher value.

The selection of a strategy can be made by the operator or by an intelligent algorithm. Depending on the selected strategy, the scheduling problem that is retrieved from the component *Job Buffer* is modified and passed to the component *Scheduler*. If for some reason jobs are not schedulable, the component *Strategy Selector* is notified that scheduling is FAILED. In that case, the error handling strategy is activated, which is defined as redefining the scheduling problem as a smaller set of scheduling jobs. If failures are repeated multiple times the operator is warned. If the jobs are schedulable, they are passed to the component *Adaptor* which communicates with the corresponding task forces to inform them about the jobs to be carried out.

The component *Scheduler* implements a *job shop scheduling* problem [21], in which multiple jobs are carried out by several task forces. The job shop problem is defined as:

$$J = ST/SR, prec/JTmin \tag{5}$$

Here, ST is a sequence of tasks, SR is a set of resources, *prec* is precedence constraints and $JTmin$ is the minimum scheduling time. The job shop problem is to schedule tasks on resources to minimize the time it takes for all jobs to be handled.

We adopt Google OR Tools [22] to implement the component *Scheduler* shown in Fig. 1. As such the scheduling problem that is generated by the component *Task to Job Transformer* must conform to the following input format of this tool:

$$j = (TF1, D1)(TF2, D2)...(TFn, Dn) \tag{6}$$

Here, each pair corresponds to a task expressed as a combination of the task force id and the duration of the corresponding task. The duration of each task is estimated based on the intensity of the corresponding emergency instance with the help of a set of rules stored in the *Emergency Handling Procedures*.

4 Case studies

Based on our proof-of-concept implementation, in the following subsections, we present two case studies to demonstrate the contribution of this chapter. The case studies are based on the example given in Section 2.1. For both case studies, the probabilities of disaster kinds *collapse, fire*, and *landslide* are assumed to be 0.5, 0.33, and 0.17, respectively. The probability of the intensity values of disasters *low, medium-low, medium, medium-high, and high* are all equal to 0.2. The probabilities of the existence of the number of persons in each building class are shown in the following table.

The first column in Table 2 lists the building classes. The second and third columns show the minimum and the maximum number of persons for each building class. For the following case studies, a number between these minimum and maximum numbers is randomly selected based on uniform distribution.

4.1 An earthquake with a minor impact

Let us assume that an earthquake occurred with an impact probability of 0.02. In our simulation model, 12 emergency instances are generated and stored in the component *Data Buffer*. The first column of Table 3 shows the unique identifiers of the emergency

Table 2 Distribution of the number of persons for each building class.

Building class	Minimum number of persons	Maximum number of persons
Residence	1	6
School	200	400
Hospital	200	400
Government building	50	150
Shopping center	200	800
Shop	5	20
Finance center	10	100
Sports center	10	1000
Religious building	2	20
Museum	20	50
Bridge	30	80
Historical building	20	50
Office building	15	50
Firefighters and security forces Facilities	50	100
Electricity, gas, and water networks	20	30

Table 3 The generated emergency instances with a 0.02 probability.

Unique identifier	Kind	Intensity	Building class	Number of persons
9	Collapse	Medium high	Electricity, gas, and water networks	27
2	Fire	Low	Residence	2
3	Fire	Medium	Residence	5
4	Fire	Medium	Residence	3
5	Fire	High	Residence	4
6	Collapse	Low	Bridge	42
7	Fire	Medium low	Residence	4
8	Fire	Low	Residence	4
10	Fire	High	Finance center	14
11	Collapse	Low	Residence	4
12	Landslide	Low	Residence	4
1	Fire	Medium	Residence	2

instances in which the first row and the last row correspond to the maximum and minimum span time in the scheduling process, respectively. This disaster case is considered minor since the available resources are assumed to be higher than the demand.

The definitions of other columns are explained in Section 3.1. This disaster case happens to represent 3 kinds of damages—*Collapse*, *Fire*, and *Landslide*—on the building types of *Residence, Electricity Gas and Water Networks, Bridge*, and *Finance Center*. Due to the probabilistic nature of the generation process, the number and attributes of these emergency instances may vary from case to case. The component *Task Generator* retrieves these instances and the rules stored in the component *Emergency Handling Procedures*, and by executing the rules, teams are generated.

Consider Table 4 which shows an excerpt of five emergency instances from Table 3. Columns 2–6 here indicate the required composition of the task forces for these emergency instances. It is assumed that as for the total number of resources, there are 50 teams for each task force. The component *Analyzer* evaluates the scheduling demand with respect to *Referenced Control Parameters* and the available resources in *Resource Database* in which all the three rules given in Section 3.5 evaluate TRUE. Table 5 gives the total number of required teams for all emergency instances.

In Fig. 2, teams for the task force *Repair Team* are numbered from 0 to 49, for the task force *Security Forces* from 50 to 99, for the task force *Medical Team* from 100 to 149, and so on. The generated tasks and the required resources are provided to the component *Scheduler* which orders the tasks according to their priorities. This figure depicts an excerpt from the scheduled tasks for the example case. Here, the rows represent the assigned task forces. The component *Scheduler* treats the teams for each task force equally. The

Table 4 Excerpt from the inferred tasks for the emergency instances shown in Table 3.

Unique identifier	Firefighters	Medical team	Repair team	Rescue team	Security forces
9	1	2	4	3	3
10	5	5	1	1	2
5	5	5	1	1	2
11	1	1	1	1	1
1	3	3	1	1	1

Table 5 The total number of demanded resources.

Firefighters	Medical team	Repair team	Rescue team	Security forces
27	30	15	16	16

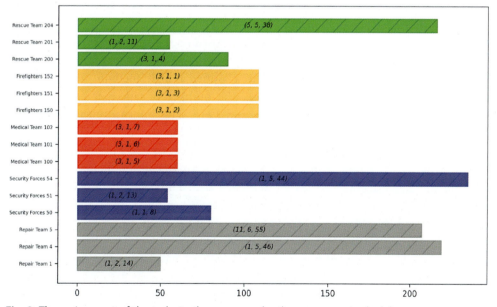

Fig. 2 The assignment of the tasks to the resources by the component *scheduler*.

horizontal bars indicate the scheduled time intervals for the corresponding tasks. In each bar, the priority, unique identifier of the emergency instance, and the task are shown respectively. The *X*-axis shows the execution time unit. We assume that the maximum required time for the tasks is considered acceptable. For simplicity, the traveling time of a resource from one location to another is neglected.

Table 6 Total number of demanded resources.

Firefighters	Medical team	Repair team	Rescue team	Security forces
407	398	524	319	371

4.2 An earthquake with a major impact

In case of a major impact, we increase the probability from 0.02 to 0.2. This results in 224 emergency instances in the component *Data Buffer*. With the help of the rules stored in the component *Emergency Handling Procedures*, 2019 tasks are generated that demand the resources in Table 6.

The available resources are assumed to be 50 teams per task force. Since the demand is much higher than the available resources, this is an earthquake with a major impact. Based on these values, the component *Analyzer* reports that the available resources are not sufficient, and as such control strategies must be adopted. In the following sections, we consider the effect of two strategies on the overall performance of the scheduled tasks.

4.2.1 Priority based scheduling

In this strategy, all the 2019 tasks are prioritized based on the attributes of *Building Class*, *Number of Persons*, and *Intensity* of the corresponding emergency instances. Here, the values in the range 11–15, 5–10, and 1–4 are classified as high, medium, and low prioritization values, respectively. For this case, the number of tasks per category high, medium, and low are respectively 31, 45, and 1943. This information is provided to the component *Scheduler*, whose output is depicted in Fig. 3.

The Y-axis in Fig. 3 shows the prioritization values of tasks on a nonlinear scale. The X-axis shows the end-time of the tasks depicted as a point in the figure. The tasks in medium and high categories are placed in the red area. The component *Scheduler* assigns first the resources to the tasks with higher priorities. When there are no more resources, tasks are suspended until the required resources are freed again. The total required time for an aid operation of a task is randomly defined. A task with a high priority, for example, aid operations for a collapsed hospital or a school, generally demand a complex set of resources compared to an empty building with a low priority. As such, one cannot conclude the exact end time of a task unless all parameters are known. Nevertheless, it can be seen from the figure that tasks with lower priorities on average require more time to complete than the tasks with higher priorities. The median value of the end-time is 585 for high-priority tasks. Fig. 4 depicts the histogram of the number of tasks versus the end-time of tasks with low priority, where the median value is 1067.

4.2.2 Scheduling based on 20% compromise

The resource allocation based on a 20% compromise is executed as follows. First, all tasks are prioritized according to the rules as adopted in the previous section. Second, starting

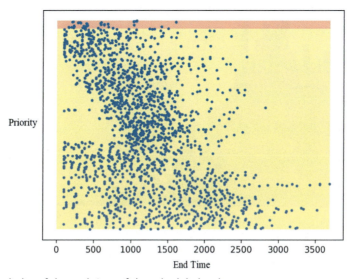

Fig. 3 Scattered plot of the end-time of the scheduled tasks.

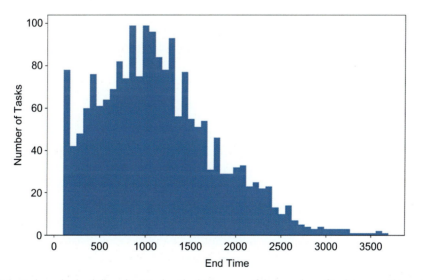

Fig. 4 Priority-based scheduling depicted as the histogram of the number of tasks versus the end-time of the tasks with low priority.

from the high to lower priority of tasks, only 80% of the demanded resources are assigned. When there are no more resources, tasks are suspended until the required resources are freed again. Starting higher to lower priority, the remaining 20% of resources are assigned in the next round of the scheduling process.

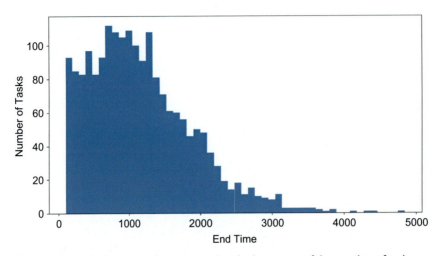

Fig. 5 Scheduling with 20% compromise depicted as the histogram of the number of tasks versus the end-time of the tasks with low priority.

We notice that in this strategy, the median value of the tasks with high priority is increased from 585 to 634. The result of the scheduling process for low priority of tasks is depicted in Fig. 5. The median value of this histogram is 1052, which is somewhat lower than the previous case. However, the completion time of some of the tasks is stretched toward the 5000-time unit.

5 Threats to validity

With respect to the objectives of this article, we observe the following threats to the validity of the assumptions presented in this chapter:

(1) **Reporting of the emergency instances is not regular**: The article assumes that all the relevant emergency instances have been stored in the resource database before the aid activities are determined. If, however, the emergency instances continue to flow in, an online scheduling algorithm can be more appropriate.

(2) **The data model of an emergency instance is not expressive**: In Section 3.1, the data model of an emergency instance is given based on several predefined attributes such as building classes, number of persons, and intensity. It may be the case that for certain occurrences of disasters, new attributes are required. To provide flexibility, extensible modeling techniques can be adopted.

(3) **Rules for emergency handling procedures are not effective**: The effectiveness of the proposed system is largely dependent on the way how emergency instances are interpreted and converted into a set of tasks. The required resources are also computed based on predefined procedures, which are formulated as heuristic rules. Such

a rule-based decision support system may suffer from several challenges: (a) Rules may be unnoticeably conflicting with each other; (b) Rules may be based on a wrong assumption; (c) Due to evolutionary circumstances, rules may eventually be obsolete; (d) Different strategies may cause unanticipated results for the same set of rules. One may try to reduce these risks with the help of experimentation and the involvement of experts.

(4) Emergency instances from neighboring locations are ignored: The generated tasks are converted into a common job in case the corresponding instances have the "same" geographical location. It might be more effective if the reported emergency instances from neighboring locations in proximity are grouped in a common job.

(5) The prioritization and/or compromising strategies are not effective: The strategies to group tasks and allocated resources are based on the expert's opinion, which may not be accurate as desired. Nevertheless, experiments can be carried out and machine learning algorithms can be adopted to improve the performance of strategies.

(6) Rerouting of resources has a major impact: In this article, for simplicity, several assumptions are made. For example, rerouting time and the location of task forces are omitted. Nevertheless, these can be added as additional tasks and be considered in the scheduling process.

(7) Scheduling of tasks is computationally too expensive: If the tasks are too complex and/or the number of tasks is large, the scheduling process may exceed the acceptable deadline. This may demand simplifications and/or parallelization of the scheduling problem. Nevertheless, in the example cases presented in this chapter, the scheduling processes are considered within acceptable limits. In the first example case of Section 4.1, the schedule is generated in 55 ms on an Intel processor of the 11th generation at 2.30 GHz with 16 GB memory running on a Windows 11 operating system. In the second example, the schedule is generated in 486,327 ms. For the last case, the scheduling process takes 532,139 ms.

6 Conclusions

In Section 2.3, three research questions are formulated as the objective of this chapter. The first research question is about designing an open architecture for emergency management. In Section 3.2 an architecture is presented that includes the necessary components *Emergency Sensor, Emergency Handling Procedures, Feedback, Analysis, Control Strategies*, and *Scheduler* for emergency management. Furthermore, the architecture is considered open to introducing new rules and modifying and removing the existing ones for emergency instances to job conversion, situation analysis, and control strategies. The second research question which is related to the conditions that indicate resource allocation conflicts is described as *Reference Control Parameters* in Section 3.5. The third research question which is about resource allocation and conflict resolution techniques is addressed in Section 3.6 by proposing prioritization, compromising, and combined strategies.

Acknowledgment

We appreciate the contributions of our students A. Türksoy, I. Yüksel, and R. Bilgücü on an earlier version of this chapter. In this version, the software implementation is rewritten, and the corresponding data is regenerated. This work has been generously supported by the TÜBİTAK BİDEB program 2232 International Fellowship for Outstanding Researchers.

References

[1] EU Commission, Communication from the commission on a European Programme for critical infrastructure protection, COM (2006) 786, 2006.

[2] J.C. Knight, Safety critical systems: challenges and directions, in: Proceedings of the 24th International Conference on Software Engineering, 2002 May 19, pp. 547–550.

[3] M.J. Fagel, R.C. Mathews, J.H. Murphy (Eds.), Principles of Emergency Management and Emergency Operations Centers (EOC), CRC Press, 2021 September 26.

[4] J.L. Wybo, K.M. Kowalski, Command centers and emergency management support, Saf. Sci. 30 (1–2) (1998) 131–138.

[5] M. Ryan, Planning in the emergency operations center, Technol. Forecast. Soc. Chang. 80 (9) (2013) 1725–1731.

[6] L.G. Militello, E.S. Patterson, L. Bowman, R. Wears, Information flow during crisis management: challenges to coordination in the emergency operations center, Cogn. Tech. Work 9 (1) (2007) 25–31.

[7] P. Toth, D. Vigo (Eds.), Vehicle Routing: Problems, Methods, and Applications, Society for Industrial and Applied Mathematics, 2014 November 24.

[8] Q. Zeng, C. Liu, H. Duan, M. Zhou, Resource conflict checking and resolution controller design for cross-organization emergency response processes, IEEE Trans. Syst. Man Cybern. Syst. Hum. 50 (10) (2019) 3685–3700.

[9] F. Fiedrich, F. Gehbauer, U. Rickers, Optimized resource allocation for emergency response after earthquake disasters, Saf. Sci. 35 (1–3) (2000) 41–57.

[10] N. Altay, Capability-based resource allocation for effective disaster response, IMA J. Manag. Math. 24 (2) (2013) 253–266.

[11] H.D. Sherali, J. Desai, T.S. Glickman, Allocating emergency response resources to minimize risk with equity considerations, Am. J. Math. Manag. Sci. 24 (3–4) (2004) 367–410.

[12] S. Basu, S. Roy, S. Bandyopadhyay, S.D. Bit, A utility driven post disaster emergency resource allocation system using DTN, IEEE Trans. Syst. Man Cybern. Syst. Hum. 50 (7) (2018) 2338–2350.

[13] K. Yin, L. Yu, X. Li, An improved graph model for conflict resolution based on option prioritization and its application, Int. J. Environ. Res. Public Health 14 (11) (2017) 1311.

[14] Y. Gao, C. Liu, Q. Zeng, H. Duan, Two effective strategies to support cross-organization emergency resource allocation optimization, Mob. Inf. Syst. 2021 (2021) 7965935.

[15] Y. Wang, Z. Wu, Model construction of planning and scheduling system based on digital twin, Int. J. Adv. Manuf. Technol. 109 (7) (2020) 2189–2203.

[16] A. Tecle, F. Szidarovszky, L. Duckstein, Conflict analysis in multi-resource forest management with multiple decision-makers, Nat. Resour. 31 (3) (1995) 8–17.

[17] Y.X. Han, X.Q. Huang, X.M. Tang, Development of a new tool for constrained conflict resolution, Proc. Inst. Mech. Eng. G J. Aerosp. Eng. 233 (5) (2019) 1683–1694.

[18] A. Valenzuela, D. Rivas, R. Vázquez, I. del Pozo, M. Vilaplana, Conflict resolution with time constraints in trajectory based arrival management, in: Second SESAR Innovation Days, 27th–29th November 2012, 2012.

[19] F. Fiedrich, An HLA-based multiagent system for optimized resource allocation after strong earthquakes, in: Proceedings of the 2006 Winter Simulation Conference, IEEE, 2006 December 3, pp. 486–492.

[20] M. Akşit, The 7 C's for creating living software: a research perspective for quality-oriented software engineering, Turk. J. Electr. Eng. Comput. Sci. 12 (2) (2004) 61–96.

[21] G. Orhan, Optimal Modeling Language and Framework for Schedulable Systems, University of Twente, 2019 October 31.

[22] L. Perron, V. Furnon, OR-Tools, Google Developers, 2022. (Internet), cited 1 June 2022. Available from: https://developers.google.com/optimization/.

PART IV

Application domains

CHAPTER 11

Urban water distribution networks: Challenges and solution directions

Miguel Ángel Pardo Picazo[a] and Bedir Tekinerdogan[b]
[a]Department of Civil Engineering, University of Alicante, Alicante, Spain
[b]Information Technology Group, Wageningen University & Research, Wageningen, The Netherlands

1 Introduction

Water utility managers and policymakers are acutely aware that water distribution networks are vulnerable to a range of natural and man-made disasters. These include water scarcity, droughts, water shortages, flooding, earthquakes, and more. In the latest report from the American Water Works Association [1], the major challenges facing the water sector are outlined in Table 1. These challenges are ranked according to their level of importance, with a value of one indicating a low level of importance and a value of five indicating high importance.

According to the report, the most pressing challenge facing the water sector is the renewal of aging infrastructure. This has been the primary concern for the past 9 years and continues to be a major challenge. The report is updated annually to reflect changes in the industry and emerging challenges.

The Environmental Protection Agency (EPA) published [2] a report considering the potential drinking water hazards. Natural disasters and hazardous releases affected drinking water systems and received the US government's attention. The potential hazards in drinking water systems are shown in Fig. 1. This report did not show any hierarchy that demonstrated the key future problems in water-pressurized networks.

One of the greatest problems identified is water stress. Today, 1.9 billion people live in arid regions with intense water scarcity, a figure that is expected to rise to 5.7 billion by 2050 [3]. According to the report "Water resources across Europe—confronting water stress: an updated assessment" [4] water stress affects 20% of Europe's territory, an area where 30% of Europe's population lives. Southern Europe is facing extreme water stress problems [5], with agriculture, urban supply, and tourism being the major consumers.

The European Federation of National Water Services Associations (EurEau), a body representing national water service providers in 30 countries, identified the main challenges facing the water sector (Table 2).

With these general ideas gained from observation of the European [6] and American situations [1,2], the aim is to find the problems of such an essential infrastructure as

Management and Engineering of Critical Infrastructures
https://doi.org/10.1016/B978-0-323-99330-2.00005-2

Table 1 Challenges facing the water sector [1].

Ranking	Challenge	Weighted average	% Ranged as critical
1	Renewal and replacement of aging infrastructures	4.58	65.2
2	Financing	4.46	57.4
3	Long-term water supply availability	4.32	52.4
4	Emergency	4.13	35.2
5	Public understanding of the value of water services and systems	4.13	38.0
6	Watershed/source water protection	4.08	36.7
7	Public understanding of the value of water resources	4.06	32.3
8	Aging workforce/anticipated retirements	4.03	38.4
9	Regulations	4.00	29.3
10	Groundwater management and overuse	3.98	30.2
12	Cybersecurity issues	3.91	24.5
13	Cost recovery	3.89	28.5
14	Water loss control	3.71	18.1

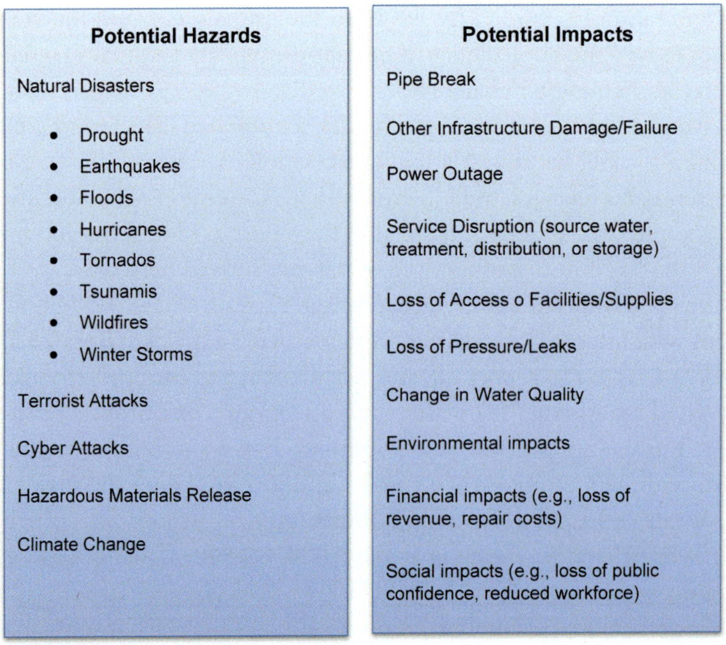

Potential Hazards

Natural Disasters

- Drought
- Earthquakes
- Floods
- Hurricanes
- Tornados
- Tsunamis
- Wildfires
- Winter Storms

Terrorist Attacks

Cyber Attacks

Hazardous Materials Release

Climate Change

Potential Impacts

Pipe Break

Other Infrastructure Damage/Failure

Power Outage

Service Disruption (source water, treatment, distribution, or storage)

Loss of Access o Facilities/Supplies

Loss of Pressure/Leaks

Change in Water Quality

Environmental impacts

Financial impacts (e.g., loss of revenue, repair costs)

Social impacts (e.g., loss of public confidence, reduced workforce)

Fig. 1 Hazards and impacts on water networks [2].

Table 2 Challenges facing the water sector [6].

Ensure the security and reliability of the water supply	Move toward efficient and climate-friendly water services
Protect water (a vulnerable resource)	Manage assets for the long term
Finance sustainable water in the long term	Increase the resilience of water supply infrastructures
Promote water in the circular economy	

drinking water distribution networks. Other approaches separate existing challenges according to their temporal distribution, such as short, medium, and long-term issues [7]. Although several works have been carried out, to the best of our knowledge, no study presents the management problems in pressurized water distribution networks, and the challenges to be addressed are often confused with the associated solutions and research. The studies indicated are often based on interviews with water sector professionals, and the specific scope (technical, economic, and safety) of the problems addressed are not delimited. In this paper, the challenges have been proposed from the perspective of water management, and problems related to economic aspects (known to concern managers) have not been considered, as they are beyond the scope.

Likewise, it should also be noted that the terminology is different. For example, in Bello et al. [7], the optimal design and the segmentation are considered (long-term) challenges. However, these two challenges are considered tools that to manage leakage. In our approach, we tried to include more generic titles for grouping the issues into only seven. These main infrastructure problems are identified from a pure infrastructure management perspective (Fig. 2). These challenges are:

1. Aging Infrastructure (1st for AWWA)
2. Leakage (20th)
3. Environmental concerns (5th and 6th)
4. Increasing demand (3rd)
5. Energy Management (2nd)
6. Cyber-security (12th)
7. Terrorism (4th)

This list indicates the order assigned according to the American Water Works Association study (the only showing hierarchy). In the environmental protection agency surveys, terrorist attacks and cyber-attacks are considered hazards, and all challenges are considered impacts of natural hazards. Environmental problems, pipe bursts, changes in quality, etc. Finally, the European viewpoint also highlights the importance of Managing assets (1. Aging Infrastructure), moving toward an efficient climate (3. Environmental concerns), Ensuring security and reliability (4. Increasing demand, 6. Cybersecurity, 7. Terrorism), etc.

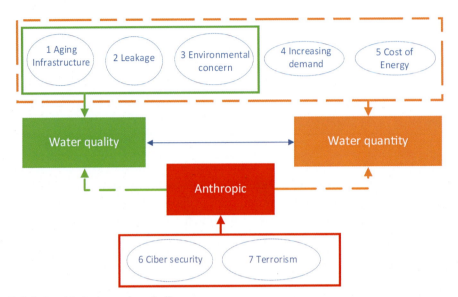

Fig. 2 Relationship between key challenges.

The key demonstrated is related to resource quantity problems to a greater or lesser extent (Note that challenges six and seven caused by a human being are also related to quantity problems). But only some of these challenges are related to quality (for example, leakage, obsolete infrastructure, and environmental concerns), and the others are unrelated to the quality of water delivered to the consumer. The final challenges to be faced by utility managers are caused by humans, cyber-security, and terrorism, the two challenges linked to quantity and quality purposes.

This chapter establishes the main challenges and offers solutions to the catchment, treatment, and distribution stages of the urban water cycle. It so analyses all steps until the water leaves the consumer's tap (domestic or industrial), being like "cradle to gate" in life cycle analysis when calculating carbon emissions. This subject can be employed by utility managers and/or practitioners working in water supply industries. Other stakeholders, such as decision-makers or politicians, should incorporate this approach to find the major issues in this industry. Policymakers can exploit future water distribution policies by taking this information into account.

This study can be used by utility managers of water supply companies (public, private, or mixed companies), decision-makers, and/or professionals. It is used to deepen the knowledge of the current problems to be faced in the management and operation processes of water networks.

The research method has comprised carrying out a bibliographic review in which most of the observed problems have been incorporated. Of course, our own experience in research on pressurized networks allowed us to find, according to our understanding, the best way to group problems by reducing their number and indicating which aspects each of them affects. It also allows a broader view of the problem to be taken when planning strategic service decisions. Other stakeholders, such as the consumers themselves (households, industries), can also incorporate this approach as a method to better understand the infrastructure that provides them with water. Finally, policymakers can shape future water policies with these guidelines. Does this work aim to answer the following questions: (i) What are the problems that supply network managers face? (ii) What are the proposed solutions to these problems? (iii) How are these problems related to each other? (iv) Why this infrastructure should be critical?

The rest of this work is organized as follows. Section 2 shows the challenges of urban water distribution infrastructures. Water quantity issues are shown in Sections 2.1 and 2.2, while water quantity issues alone are shown in Section 2.2. Human-induced problems are described in Section 2.3. The lines of research are shown in Sections 3.1, 3.2, and 3.3. Lastly, the discussion section is in Section 4 and the key conclusions are depicted in Section 5.

2 Challenges in water pressurized networks

The most significant challenges facing water distribution systems are presented below. An extensive listing of the challenges has been attempted without indicating hierarchy among them, as each utility has different priorities. The seven problems to be addressed have been distributed into problems related to water quality, the quantity of the resource (water) available and human-induced problems. The relationship between all of them is shown in Fig. 2 and they are outlined here.

2.1 Water quantity

The European Commission is preparing a Drinking Water Directive to make sure better protection of human health and to meet sustainable development goal six (Water). Agriculture is the largest freshwater consuming sector (59%) of total water consumption in Europe in 2017. The primary cause is agriculture in Southern European countries, which accounted for 64% of total surface runoff water abstraction and 24% of groundwater abstraction.

2.1.1 Energy consumption

Water has been regarded as an energy-producing resource. However, hydropower has reached its peak in developed countries today, large installations have already been built. Currently, a greater sensitivity to environmental disturbances appeared and water is studied not as an energy generator, but as an energy-hungry procedure, as huge amounts of energy are needed to transport water and to restore it to its former quality. The energy consumed in these steps is no small matter, so much so that 19% of the energy consumed in California [8] is used for water. Energy is also an indispensable and limited resource, which varies in price with an upward trend and whose price influences decisions such as the renewal of pipes, implementation of demand reduction policies, pressure reduction, etc. The latest energy crisis, because of climate change, is having a much greater social impact than earlier crises and is putting water and energy at the top of the political and scientific agendas.

The International Energy Agency [9] assessed the energy used in the water industry at 4% of overall consumption. IEA also stated that energy consumption would increase by 30% by 2040 [10]. In Europe, the water sector accounts for 3% of global electricity expenditure [11]. The energy footprint in the urban water cycle ranges from $0.21-4.07 \, kWh/m^3$ [12], depending on water origin, topography, or treatment technology, among others. Given the evidence that aging pipes are still in use and then, the global level of leakage represents 10%–35% in the world [13], and 24% in Europe [14].

2.1.2 Increasing demand for potable water (water scarcity)

Water consumption in 2014 was estimated at four trillion m^2 [14] or $3880 \, km^3/year$ [15]. The distribution of this resource is so unequal that over one million inhabitants do not have access to drinking water. Without a doubt, freshwater availability fluctuates on an annual and seasonal basis. Therefore, periods of scarcity generate great pressure on resources and economic activities. Agriculture—the most water-hungry sector, with 70% of total water use in Europe in 2017 [16]—and to a lesser extent urban water supply exploit groundwater resources in spring and summer, while other sectors such as refrigeration in power generation exert great pressure on surface water in autumn and winter.

Today's high-water consumption poses a risk of overexploitation of water resources and, if left unchecked, can lead to their depletion with irreversible consequences. It is therefore essential to develop policies that improve water efficiency, as water is a limited resource and increasing supply is almost impossible. Faced with this increase, demand management policies are advocated in the first place, which seek to reduce consumption by the user and reduce losses at all stages of the water cycle before the final use of water.

Society requires the search for other supply alternatives to make sure supply. Water for irrigation does not have to meet quality requirements as restrictive as drinking water—Directive 98/83/EC Directive and Commission Directive (EU) 2015/1787—Therefore, alternative water sources to irrigation water consumption are more likely to be found. In this context, groundwater can supply irrigation consumption and keep a balance between water supply and demand in urban areas [17].

Water scarcity must be anticipated while identifying opportunities to increase the sustainability in the overall supply of the resource. Water restrictions are a valid short-term policy action, long-term guidelines are required, ranging from increasing water supply (water of different qualities), better management of the resource (increasing resource efficiency) and maintaining quality standards. According to the SOTWI survey (2021), 44%

of utilities have drought contingency programs [1]. Europe has developed policies to reduce household water demand between 2000 and 2017 (−17%) with price increases, water-saving technologies, and awareness-raising campaigns, despite Europe's growing population.

2.2 Water quality

The directive on the quality of water for human consumption (Council Directive 98/83/ EC of 3 November 1998) can be found in the Directory of European Union consolidated legislation (EU) 2015/1787 of 6 October 2015. The Directive fixed the quality standards at the EU level. A total of 48 microbiological, chemical and indicator parameters must be monitored and analyzed. On 19 January 2022, two endocrine-disrupting substances (beta-estradiol and nonylphenol) were added to the first watch list, meaning that drinking water across the EU has to be more strictly monitored for its potential presence.

Source water protection involves reducing potential risks and their impacts. Protection programs can be very cost-effective in maintaining, safeguarding, and improving the quality and quantity of drinking water.

Governments enforce compliance with requirements affecting water protection under existing laws. Initiatives are therefore established to give technical and financial support to water utilities to carry out source water protection tasks. Common water quality problems are because of biological and physicochemical phenomena. Microbiological problems are dangerous with human consumption (for example, bacterial regrowth, nitrification and waterborne diseases caused by microbial pathogens).

The key drivers of water quality are agricultural and livestock activities (fertilizers, pesticides), waste and sewage discharges, fuels, runoff, dissolved oxygen, pH, global warming (temperature), etc. These are influenced by limits that cannot be changed by the network manager, such as rainfall, climate, soil, vegetation, geology, and groundwater. The network manager can only control human activities that interfere with water quality. The causes are many, such as cholera, diarrhea, and hepatitis A. Estimated that at least five million people die each year from drinking water that does not meet quality standards.

2.2.1 Aging infrastructure

The environmental protection agency (EPA) showed that the total length of pipes for public water systems is equal to 2,231,710 miles (3,591,589 km; [18]). In Europe, this length is 4,225,527 km [6].

Water distribution networks need maintenance and investment to make sure service standards and offer clean, healthy water. The EPA conducts reports to find water network needs, and the 2018 report indicates a total 20-year capital improvement need of 472.6 billion US$ [18]. While in Europe, the investment is quantified as 45 billion €/year [6] to rejuvenate the infrastructure, reduce costs and protect the environment. Moreover, seven million kilometers of the EU's pipelines have been in operation for over 100 years, and any big city in the United States will have aged cast-iron pipes older than 100 years.

Investment in water infrastructure is financed through tariffs (water bills), taxes, and transfers (from EU funding schemes or loans from other countries). Aging infrastructure means a loss in the quality standards of water supply services and a waste of scarce resources. Although one of the main reasons for burst and leaking pipes is aging, not the only one because fluid pressures and the time spent seeking the leak and repairing it (after it has been found) also play a role. Many pipes have been installed for a long time, with a high number of breaks and the consequent risk of pathogenic intrusion, insufficient service, etc.

2.2.2 Leakage
The reduction of leaks in the networks contributes not only to satisfying demand but also avoids various problems of system management, quality, overpressure, etc. Water leakage consumes available water resources with no benefit (water escaping from the pipes) for utility managers. In addition, all the earlier stages in the water cycle (catchment, treatment, and distribution) involve asset consumption for water that escapes through breaks.

Other studies estimate the average leakage rate of the water pressurized networks in Europe is 23% of all water distributed [6]. Although, water utilities' efficiency may range between 10% and 72% [19]. Statistical modeling of pipe burst risk has been attempted [20], and failure prediction systems have been developed to overcome the main problem, the absence of historical burst data [21].

2.2.3 Environmental concerns
Although many countries are committed to meeting net zero emissions targets. If implemented early and fully, they bend the global emissions curve downwards. Low-emission power generation sources have been developed to reduce carbon footprint. In this new scenario, solar photovoltaic and wind power are estimated to produce 500 gigawatts (GW) by 2030. The final aim pursued would be for CO_2 emissions related to energy consumption to fall by 40% by the year 2050 [22].

The water-energy nexus interacts with climate change. On the one hand, because energy expenditure increases greenhouse gas (GHG) emissions and, on the other, in compliance with the forecasts of the Intergovernmental Panel on Climate Change (IPCC), in many geographical areas of the Mediterranean, water resources availability will decrease, accentuating the stress they already suffer today. Frequent droughts and the challenge of climate change need higher levels of efficiency in these two resources, especially because their interdependence amplifies the problem. Global warming, the greatest evidence of climate change, decreases water availability in the most water-stressed areas, which also almost always coincides with those with the highest demand. These forces use alternative, more consumptive energy sources and the abstraction of groundwater from greater depths. But the interdependence does not finish there. On the one hand, less water means less hydropower production, and on the other, higher temperatures increase the domestic energy consumption required for heating and cooling homes. All of this leads to greater use of non-renewable energy, promotes climate change, and feeds a vicious cycle that is difficult to control.

2.3 Anthropic problems

2.3.1 Cyber-security

The cyber-security was defined as "the process of protecting information by preventing, detecting, and responding to attacks" [23]. In the last decade, urban water distribution infrastructures have shifted from being isolated control systems to communicating in a network of collective systems. The relationships between control and data acquisition software (SCADA) and the control hardware characteristic of each infrastructure are produced in centralized control systems, common in cities, using IoT technologies to create the smart city's philosophy. Detecting these attacks in WPNs resulted in a key topic in the water industry, as proven by many approaches published for this purpose [24].

The migration toward common control and operation systems was driven by cost reduction and ease of use. But these benefits also mean increased vulnerability to cyber threats. Control systems are now remotely operated (via intranet, internet, or any wireless connection), so these signals can be accessed by unauthorized users. While water networks are not a major concern, the water industry relies on reliable communication systems that make sure that no external agents have influenced it. This is sensitive when dispensing chemicals and treating contaminants. Water management companies have changed privacy and cybersecurity programs to follow national regulations.

The number of recorded cyber-attacks has increased. Among them, are attacks on water infrastructures that have been made public in Israel, the United States [25] and Australia [26]. It has been checked that sometimes cyber-attacks come from hackers/ Nation attacks but also ex-employees who used stolen credentials to access the district's computer systems.

In the United States, an official agency (Industrial Control Systems Cyber Emergency Response Team, ICS-CERT) recorded cybersecurity incidents in the water sector. It accounted for 25 cybersecurity incidents in the water sector in 2015 [27]. Similar numbers 15 and 24 were achieved by Hassanzadeh et al. [28] and Panguluri et al. [29]. The latest also showed that the water industry was ranked as the fourth industry on this list (behind power, petroleum, and transportation sectors).

2.3.2 Terrorism

The EU definition of terrorism is: "Terrorist offences are acts committed with the aim of seriously intimidating a population, unduly compelling a government or international organisation to perform or abstain from performing any act and/or seriously destabilising or destroying the fundamental political, constitutional, economic or social structures of a country or an international organisation" The Environmental Protection Agency (EPA) stated, "water security is defined as prevention and protection against contamination and terrorism."

The need for water consumption to make sure human life means that water distribution networks are a critical target for military, political, and terrorist actors [30]. Terrorist threats and the possibility that terrorists may infiltrate critical infrastructure management mean that the water industry must consider how to respond to these potential acts. A state is created in which anticipation of potential attacks becomes essential to follow current regulations. In addition, as the water sector is not the first target of terrorism, solutions already developed in other industries (financial services, large-scale electricity sector, etc.) can be employed in this industry.

3 Research directions and actions

In this section, the aim is to outline the lines of research and policies that are being carried out to solve each of the challenges outlined in the earlier section. Therefore, the tools are shown in the same order as they were presented, that is the answer to Section 2.1.1 is shown in Section 3.1.1.

3.1 Water quantity solutions

3.1.1 Energy management

The first stop in terms of innovations in the integral water cycle is water abstraction. In many regions, because of natural watercourses' absence, the main source of water resources has been groundwater abstraction or even desalination (very energy-hungry procedures). These facilities have driven the implementation of innovative technologies in search of greater efficiency. Some, such as powering well pumping using solar energy, have made their way, and are becoming commonplace. In this current scenario, solar energy appeared as a "green" source with zero emissions. Photovoltaic production has increased worldwide for many reasons, high irradiation values in (latitudes between 40N and 40S) [31], an increase in the price of electricity from 0.09 Euros/kWh in 2004 [32] and 0.4 Euros/kWh in 2022 (this data was gathered in September 2022), changes in legislation, taxes elimination, the increase in oil prices, etc. However, the falling cost of solar PV modules (30%–60% in 10 years; [33]) is the most important factor driving the technology. These have made solar PV production a cost-effective alternative that is increasingly being considered by engineers, researchers, and various professionals in the water industry.

3.1.2 Politics for reducing freshwater consumption

Many programs are proposed to reduce the final consumption of drinking water. Demand reduction does advocate resource conservation and efficient water use. This is meant any activity aimed at reducing the amount of water withdrawn from the environment while maintaining the level of service quality. Demand management programs are not only information or awareness-raising campaigns but also encompass certain actions aimed at reducing the volume consumed by end-users (encouraging final users to take shorter showers, not to leave taps running, to use eco cycles in machines or mechanisms as aerators to reduce water consumption in faucets, etc.). Demand management has several advantages for both suppliers and consumers, although the main one is from a sustainability standpoint.

A new research line comes from increasing water offers by using reclaimed water. It can be used for irrigation, flushing toilets, washing houses and cars, and reducing potable water consumption [34]. It can also be used in industrial processes, such as cooling and dust control systems on construction sites.

Other uses include industrial processes, water cooling in manufacturing and dust control on construction sites. There may also be industrial uses, such as water cooling in manufacturing and dust control on construction sites. Many investigators fixed the limits of the quality of reclaimed water [35], and its affection for crops [36]. The issue is so hot that regulation (EU) 2020/741 on minimal conditions for water reuse [37] has been published. This regulation determines the least water quality requirements and proposes procedures to be followed to get recycled water suitable for the environment, and human and animal health.

3.2 Water quality solutions

3.2.1 Replacement and renovation

Small municipalities have fewer possibilities of specialization and cannot bear the costs of the infrastructure required for such systems. They often turn to incomplete solutions far from the high-quality standards required by a public service of such importance, vital for the human, social and economic development of a population. With the recent growth of Infrastructure Asset Management, the shortcomings of traditional systems have become clearer. There is no complete inventory of network elements, no coordination between GIS and mathematical models, databases store inconsistent information, etc. In these circumstances, analysis, forecast of future needs and planning of urban water networks becomes a complicated task. This results in a need to discuss this complex reality, analyze the needs of a water supply service provider and structure their information systems to make sure efficient management, as well as to contribute to the increase of the service's quality standards. In this regard, the information systems of the company have been organized into five major sections: the inventory, the clients' database, the customer service, the geographic information system, and the remote-control system of the infrastructure.

Infrastructure asset management (IAM) is a planning process that brings together multidisciplinary tools and methods to provide an integrated approach to provide a decision support system to manage tangible physical assets in such a way as to guarantee tangible physical assets in such a way as to guarantee a level of service with the greatest possible efficiency and at the lowest possible cost. It is not, therefore, a tool that is limited to the traditional business concept of (physical) asset management traditional business concept of asset management (physical and financial), which seeks to optimize the life cycle of these assets to obtain the highest possible efficiency and the lowest possible cost life cycle of these assets to obtain the highest possible return on investment.

The deterioration that all infrastructure undergoes over time, together with the need to offer a certain level of service with a certain number of resources, is the main reason for this a certain level of service with limited economic resources and demanding external constraints (legal, regulatory, etc.) increasingly demanding external conditions (legal, environmental, etc.), make it necessary to resort to new techniques for analyzing the problem. new techniques for analyzing the problem. IAM offers a method that systematizes such an analysis, considering three fundamental axes of study (cost, risk, and performance). In this way, the company ensures not only that the best value is obtained from each of the assets, but also that the best value is obtained from each of the assets and that the financial resources are available to rehabilitate and replace them when necessary, but also that the company and replace them when necessary, but also to increase the efficiency of the system's operation, satisfying at all times the system, satisfying at all times the present and future demands on the infrastructure under analysis.

3.2.2 Leakage

Leakage poses a phenomenal challenge for managers of drinking water distribution companies because the quality of the water delivered to end-users must be assured. Many proposals in recent years investigate the data collected from pressure transducers and flowmeters along pressurized water networks [38], the relationships between pressure and leakage [39], on the influence of pipe material [40] and the energy lost in leakage [41].

Many studies address the problem of leakage and its control strategies [42], among them, the Water Loss Task Force of the International Water Association (IWA) proposes four strategic lines of action to reduce and control the level of losses. These actions are described in Fig. 3. Two concepts are also defined: the *Economic level of leakage* (ELL) and the *Unavoidable Annual Real Losses* (UARL). The former has focused on the economic effect of active leakage control, using an approach of minimizing total economic costs. Both the cost of leaked water (proportional to leakage) and the cost of active leakage control (influenced by the frequency of interventions) are considered. This results in a total cost curve (sum of both the cost of leakage and the cost of active leakage control interventions) that has a minimum value corresponding to the economic level of leakage in the short term (ELL).

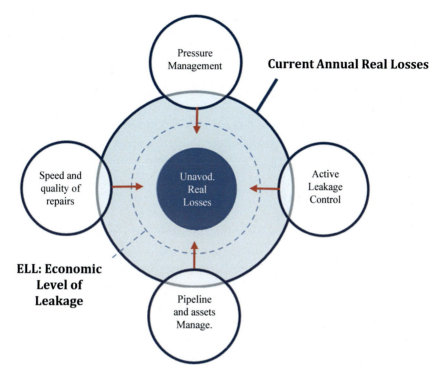

Fig. 3 Action strategies on leakage [43].

The second concept shows a limit to the minimization of leakage and refers to background leakage (water losses below the detection level with the current technology). The leaks appear during the search and repair time and are so always in the pressurized water distribution network.

The four strategies developed to reduce leakage are shown in Fig. 3. Of these, infrastructure renewal has been discussed in Section 3.2.1, pressure management in Section 3.2.2.1 and active leakage control is discussed in Section 3.2.2.2. We have highlighted two Active Leakage Control techniques such as segmentation and night-flow analysis, as they have deserved much attention from researchers in the latest years. Finally, the fourth strategy (speed and quality of repairs) has not been described here, as understood as a technique for management.

3.2.2.1 Pressure management to reduce leakage

The pressure-leakage relationship is a well-known concept when operators manage water distribution networks. The primary aim has been quantifying water savings by reducing overpressure [44] during night periods of lower water consumption. Other studies focus on the benefits of pressure management by reducing the number of breaks [39], new methods of designing district metering areas, or studies on the relationship between consumption and pressure. The energy influence of leakage has also been studied in water distribution systems [45] and also in irrigation networks [41]. The influence of installing a Pressure Reducing Valve (PRV) to reduce leakage demonstrates energy savings are also a nice consequence because of the water-energy nexus.

Most efforts to mitigate leakage occurred after occurring a break. When it became clear that significant fluctuations in pipe pressure within a water distribution system were the main cause of breaks and leakages, research was conducted to improve water pressure monitoring; the resulting improvements allow a system to reduce its risk factor, protect system components prone to leakage, and offer safer and cleaner potable water.

3.2.2.2 Active leakage control

To avoid wasting water resources, one of the most widely used policies is Active Leakage Control [46], which aims at early detection, location and repair of pipe breaks, thus reducing damage to third parties, reducing unplanned works and the volume of water lost. It is a manual process that requires specialized operators and using acoustic measuring equipment and correlators to inspect the distribution system segmentation or continuous night-time flow measurements [47] and to locate the leak.

• Segmentation as a leak detection technique

Segmentation of a drinking water distribution network is a technique used in the control and global localization of the uncontrolled volume of leakage and comprises dividing the network into several smaller sub-networks, called sectors or District Metered Areas (DMAs) [48], on which all the volume entering and leaving the defined sector delimitation is measured. It is thus a strategic technique for the delimitation, detection and monitoring of irregular/atopic consumption or uncontrolled leakage volumes that occur along drinking water distribution networks.

Segmentation is a strategic choice that reduces the inspection area for the detection and location of anomalies such as breaks, leaks or pressure deficiencies. If this technique is combined with a control system, it could improve managing the overall operation of the network, optimizing the pressure in each sector [49].

• Night flow analysis as a leakage detection technique

Night-time flow analysis is performed in situations where enough network data is not available. Consumption during night-time periods is limited to sporadic use in households and industry. The former are small-scale consumptions that can be quantified, while the latter can be known for proper metering. In short, at certain times of the night, the volume of leakage can be known from the difference between the volume injected and the volume consumed. The nocturnal flow rate can be a reliable indicator that an anomaly has occurred in the system (a break, consumption in fire hydrants, etc.). In this line of work, methods for the detection and monitoring of leaks, such as minimum night-time flow rates, are used [50]. The minimum night flow is broken down into the flow delivered to consumers and losses in the distribution network; the flow delivered to consumers is composed of leakage in indoor installations (of long duration and constant size) and intentional consumption (of short duration and variable size). Small tanks (cisterns) distort night-time consumption and, although typical of insufficient networks, a common problem in many supplies.

3.2.3 Environmental concerns

Since a direct relationship between energy management policies and the emissions produced, knowing that all policies that save energy from burning fossil fuels will save emissions. And, climate change, will cause greater amounts of resources (water). These relationships are shown in Fig. 4.

To reduce emissions, the water utility has focused on incorporating new ways of power sources (solar PV production, wind production, etc.). Other long-term policies to reduce emissions in any activity (industrial, transport, agricultural, etc.) have focused on quantifying emissions related to the useful life of any infrastructure, the Life Cycle Assessment (LCA). It has been calculated that the largest proportion of emissions is related to the manufacture of the materials that make up the pipe network, and to repairs and maintenance. Finally, the end of life also involves a high number of emissions, so the material also influences this stage of the infrastructure's life.

3.3 Solutions to anthropic problems

3.3.1 Cyber-security

This challenge relates to the security that ensures the proper interaction between the physical and cyber elements of water systems. Hence, work is underway to design tools

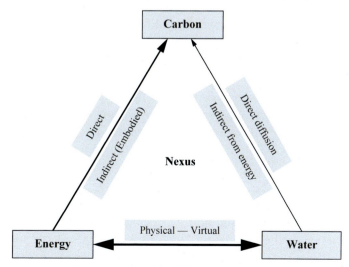

Fig. 4 Water, energy and emissions relationship [51].

to simulate deliberate pollution attacks on water networks, together with cyber-attacks on the pollution warning system.

To discuss the challenge of ensuring cybersecurity, information sharing has been promoted, and legislative frameworks have been established that define decision-making processes [27]. These actions have led to a prioritization of cyber security to reduce the potential risks from cyber-physical attacks. But most of the practical recommendations found in the literature focus on basic hardware security concepts (patching, software updates, anti-virus software, etc.). Such choices can lead to limitations in water system operation, such as offline procedures when updating security patches and installing software, the lack of real-time decision-making processes, and the incompatibility of legacy control systems [52].

Preventing attacks against IT systems concentrates on the three pillars of security: integrity, confidentiality, and availability. Integrity is protection against improper modification or destruction of data, adding to the authenticity of the information. Confidentiality safeguards that data are not published to any unauthorized body, process, or device. Finally, availability refers to fast and reliable access to data for allowed entities.

3.3.2 Terrorism

The major risk would be contamination of water in distribution networks with the later contamination of the population. Therefore, research work focuses on early warning systems that minimize the reaction time of utility managers. These tools should handle all real-time monitored signals, using pollutant dispersion models (obtained from hydraulic simulation software) performing transport and diffusion. Finally, the network manager must know the location and density of the population that may be affected and set up contingency plans (disinfection stations or others) to minimize the effect. These protection systems are valid for malicious actions and natural disasters and any other unusual phenomena, such as fires.

Water utilities did not take any action by themselves to avoid terrorism and the key lines followed are defined by their government and institutions. However, many researchers have worked on ways to reduce the impact of terrorism on water networks. For instance, critical locations for potential terrorist actions were identified [53], vulnerability index for drinking water distribution systems were proposed [54] and attack models were characterized [55].

4 Discussion

This paper has identified several significant challenges that are faced by infrastructures for supplying water to the population, and it has highlighted the interrelationship among these challenges. The interrelationships among the seven challenges are shown

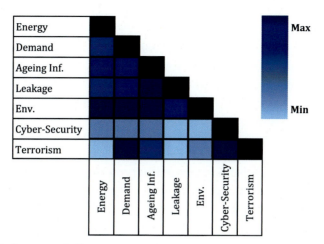

Fig. 5 Relationship between challenges.

in Fig. 5, with color coding to indicate the strength of each relationship. The relationships between aging water distribution networks, leaks, and environmental problems due to resource wastage, as well as water demand, are strong. Meanwhile, the relationships with other parameters such as leakage and terrorism are intermediate, and the relationship with cyber security is weak. It is also noted that the two man-made challenges, cyber-attacks and terrorism, are only related to each other, as the remote-control part of the network is separate from the physical part of the distribution network.

The interrelationship among challenges emphasizes the need for policies that consider the impact of multiple challenges, not just one. Furthermore, it is essential to recognize that there is no one-size-fits-all solution that can address all network problems, and constant analysis of the variables is necessary to maintain optimal performance. Although the challenges of protecting against cyber-attacks and terrorism are less developed than other more common problems, it is crucial to acknowledge their potential impact and incorporate measures to mitigate them.

Although it has been commented that aging infrastructure was the most prominent problem [1], perhaps the new socio-political landscape with tremendous energy dependence with energy prices surpassing record highs week after week (this work is done in 2022), with geopolitical uncertainty across Europe and with great concern about the temperatures that the summer has caused, is a large change in the water industry's perception of the challenges to be faced. In such a changing time, the challenges are not ranked, but the general uncertainty makes it impossible to make an accurate prediction.

5 Conclusions

Water is a critical resource, and its distribution involves a significant responsibility for practitioners to ensure that water is safe for consumption. As such, managing water distribution networks is a highly critical infrastructure.

The primary objective of this work is to provide insights to utility managers who operate water distribution networks or are involved in strategic planning. The paper identifies seven key challenges that drinking water distribution network managers need to address, with two related to human actions and the other five pertaining to the water distribution process. These challenges range from ensuring water quantity to maintaining water quality standards, and the paper outlines potential solutions for each challenge.

Although several approaches and solutions exist, the paper focuses on the most promising options available today. These solutions include incorporating renewable energy sources to meet high energy costs, implementing water reduction programs, utilizing reclaimed water to address increasing water demand, and deploying infrastructure asset management tools to manage pipeline renovation and active leakage control. The paper also highlights the importance of addressing environmental concerns, such as focusing on the life cycle assessment process.

The challenges identified in the paper are interconnected, with those related to water quality and quantity significantly impacting one another. This complexity presents a significant challenge to managers who must handle these issues with limited budgets and political constraints that are inherent in public infrastructure. Additionally, the paper highlights man-made challenges such as cyber-attacks and terrorism that are less connected to the other challenges but pose significant risks to the water distribution network's operation.

Acknowledgments

The authors have no conflicts of interest to declare. This work was supported by the research project "DESENREDA" through the 2021 call "Estancias de movilidad en el extranjero Jose Castillejo" of the Ministerio de Universidades (CAS21/00085).

References

[1] AWWA, The state of the water industry-2021, in: Water Encyclopedia, 2021, https://doi.org/10.1002/047147844x.mw572.
[2] Environmental Protection Agency, Systems Measures of Water Distribution System Resilience, 2015.
[3] UN, 2018 UN World Water Development Report, Nature-based Solutions for Water, 2018.
[4] European Environment Agency, Water Resources Across Europe—Confronting Water Stress: An Updated Assessment, 2021.
[5] T.I.E. Veldkamp, Y. Wada, J. Aerts, P. Döll, S.N. Gosling, J. Liu, Y. Masaki, T. Oki, S. Ostberg, Y. Pokhrel, et al., Water scarcity hotspots travel downstream due to human interventions in the 20th and 21st century, Nat. Commun. 8 (2017) 1–12.

[6] EurEau, Europe's Water in Figures—An Overview of the European Drinking Water and Waste Water Sectors, The European Federation of National Associations of Water Services, 2017.

[7] O. Bello, A.M. Abu-Mahfouz, Y. Hamam, P.R. Page, K.B. Adedeji, O. Piller, Solving management problems in water distribution networks: a survey of approaches and mathematical models, Water 11 (2019), https://doi.org/10.3390/w11030562.

[8] CEC, California's Water—Energy Relationship, 2005.

[9] IEA, World Energy Outlook 2019, 2019. París.

[10] IEA, World Energy Outlook 2017, 2017. Paris, France.

[11] M.A.D. Larsen, M. Drews, Water use in electricity generation for water-energy nexus analyses: the European case, Sci. Total Environ. 651 (2019) 2044–2058.

[12] M. Wakeel, B. Chen, Energy consumption in urban water cycle, Energy Procedia 104 (2016) 123–128.

[13] Á.-F. Morote, M. Hernández-Hernández, Unauthorised domestic water consumption in the city of Alicante (Spain): a consideration of its causes and urban distribution (2005–2017), Water 10 (2018) 851.

[14] IEAIE, Water Energy Nexus; Excerpt From the World Energy Outlook 2016, 2016. Paris.

[15] K. Ozano, A. Roby, A. MacDonald, K. Upton, N. Hepworth, C. Gorman, J.H. Matthews, K. Dominique, C. Trabacchi, C. Chijiutomi, et al., Groundwater: Making the Invisible Visible: FCDO Briefing Pack on Water Governance, Finance and Climate Change, 2022.

[16] Y. Wada, L.P.H. Van Beek, M.F.P. Bierkens, Nonsustainable groundwater sustaining irrigation: a global assessment, Water Resour. Res. 48 (2012), https://doi.org/10.1029/2011WR010562.

[17] V.K. Kandiah, E.Z. Berglund, A.R. Binder, Cellular automata modeling framework for urban water reuse planning and management, J. Water Resour. Plan. Manag. 142 (2016) 4016054.

[18] EPA, Drinking Water Infrastructure Needs Survey and Assessment. System, 2018.

[19] E.C, Resource and Economic Efficiency of Water Distribution Networks in the EU Final Report, 2013.

[20] T.G. Mamo, I. Juran, I. Shahrour, Urban water demand forecasting using the stochastic nature of short term historical water demand and supply pattern, J. Water Resour. Hydraul. Eng. 2 (2013) 10–92.

[21] R. Jafar, I. Shahrour, I. Juran, Application of artificial neural networks (ANN) to model the failure of urban water mains, Math. Comput. Model. 51 (2010) 1170–1180.

[22] IEA, World Energy Outlook 2021, 2021. Paris, France.

[23] K. Stouffer, K. Stouffer, T. Zimmerman, C. Tang, J. Lubell, J. Cichonski, J. McCarthy, Cybersecurity Framework Manufacturing Profile, US Department of Commerce, National Institute of Standards and Technology, 2017.

[24] T. Riccardo, G. Stefano, Deep-learning approach to the detection and localization of cyber-physical attacks on water distribution systems, J. Water Resour. Plan. Manag. 144 (2018) 4018065, https://doi.org/10.1061/(ASCE)WR.1943-5452.0000983.

[25] L.L. Thomson, Insecurity of the internet of things, Scitech Lawyer 12 (2016) 32.

[26] N. Tuptuk, P. Hazell, J. Watson, S. Hailes, A systematic review of the state of cyber-security in water systems, Water (2021), https://doi.org/10.3390/w13010081.

[27] R.M. Clark, S. Panguluri, T.D. Nelson, R.P. Wyman, Protecting drinking water utilities from cyberthreats, J. Am. Water Work. Assoc. 109 (2017) 50–58.

[28] A. Hassanzadeh, A. Rasekh, S. Galelli, M. Aghashahi, R. Taormina, A. Ostfeld, K. Banks, A review of cybersecurity incidents in the water sector, 2020. arXiv Prepr. arXiv2001.11144.

[29] S. Panguluri, W. Phillips, J. Cusimano, Protecting water and wastewater infrastructure from cyber attacks, Front. Earth Sci. 5 (2011) 406–413.

[30] P.H. Gleick, Water and terrorism, Water Policy 8 (2006) 481–503, https://doi.org/10.2166/wp.2006.035.

[31] L. Wald, Basics in Solar Radiation at Earth Surface, 2018.

[32] Mincotur, Precio neto de la electricidad para uso doméstico y uso industrial Euros/kWh, 2018. WWW Document https://www.mincotur.gob.es/es-ES/IndicadoresyEstadisticas/DatosEstadisticos/IV.Energíayemisiones/IV_12.pdf. (Accessed 20 September 2004).

[33] A. Closas, E. Rap, Solar-based groundwater pumping for irrigation: sustainability, policies, and limitations, Energy Policy 104 (2017) 33–37.

[34] M.A. Pardo, A. Pérez-Montes, M.J. Moya-Llamas, Using reclaimed water in dual pressurized water distribution networks. Cost analysis, J. Water Process Eng. (2020), https://doi.org/10.1016/j. jwpe.2020.101766. 101766.

[35] N. Voulvoulis, Water reuse from a circular economy perspective and potential risks from an unregulated approach, Curr. Opin. Environ. Sci. Heal. (2018), https://doi.org/10.1016/j.coesh.2018.01.005.

[36] S. Toze, Reuse of effluent water—benefits and risks, Agric. Water Manag. (2006) 147–159, https:// doi.org/10.1016/j.agwat.2005.07.010.

[37] EU, Regulation (EU) 2020/741 of The European Parliament and of the Council on Minimum Requirements for Water Reuse, 2020.

[38] K. Adedeji, Y. Hamam, B. Abe, A. Abu-Mahfouz, Leakage detection and estimation algorithm for loss reduction in water piping networks, Water 9 (2017) 773.

[39] J. Thornton, A. Lambert, Progress in practical prediction of pressure: leakage, pressure: burst frequency and pressure: consumption relationships, in: Proceedings of IWA Special Conference' Leakage, 2005, pp. 12–14.

[40] G. Germanopoulos, P.W. Jowitt, Leakage reduction by excess pressure minimization in a water supply network, Proc. Inst. Civ. Eng. 87 (1989) 195–214.

[41] M.A. Pardo, J. Manzano, E. Cabrera, J. García-Serra, Energy audit of irrigation networks, Biosyst. Eng. 115 (2013) 89–101, https://doi.org/10.1016/j.biosystemseng.2013.02.005.

[42] D. Giz, A. Vag, Guía para la reducción de las pérdidas de agua-un enfoque en la gestión de la presión, 2011.

[43] A. Lambert, What do we know about pressure-leakage relationships in distribution systems, in: IWA Conf. Systems Approach to Leakage Control and Water Distribution System Management, 2001.

[44] M.A. Pardo, A. Riquelme, J. Melgarejo, A tool for calculating energy audits in water pressurized networks, AIMS Environ. Sci. 6 (2019) 94–108, https://doi.org/10.3934/environsci.2019.2.94.

[45] A.F. Colombo, B. Karney, Energy and costs of leaky pipes: toward comprehensive picture, J. Water Resour. Plan. Manag. (2002), https://doi.org/10.1061/(ASCE)0733-9496(2002)128:6(441).

[46] L. Berardi, D. Laucelli, A. Simone, G. Mazzolani, O. Giustolisi, Active leakage control with WDNetXL, in: Elsevier (Ed.), 12th International Conference on Hydroinformatics, HIC 2016. Incheon, South Korea, 2016, pp. 62–70, https://doi.org/10.1016/j.proeng.2016.07.420.

[47] R. Puust, Z. Kapelan, D.A. Savic, T. Koppel, A review of methods for leakage management in pipe networks, Urban Water J. 7 (2010) 25–45.

[48] V. Tzatchkov, V. Alcocer Yamanaka, V. Bourguett Ortiz, F. Arreguín Cortés, Avances en la hidráulica de redes de distribución de agua potable, 2014. http://repositorio.imta.mx/handle/20.500. 12013/1610.

[49] G. Kunkel, R. Sturm, Piloting proactive, advanced leakage management technologies, J. AWWA 103 (2011) 62–75, https://doi.org/10.1002/j.1551-8833.2011.tb11402.x.

[50] J.M.A. Alkasseh, M.N. Adlan, I. Abustan, H.A. Aziz, A.B.M. Hanif, Applying minimum night flow to estimate water loss using statistical modeling: a case study in Kinta Valley, Malaysia, Water Resour. Manag. 27 (2013) 1439–1455, https://doi.org/10.1007/s11269-012-0247-2.

[51] F. Meng, G. Liu, S. Liang, M. Su, Z. Yang, Critical review of the energy-water-carbon nexus in cities, Energy 171 (2019) 1017–1032, https://doi.org/10.1016/j.energy.2019.01.048.

[52] S. Shin, S. Lee, S.J. Burian, D.R. Judi, T. McPherson, Evaluating resilience of water distribution networks to operational failures from cyber-physical attacks, J. Environ. Eng. 146 (2020) 4020003.

[53] S.A. Patterson, G.E. Apostolakis, Identification of critical locations across multiple infrastructures for terrorist actions, Reliab. Eng. Syst. Saf. 92 (2007) 1183–1203, https://doi.org/10.1016/j. ress.2006.08.004.

[54] M. Maiolo, D. Pantusa, Infrastructure vulnerability index of drinking water systems to terrorist attacks, Cogent Eng. 5 (2018), https://doi.org/10.1080/23311916.2018.1456710.

[55] I.N. Fovino, M. Masera, A. De Cian, Integrating cyber attacks within fault trees, Reliab. Eng. Syst. Saf. 94 (2009) 1394–1402.

CHAPTER 12

Critical infrastructure security: Cyber-threats, legacy systems and weakening segmentation

William Hurst[a] and Nathan Shone[b]
[a]Information Technology Group, Wageningen University & Research, Wageningen, The Netherlands
[b]School of Computing and Mathematics, Liverpool John Moores University, Liverpool, United Kingdom

1 The emerging cyber-threat

Critical infrastructures have seemingly invisible boundaries, inherently intertwined, interconnected and codependent. Initially, this was a physical overlap, where one service relied on the direct output or production of another. More recently, the boundaries have become increasingly fuzzy—cyber-based—as a result of the societal move toward smart city innovations, which replace manual (labor-intensive) tasks with automation [1] by means of sophisticated digital technologies. Their digital interconnectivity is now complex, and this technical switch has produced a new frontier for security, where threats to the health and wellbeing of critical infrastructures are no longer just weather-related (climate change [2], extreme weather [3]), physical (terrorism [4], drone [5]) or blockade-based [6] but now also cyber. Displayed in Fig. 1, a simplified network of external threat vectors is based on a compilation of research articles that focus on varied domains of critical infrastructure protection (specifically [2–7]).

For visualization purposes, Fig. 1 arranges the threat vectors into groups, Weather, Physical and Cyber. Yet each is multifarious resulting in three issues: (i) The nature of the impact of a successful attack is somewhat unpredictable. For instance, impacts have the potential to range from severe (e.g., nuclear meltdown, flooding, blackouts, economic shutdown, etc. [8]) to minor (individual loss of personal data, minor delays to service provision); (ii) The threat source is diverse, where security solutions (physical or cyber) in one domain may have no or little benefit for others; and (iii) Interdependency, while a major benefit, also has the potential to spread the damage exponentially (referred to as *cascading*) across differing critical infrastructure types.

1.1 Research trend

These challenges have created a spate in research over the last 20 years into cascading failure analysis, protection plans and simulations of interconnectivities. Yet, what we see is

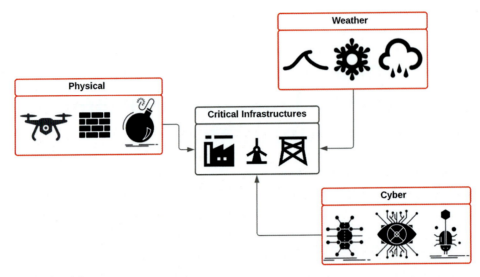

Fig. 1 External threat vectors.

also a shift toward cyber-security research, as indicated in Fig. 2 displaying the research focus trend from 1922 journal articles on the Scopus digital library over a 20-year period between January 1, 2001 and December 31, 2021.

While now a prominent topic in the research community, for most this cyber-threat is invisible and unseen, despite some staggering statistics. For instance, Kaspersky solutions blocked 1,686,025,551 cyber-attacks from online resources across the globe during the second quarter of 2021, documented in a report available in Ref. [9]. Yet, recent high-profile examples (in varied critical infrastructure domains) have found prominence in the wider media, pushing critical infrastructure security more into the mainstream. One of particular prominence being WannaCry.

1.2 WannaCry

WannaCry simultaneously established a foothold throughout multiple continents and organizations in May 2017 [10]. WannaCry is classified as a type of an attack known as *ransomware*, involving the use of malware (malicious software) to encrypt files on a target device, rendering it inoperable (Fig. 3). Perpetrators of ransomware demand payment (ransom) for the decryption of the device, thus granting access to the files. Ransomware is common for use in small (i.e., personally-directed) attacks. Kaspersky defeated 97,541 ransomware attacks for their users documented in Ref. [9]. Yet, on this scale, with a critical infrastructure target, the use of ransomware was almost unprecedented prior to WannaCry. It is now known to have spread to over 150 countries and did so within a matter of a few hours [11].

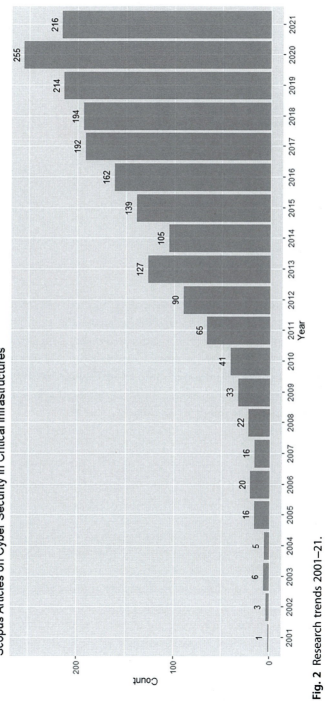

Scopus Articles on Cyber Security in Critical Infrastructures

Fig. 2 Research trends 2001–21.

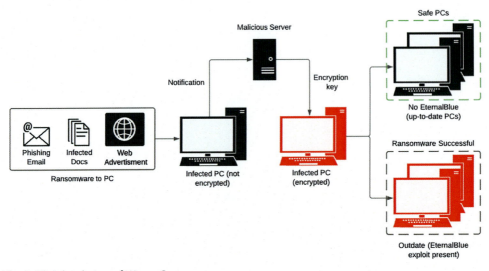

Fig. 3 High-level view of WannaCry.

Ghafur et al. document the impact of WannaCry on the United Kingdom's National Health Service (NHS), by means of a retrospective analysis in Ref. [10]. Despite WannaCry infecting around 600 organizations nationwide, the main documentation in the media was relating to the 34 infected hospital trusts. Ghafur et al. detail in their findings that, specifically, fiscal costs were the main impact of WannaCry. For instance, £4m in lost inpatient admissions and £1.3m from canceled outpatient appointments. However, no difference in mortality was noted [10].

WannaCry had further impact after 2017 with The Hacker News media outlet documenting that WannaCry briefly resurfaced in 2018 as a new variant, causing the Taiwan Semiconductor Manufacturing Company (a producer of microchips) to shut down several factories temporarily as a result of 10,000 machines becoming infected [12].

1.2.1 Cryptoworm

WannaCry is known as a *ransomware cryptoworm*, specifically designed to target users working with the Microsoft Windows Operating System (OS). The cryptoworm works by encrypting data and then, as with other ransomware approaches, demands payment for decryption (in this instance in cryptocurrency). The cryptoworm propagated through an exploit (a software which takes advantage of a vulnerability within another software) known as EternalBlue. EternalBlue was an exploit allegedly developed take advantage of a vulnerability in Microsoft's Server Message Block (a communication protocol created to share access to files or printers on a shared network). EternalBlue's designs were leaked by hackers. However, the deployment of WannaCry also required coupling

EternalBlue with a backdoor implant tool known as DoublePulsar, which allowed WannaCry to install and execute a copy of itself [13].

1.3 ThreatNeedle

However, Windows OS is not alone in facing high-profile threats, macOS also finds itself a target, notably within the defense industry critical infrastructure grouping. During 2020, a spate of spyware attacks from using the ThreatNeedle cluster on defense enterprises in over 12 countries were documented in a report by Kaspersky [14,15]. ThreatNeedle is an advanced cluster of malware known as NukeSped, which is a trojan designed to target macOS devices, allegedly linked to the North Korean government [16]. Cybersecurity expert Patrick Wardle provides a clear breakdown of NukeSped's functionality on his blog *Objective-See* found in Ref. [17]. The trojan works by posing as a cryptocurrency application and subsequently gains persistence on the host device. The infected device then contacts a server for the second-stage malicious payload [16].

1.4 Stuxnet

ThreatNeedle is not unique as a state-backed cyber-threat. Possibly, the most well-known state-based cyber-attack example is Stuxnet. Stuxnet was first uncovered in 2010 and immediately gained prominence within cyber-security research communities and the mainstream media for both its level of sophistication and target. Stuxnet was designed specifically to damage the Programmable Logic Controllers (PLC) at a nuclear power plant in Natanz, Iran [18]. PLCs are used to automate mechanical processes, such as gas centrifuges. Stuxnet was able to spread toward its target using a number of different mechanisms, namely: USB flash drives, Siemens SIMATIC WinCC (interface to SCADA systems), Siemens SIMATIC Step7 (industrial control projects) and network shares.

1.4.1 Zero-day

Stuxnet exploited zero-day flaws to deploy its payload which, as Bilge et al. discuss, are vulnerabilities that have not been disclosed publicly or to those who may directly be invested in knowing the vulnerabilities exist [19] (i.e., security teams). Typically, as vulnerabilities remain unknown, there is no defense against a zero-day exploit, with anti-virus solutions (and traditional intrusion detection systems) unable to provide patches to attack signatures that are yet unidentified [19] and only understood after an attack has taken place. A further nature of a zero-day exploit is that the attacks are targeted. For instance, the attackers are specifically aware of a vulnerability in a particular system of their target and the attack is adjusted as such. For example, Stuxnet exploited several Microsoft zero-day flaws, two of which are labeled MS10-061 (Print Spooler Service) and MS08-067 [20] (server service allowing remote code execution in Windows XP [21]).

To date, much research has been conducted on detecting zero–day exploits, for example, in Refs. [22–24], with a comprehensive review by Abri et al. in Ref. [25]. Effective solutions tend toward proactive security models, where the detection of an attack relies upon identifying when there are instances (outliers) that vary from the norm (benign traffic). Approaches to achieve this can be grouped using subfields of artificial intelligence, for instance (i) supervised learning [26] (involving training autonomous security solutions on known benign traffic and outliers), (ii) unsupervised learning (where the AI algorithm does not require pretraining to detect outliers), (iii) a combination of both, where a hybrid approach leverages the benefits of the detection accuracy provided by supervised learning and also the flexibility and adaptability of unsupervised [27]; or, more recently, (iv) deep learning, using techniques known as autoencoders, able to detect outliers in traffic patterns with high success [28].

1.4.2 Honeypots

In addition to the use of proactive security systems with integrated AI, one possible solution to overcome zero-day attacks is to find the vulnerabilities using honeypots. Honeypots involve isolating traffic, deceiving the attacker into assuming they have breached the system and generate attack signatures [29] based on monitoring the behaviors within a controlled environment.

In short, the goal of a Honeypot is to let the attacker believe they are successfully breaching the system, but in fact are being studied in an artificial environment (e.g., a virtual machine) where no actual damage can be caused. The Honeypot concept is depicted in Fig. 4. Critical infrastructures adopt a *defense in depth* model, firstly involving an external perimeter consisting of an advanced intrusion detection system and firewall. Beneath this, different levels of access rights are provided depending on the staff role within the organization. The high-level layer, for example, would provide access to vital systems and would not be accessible to those with low security clearance. Between each layer, further intrusion detection systems and firewalls are in place. The Honeypot will mimic this structure, for instance as documented in many works over the last 20 years, for example, in Refs. [30–33].

1.4.3 Duqu

Related to Stuxnet, Duqu is a collection of malware which was discovered in 2011 and received far less publicity despite using much of the Stuxnet code. Duqu derives its name from the temporary files it creates on the infected device, which all start with ~DQ. Similar to Stuxnet, the malware also functions by exploiting Windows zero-day vulnerabilities [34]. A full report on Duqu is published by the Laboratory of Cryptography and System Security (CrySyS) at Budapest University of Technology and Economics found at Ref. [34].

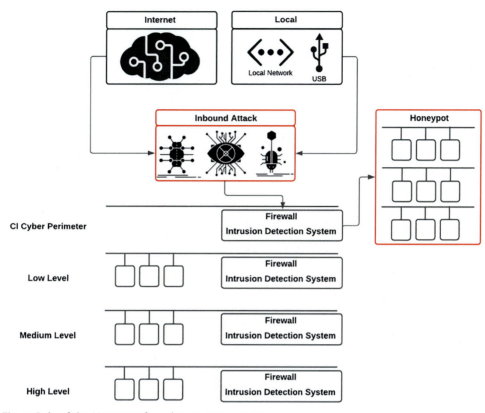

Fig. 4 Role of the Honeypot for cyber-security support.

1.5 Flame

At the time of writing this manuscript, Flame (discovered in 2012 and also recognized under the name Flamer) is documented as potentially the most complex cyber–attack ever created. Similar to Stuxnet, Flame, in the form of a trojan, has the ability to spread over a local area network (LAN) or via USB, targeting vulnerabilities within Microsoft OSs. The European Network and Information Security Agency (ENISA) document the threat of Flame in Ref. [35] and its mission as an information stealer, particularly targeted at the Middle East. Once deployed, Flame has a compilation of spying tools (documented by ENISA [35]) for listening to microphones, exfiltrating documents, tapping phones and online chats, Bluetooth scanning, taking screenshots, recording keyboard activity and intercepting Internet traffic.

Similar to Stuxnet, Flame is specifically a targeted attack, and, therefore, has infected relatively few PC devices (estimated between 100 and 1000 devices). Yet, differing from Stuxnet, Flame was designed for espionage rather than damage [36,37]. In terms of its

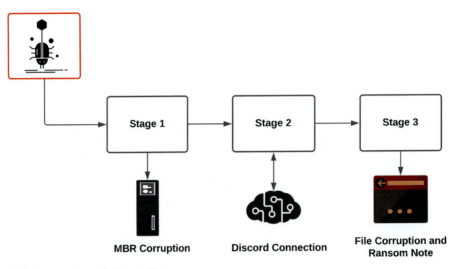

Fig. 5 High-level view of WhisperGate.

construction, Flame is uniquely larger than other viruses (approximately 20 megabytes in size) and, in result, it masquerades as a legitimate Windows update.

1.6 WhisperGate

In early 2022, a destructive malware was discovered targeting Ukrainian organizations [38]. At the time of writing this manuscript, relatively little is known about WhisperGate compared to the other aforementioned high-profile attacks given its infancy; and most information regarding the attack is found in the mainstream media and security blogs such as in Refs. [38–40] (Fig. 5).

What is known is that WhisperGate is aimed to be a destructive malware, with a technical analysis provided by the INSIKT Group at Ref. [39]. According to their article, WhisperGate is deployed in three stages (which must be executed before the PC reboots to take effect): (i) corruptions of the Master Boot Record (MBR), which in turn corrupts other drives on reboot. The MBR is then overwritten with a ransom note demanding payment and containing the details for a Bitcoin wallet; (ii) download of the malware (malicious file corrupter) needed for stage 3 which are hosted on a Discord channel; (iii) written in .NET, corruption of files on systems and network drives takes place.

1.7 Cyber-threat landscape

The emergence of the aforementioned media-documented attacks is depicted in the timeline in Fig. 6, with a summary provided in Table 1. All have specific goals and in some cases, direct targets.

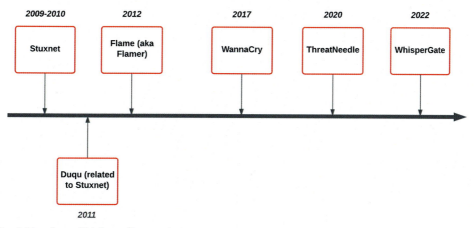

Fig. 6 Timeline of high-profile attacks.

Table 1 Comparison of high-profile attacks.

Name	Goal	Operating system
Stuxnet	Damage	Windows
Duqu	Damage	Windows
Flame	Espionage	Windows
WannaCry	Ransom	Windows
ThreatNeedle	Espionage	MAC
WhisperGate	Damage and ransom	Windows

Despite the developing landscape with varied sophistication, cyber-based security threats can still be assigned to 1 of 7 categories (Table 2), as defined in various gray literature sources, such as Refs. [42–45].

To offer a quantitative understanding of the division of these attack types within the critical infrastructure sectors, IBM Security X Force Threat Intelligence provide a detailed account of the percentages in Refs. [46,47]. Key findings from the report outline that ransomware malware accounted for 1 in 5 cyber-attacks (worldwide) in 2021 and Phishing was the top infection method, particularly in the United Kingdom against businesses where it accounted for 63% of incidents [46]. Globally, manufacturing was the most attacked infrastructure (23% of attacks), with ransomware as the main attack type. In the United Kingdom, the energy critical infrastructure sector was the main focus, with 24%, with manufacturing and finance second with 19% [46].

With this growing threat landscape, it is essential to consider how attack success continues to be possible and what methods are appropriate for securing critical infrastructures in a time of constant change. In the following section, the focus is on legacy systems, the

Table 2 Cyber-threat categories.

Cyber-threat	Description
Malware	Malware is software developed for malicious purposes, with further subcategories (spyware, ransomware, viruses and worms)
Emotet	Specifically a banking trojan, which is a type of malware but given its own unique category
Denial of Service	A process involving flooding computer networks with illegitimate requests, thus blocking legitimate access requests
Man in the Middle (MITM)	When attackers insert themselves into a two-party transaction
Phishing	Fake communication-based attack, namely using fake messages (such as emails) which seem legitimate to the user in order to request information or follow instructions which lead to the loss of information (such as a credit card number)
Structured Query Language (SQL) Injection	Involves inserting malicious code into a server that uses SQL to store data (such as customer details). In result, the server can be manipulated into releasing sensitive information
Password Attacks	Social engineering to ascertain data which can be used to trick people into providing information about their password. For instance, trending posts on social media asking for "first car," "pet names," etc. This could also include a brute-force attack, where the attack type relies on the computing power to find the correct combinations of usernames and passwords [41]

Industrial Internet of Things (IIoT), weakening segmentation and how legacy and aging systems are a considerable road block for holistic cyber-attack prevention.

2 ICS attack success

Almost all critical infrastructure sectors rely upon Industrial Control Systems (ICSs) as a core part of their operations. Unfortunately, due to the visibility and significant impact of any successful attacks, they are highly-valued targets for not only cyber-criminals but malicious state actors and politically-motivated adversaries.

Statistics published by Kaspersky highlight the severity of this issue, by showing that in the first half of 2021, 33.8% of ICS computers were attacked, which is an increase of 0.4% from the second half of 2020 [48]. This is a clear indicator that ICS-specific threats are growing. Nozomi Networks estimate that ICS shutdowns resulting from attacks cost between \$225 K and \$670 M [49].

There are various key themes emerging, which can help to explain why attacks against ICSs are successful, the most pertinent are summarized in this section.

2.1 Industrial internet of things

Industry 4.0 is considered to be the next industrial revolution, brought about by the integration of smart and connected systems to increase the level of automation in manufacturing and industrial processes. The integration of IIoT devices into modern critical infrastructure networks can offer various performance, cost and operational benefits. The improved connectivity facilitates increased levels of functionality, accessibility and communicability for IIoT devices. In turn, this poses significant security challenges by increasing both the security demands and breadth of the network's attack surface.

Numerous IIoT devices and their corresponding applications have been developed in existing ICS networks, which inherit or coexist with current impaired security practices and assumptions. Given the rapid evolution of such devices, it is unsurprising that for some manufacturers, security is not always a priority. This means that such devices are attractive targets for attacks and can introduce an element of uncertainty, which is not desirable in a sector renowned for its low-risk appetite. Increasingly, security concerns are being raised regarding the integrity of IIoT device supply chains and the vulnerabilities this may pose, from both software and hardware perspectives.

Many IIoT devices have limited computational power; therefore, data is commonly passed to a local or cloud-based controller for processing. Depending on the nature of the IIoT device, the sheer volume of data traversing the network or awaiting processing may introduce additional overheads. Ensuring the confidentiality and authenticity of any data transmitted is critical for avoiding common attacks (e.g., eavesdropping, replay or man-in-the-middle attacks), but the computational limitations also restrict the protocols supported for securing communications.

The use of cloud-based controllers requires the use of a IIoT gateway, which has been identified as a notable target for attacks and has garnered significant research interest [50,51]. It has been suggested that some of the security responsibilities regarding assuring device authenticity could be offloaded to a cloud-based platform. However, safety and real-time responsiveness are the most desired characteristics in an ICS. Therefore, any such security platform must be capable of real-time responses, which rules out cloud-based platforms on account of the unacceptable latency that would be introduced.

The use of cloud-based controllers for data processing and storage also raises many security questions relating to confidentiality, control and locality. This is especially important given the potential sensitivity and criticality of the data involved in ICSs.

2.2 Weakening segmentation

ICSs can be split into two distinct segments: Information Technology (IT)—hardware and software used for managing and utilizing data, and Operational Technology (OT)—hardware and software used to monitor and control industrial equipment. Historically, there has always been physical segmentation between the IT and OT segments of an ICS, known as an *air gap*. However, the continued adoption of IIoT and the shifting of applications, data and infrastructure to cloud providers has eroded network perimeters and this separation. This has resulted in the increasing convergence of IT and OT network segments, sometimes unbeknownst to operators. The exposure of ICS protocols to the internet can provide new attack surfaces and can lead to unusual and unwanted behavior. Shodan's ICS Radar crawls the Internet and detects direct access to ICSs [52], at the time of writing, there were over 55,000 systems accessible. The most common core ICS protocols supported by these accessible systems were ModBus, Siemens S7 and DNP3.

Some ICSs are embracing the convergence by blending technologies between the segments, e.g., using IT-side database technologies for OT. Unfortunately, many ICSs security practices have not been modernized to accommodate this. There is an over-reliance on proprietary protocols as a form of protection, this is an example of security-through-obscurity, which is not an effective security strategy.

Many ICSs are not prepared for defense against malware, and they rely on the OT practices of old and do not embrace modern cyber-security approaches. Vulnerable to many attacks such as eavesdropping, interception, replay and DoS. An increasingly common attack pattern is for the initial attack vector to reside within the IT segment and the attacker to pivot into the OT segment [53], notable examples include EKANS and Havex. The significance of this convergence is emphasized by the initial stage of the ICS Cyber Kill Chain, which is focused on intelligence gathering [54]. The top 3 initial attack vectors for ICSs are external remote services, exploiting public-facing applications and internet accessible devices [55].

The convergence has exposed ICS protocols designed for use in air gapped systems, being introduced to the wider internet. Air gaps have previously been depended upon as a defensive strategy but even in traditional networks, there were still situations where the gap was bridged. Examples of this include the transfer of config files and software patches from IT to OT (whether physically or digitally).

2.3 ICS protocol weakness

Many of the protocols utilized in ICSs are proprietary and were originally designed based upon the security assumptions of older serial-based and/or air gapped systems. Therefore, the majority of original protocols were focused on functionality and efficiency, whereas security was not factored into their designs [56]. The inherent and fundamental security issues associated with such protocols [57], especially when combined with poor documentation and maintenance practices, significantly increases the risk faced by ICSs. With many of these protocols still in use [58], it emphasizes the concern over the IT/OT convergence. This is further evidenced by Bratus et al.'s LZFuzz security fuzzing tool, which proved most successful when applied to proprietary and poorly documented protocols.

The most common core ICS protocols currently in use are Modbus, EtherNet/IP and S7comm [53]. Therefore, these three have been focused upon to examine their security attributes.

2.3.1 Modbus

The Modicon Communication Bus (Modbus) protocol is one of the oldest and widely used protocols. It was originally designed in 1979 for RS-232 or RS-485 serial networks between computationally limited microcontrollers [59]. In recent times, the protocol has migrated to being ethernet-based (Modbus TCP). Both versions of this protocol are based on the assumption that there would be limited physical access to the network, therefore negating security requirements. The resulting simplicity and efficiency of the protocol has made it popular.

However, as networking has evolved, there are increasing requirements for communications to be made from within LANs or WANs. This poses a significant problem, as Modbus was never designed for this type of use and appropriate security mechanisms are not provided. Furthermore, due to required backwards compatibility, this is not something that can or will be easily changed (although the Modbus Security Protocol has been introduced as an alternative solution). Some of the security concerns with the TCP-based variant are summarized in Table 3.

2.3.2 EtherNet/IP

EtherNet/IP (here IP stands for Industrial Protocol, not Internet Protocol) is an adaptation of Rockwell's Common Industrial Protocol (CIP), providing networking through the use of standard Ethernet frames. There are two types of communication in EtherNet/IP, implicit and explicit [60]. Implicit communication is used for I/O data transfer, and it is performed over UDP to leverage the speed and latency benefits, allowing the use of producer-consumer model. Explicit communication is used for nontime critical data and is performed over TCP.

CIP uses three object classes to define characteristics of a device: Required Objects (defines attributes e.g., identifiers, manufacturer and serial number), Application Objects (defines input and output for devices) and Vendor-specific objects (defines proprietary attributes). Its functionality can be considered similar in some respects to SNMP. To increase interoperability and cross-vendor support, objects (except Vendor-specific Objects) are standardized by device type and/or function. Some of the security concerns relating to the base protocol (i.e., not using CIP Security) are clarified in Table 4.

The existing CIP standards have not been updated to accommodate increased security requirements, in order to maintain compatibility. Instead, CIP security has been introduced as an alternative protocol that addresses most of the security limitations of the original protocol.

Table 3 Modbus TCP security concerns.

Concern	Description
Confidentiality	Modbus communications are all transmitted in clear text, offering no protection against eavesdropping, command injection or MiTM attacks
Integrity	Modbus does not support any integrity checks of the transmitted data, such as message checksums
Authentication	Modbus operates using a master/slave model, requiring only a valid address to communicate, there is no authentication used in the protocol. Therefore, there is no mechanism to verify the authenticity of devices, which facilitates various attacks such as identification of slave devices, DoS and spoofing attacks. Similarly, there is no session management used within Modbus, making replay attacks possible
Authorization	Modbus does not perform any authorization checks to ensure that privileged actions can only be performed by specific entities. Master and slave components will action valid commands received, without any security checks. Therefore, unauthorized/malicious commands can be executed by attackers
Accountability	As Modbus does not support authentication, there is no facility to maintain an audit trail
Simplicity	Although seen as a positive attribute for ease of implementation, the simplicity of the protocol and its structure makes reconnaissance activities by attackers much easier
Purpose	Part of the protocol's purpose is the programming of controllers; therefore, the repercussions of ineffective security can be severe

Table 4 EtherNet/IP security concerns.

Concern	Description
Confidentiality	Communications are transmitted in clear text, offering no protection against eavesdropping attacks or MITM attacks
Integrity	No network-based checks are performed to validate integrity, thus allowing data tampering to occur
Authentication	The lack of verification of users/devices allows for disruptive changes (e.g., change device's IP) to be made by untrusted parties [61] or the launching of replay attacks. Furthermore, the protocol is based on UDP, which is stateless and offers no transmission control mechanism to help in preventing spoofing
Availability	Various implementation weaknesses have been exposed by researchers, which facilitate DoS attacks. For example, weak session ID generation and improper TCP timeout [62]
Multicast traffic	Multicast traffic is utilized within the protocol to reduce network traffic load. However, this also lacks transmission control mechanisms and is unable to prevent transmission path manipulation using injected IGMP packets
Standardization	Unfortunately, standardizing Required and Application Objects has made common devices much easier to identify and exploit during attacks

Table 5 S7Comm security concerns.

Concern	Description
Confidentiality	The protocol provides no encryption functionality, so it is unable to prevent eavesdropping or MITM
Integrity	The protocol doesn't offer any integrity mechanisms, meaning it cannot prevent data/command modification or replay attacks [63]
Authentication	Some PLCs can be configured to utilize basic password protection for certain operations. However, this offers limited protection as password lengths are constrained. Passwords are supplied for authentication as hashes but are sent in clear text. When this is combined with the lack of confidentiality and integrity, it enables passwords to be replayed and rainbow table attacks
Authorization	Passwords can be used to prevent unauthorized actions, but various attacks are possible to bypass any protection afforded [64]
Documentation	There is limited documentation available for the protocol [65], suggesting security through obscurity is utilized as a form of protection, which is widely considered poor practice

2.3.3 Siemens S7Comm

S7 Communication (S7Comm) is a proprietary protocol from Siemens first introduced in 1994 for data transfer and programming of its S7 family of PLCs. There are three versions of the protocol: S7Comm, S7CommPlus v1 and S7CommPlus v2.

Similar to the previous protocols, S7Comm was deployed with inadequate security protection, which was infamously exploited by the Stuxnet worm. Some of the main concerns of this protocol are summarized in Table 5.

In response to the vulnerabilities exploited by Stuxnet, S7CommPlus v1 was released, which utilizes session IDs to prevent replay attacks. However, researchers identified that the mechanism used to generate the IDs was too simplistic and could be exploited and sessions stolen [64].

S7CommPlus v2 improves upon the vulnerabilities of the previous versions by utilizing encryption as well as improved antireplay mechanisms and an integrity check mechanism. However, researchers have demonstrated that these improvements can be broken too [66].

2.3.4 Weakness trends

As evidenced throughout this subsection, the flawed and outdated assumptions of physical isolation that underpin many ICS protocols are the primary cause of most security issues. Compatibility is commonly cited as the reason behind not improving existing protocol standards to meet basic security goals. Although the three protocols discussed all have security enhanced versions of the original protocols to overcome these issues, many live implementations have not, or are unable to, adopt these enhanced versions. Unfortunately, this seems to be a representative picture across the sector, and the evolution of

ICSs will continue to expose the security weaknesses and vulnerabilities of these protocols.

2.4 Legacy/aging systems

ICS systems are intended to have significantly longer operational lifespans (typically 20–25 years), than commodity systems [67], so components will inevitably be considered as legacy at some point. Therefore, the standards, principles, security assumptions and anticipated threats originally used to devise architectural designs could be significantly outdated. Similarly, cutting-edge protocols and software used for implementation at the time will rapidly become outdated given the pace of technological evolution. For instance, there are numerous examples of End-Of-Life (EOL) operating systems such as Windows XP still being used [68].

Future-proofing systems decades in advance cannot be accurately accomplished. So it is unsurprising that such systems are inadequately equipped and secured for modern requirements. Furthermore, given the critical nature of these systems and their required stability, continuous development (as seen with commodity systems) is not a desirable characteristic.

One crucial example of this is the inherent security limitations of ICSs, original designs accommodated the differing priorities and needs of the IT and OT segments. As OT systems were typically air gapped, availability was prioritized over confidentiality and integrity [69] (e.g., availability of critical sensor readings is far more important than their confidentiality).

Facilitating the modern connectivity requirements of ICSs often requires legacy systems to be used or configured in ways that they were not originally designed for, or the retrofitting of additional components. Such alterations deviate from the original design and can result in unexpected behavior, as well as the introduction of new weaknesses and vulnerabilities. The continued blurring of boundaries that used to exist between IT and OT is increasing the visibility of basic security shortcomings in the OT segments, e.g., lack of encryption, poor/no authentication, weak/default passwords [70]. A common example of this is the migration from serial-based to IP-based communications, the level of internet exposure and attack surface expansion are often not fully considered.

2.5 ICS summary

As the modern connectivity requirements of ICSs continue to shift toward greater internet accessibility, the exposed surface of these systems will become increasingly scrutinized for credible attack vectors. Throughout this section, only a few of the reasons behind successful ICS attacks have been covered, there are many more contributory factors. However, it is clear that the probability of security problems occurring and the severity of associated impacts are only going to increase. There is therefore a need to focus upon factoring security into the design of an ICS (for both new and existing systems), rather than the continued approach of retrofitting. Similarly, ICS security needs to orient toward a holistic approach, which encompasses both IT and OT segments, rather than adopting a siloed approach with separate expertise.

3 Conclusion

Future cyber-security focus is orienting toward IIoT security, smart home appliances and data storage devices, where the pervasiveness of personal information stored across multiple devices (laptops, PCs, smart phones, etc.) has increased the entry point of hackers [42]. Further, IBM details that in 2021 the global shift in work habits as a result of the Covid-19 pandemic forcing many to work from home, created opportunities for threat activators to infiltrate organizations [47].

Ensuring the security of critical infrastructure systems poses an ongoing challenge, with failings having potentially catastrophic consequences. A significant hurdle is the ever-increasing number of critical infrastructure components that are now directly or indirectly exposed to the internet, causing potential attack surfaces and associated risks to grow. Similarly, the heterogeneity of such devices also contributes to increasing the attack surface and adds additional complexity when coordinating defensive efforts.

There are three emerging attack patterns of significance, which need to be adequately considered to ensure the future security of critical infrastructures. These are:

1. **Supply chain attacks**—Attacks are no longer directly launched directly against the intended organization or system. Instead, weaker elements of the supply chain are targeted offer greater opportunity to inflict widespread damage. A notable global example of this was the Solar Winds attack in 2020. Usually, targets for this type of attack will be third-party vendors or suppliers, with poor cyber-security. Some of the example techniques used include software weaknesses, firmware backdoors, preinstalled malware installation and certificate theft.

2. **Top-down infrastructure attacks**—Traditionally, attacks have followed a bottom-up approach due to the perimeter security model adopted by many. Here, the exposed elements of a system (usually lower value assets) are exploited, and attackers work to elevate their access to the higher value targets afforded greater protection. However, with the shift to cloud-based services, assets that were typically well-protected from internet exposure (e.g., Active Directory servers) are now increasingly accessible online. Hence, attackers can now directly target these high-value cloud-based administrative/management systems, instead of having to work to elevate themselves to this point.

3. **Ransomware**—Unfortunately, this type of attack is common throughout many types of networks. However, a recent Claroty report estimated that around 80% of critical infrastructures had been subject to ransomware and around 60% had paid

the ransom [71]. The very nature of critical infrastructure means that any impairment of functionality can have devastating consequences; hence, targets are far more likely to pay. Given the likelihood of success, and the increased attack surface created by IT and OT mergence, it is inevitable that attackers will be prioritizing attacks against these systems.

Thus, *could one solution be for better use of AI?* For instance, IBM details that organizations who have a fully deployed security-based AI and automation had lower costs associated with breaches ($2.9million vs $6.71 million) compared to organizations without AI [72]. Further, as indicated in current scientific literatures, the level of preparedness is dependent on two factors: (i) the use of more effective intrusion detection processes; (ii) the types of security techniques. Panagiotis et al. provide a comprehensive list in their review article in Ref. [41], discussing some of the most well-known algorithms and models integrated into security systems for better overall resilience, focusing primarily on machine learning and deep learning. Within this domain, AI-based approaches (specifically machine learning techniques) such as K-means, Naive Bayes, support vector machines, decision trees and density-based spatial clustering of applications with noise have all been used with effect in research-based approaches [73–75]. Building on this, deep learning models are becoming increasingly employed, namely the use of autoencoders, long short-term memory networks (LSTMs), recurrent neural networks (RNNs) and convolutional neural networks (CNNs) [76–78]. Research demonstrates that AI-driven approaches are the way forward to a holistic security system. The challenge remains—translating research findings into real-world implemented systems to counter the level of sophistication discussed in Section 1.

References

[1] I. Ghafir, J. Saleem, M. Hammoudeh, et al., Security threats to critical infrastructure: the human factor, J. Supercomput. 74 (2018) 4986–5002.

[2] G.M. Karagiannis, Z.I. Turksezer, L. Alfier, L. Feyen, E. Krausmann, Climate Change and Critical Infrastructure—Floods, European Commission, Science for Policy Report, Brussels, 2019.

[3] K.V. Ruiten, T. Bles, J. Kiel, EU-INTACT-case studies: impact of extreme weather on critical infrastructure, in: European Conference on Flood Risk Management (FLOODrisk 2016), Lyon, 2016.

[4] B. Bennett, Understanding, Assessing, and Responding to Terrorism: Protecting Critical Infrastructures, Wiley, 2018.

[5] Z. Xindi, C. Krishna, Critical infrastructure security against drone attacks using visual analytics, in: Computer Vision Systems, Lecture Notes in Computer Science, vol. 11754, 2019.

[6] B. Kai, C. Charmaine, The counter sovereignty of critical infrastructure security: settler-state anxiety versus the pipeline blockade, Antipode (2021) 1–23.

[7] M. Rong, C. Han, L. Liu, Critical infrastructure failure interdependencies in the 2008 Chinese Winter Storms, in: International Conference on Management and Service Science, Wuhan, 2010.

[8] P. Wagner, Critical Infrastructure Security, 2021, Available at SSRN: https://ssrn.com/abstract=3762693.

[9] ICS Cert, IT threat evolution in Q2 2021. PC statistics, Kaspersky (2021). 12 August. Available from: https://securelist.com/it-threat-evolution-in-q2-2021-pc-statistics/103607/. (Accessed 20 January 2022).

[10] S. Ghafur, S. Kristensen, K. Honeyford, A retrospective impact analysis of the WannaCry cyberattack on the NHS, NPJ Digit. Med. 98 (2) (2019).

[11] Z. Whittaker, Two years after WannaCry, a million computers remain at risk, TechCrunch (2019). 12 May. Available from: https://techcrunch.com/2019/05/12/wannacry-two-years-on/. (Accessed 20 January 2022).

[12] M. Kumar, TSMC chip maker blames WannaCry malware for production halt, The Hacker News Media (2018). 7 August. Available from: https://thehackernews.com/2018/08/tsmc-wannacry-ransomware-attack.html. (Accessed 14 February 2022).

[13] D. Goodin, >10,000 Windows computers may be infected by advanced NSA backdoor, ars Technica (2017). 21 April. Available from: https://arstechnica.com/information-technology/2017/04/10000-windows-computers-may-be-infected-by-advanced-nsa-backdoor/. (Accessed 8 March 2022).

[14] ICS CERT, PseudoManuscrypt: a mass-scale spyware attack campaign, Kaspersky (2021). 16 December. Available from: https://securelist.com/pseudomanuscrypt-a-mass-scale-spyware-attack-campaign/105286/. (Accessed 19 January 2022).

[15] ICS CERT, Lazarus targets defense industry with ThreatNeedle, Kaspersky (2021). 25 February. Available from: https://securelist.com/lazarus-threatneedle/100803/. (Accessed 20 January 2022).

[16] New Jersey Cybersecurity & Communications Integration Cell, NukeSped, NJCCIC Threat Profile (2019). 10 December. Available from: https://www.cyber.nj.gov/threat-center/threat-profiles/macos-malware-variants/nukesped. (Accessed 20 January 2022).

[17] P. Wardle, Lazarus Group Goes 'Fileless', Objective-See, 2019. 3 December. Available from: https://objective-see.com/blog/blog_0x51.html. (Accessed 20 January 2022).

[18] D. Kushner, The real story of stuxnet, IEEE Spectr. 50 (3) (2013) 48–53.

[19] L. Bilge, T. Dumitras, Before we knew it: an empirical study of zero-day attacks in the real world, in: ACM Conference on Computer and Communications Security, Raleigh North Carolina, 101, 2012.

[20] P. Mueller, B. Yadegari, The stuxnet worm, in: Political Science, University of Arizona, Arizona, USA, 2012, pp. 1–12.

[21] Microsoft, Microsoft security bulletin MS08-067—critical, Microsoft (2019). 12 March. Available from: https://docs.microsoft.com/en-us/security-updates/securitybulletins/2008/ms08-067. (Accessed 22 February 2022).

[22] X. Sun, J. Dai, P. Liu, A. Singhal, Using Bayesian networks for probabilistic identification of zero-day attack paths, IEEE Trans. Inf. Forensics Secur. 13 (10) (2018) 2506–2521.

[23] N. Sameera, M. Shashi, Deep transductive transfer learning framework for zero-day attack detection, ICT Express 6 (4) (2020) 361–367.

[24] V. Sharma, J. Kim, S. Kwon, I. You, K. Lee, K. Yim, A framework for mitigating zero-day attacks in IoT, in: CISC-S'17, Sinchang-Asan, 2017.

[25] F. Abri, S. Siami-Namini, M.A. Khanghah, F.M. Soltani, A.S. Namin, The performance of machine and deep learning classifiers in detecting zero-day vulnerabilities, in: IEEE BigData, 2019. arXiv:1911.09586v1.

[26] M. Alazab, S. Venkatraman, P. Watters, M. Alazab, Zero-day malware detection based on supervised learning algorithms of API call signatures, in: Australasian Data Mining Conference, Ballarat, 2011.

[27] P.M. Comar, L. Liu, S. Saha, P.-N. Tan, A. Nucci, Combining supervised and unsupervised learning for zero-day malware detection, in: Proceedings of IEEE INFOCOM, Turin, 2013.

[28] H. Hindy, R. Atkinson, C. Tachtatzis, J.-N. Colin, E. Bayne, X. Bellekens, Utilising deep learning techniques for effective zero-day attack detection, Electronics 9 (2020) 1684.

[29] C. Musca, E. Mirica, R. Deaconescu, Detecting and analyzing zero-day attacks using honeypots, in: International Conference on Control Systems and Computer Science, Bucharest, 2013.

[30] Y.M.P. Pa, S. Suzuki, K. Yoshioka, T. Matsumoto, T. Kasama, C. Rossow, IoTPOT: a novel honeypot for revealing current IoT threats, J. Inf. Process. 24 (3) (2016) 522–533.

[31] M.L. Bringer, C.A. Chelmecki, H. Fujinoki, A survey: recent advances and future trends in honeypot research, Int. J. Comput. Netw. Inf. Secur. 10 (2012) 63–75.

[32] M. Nawrocki, M. Wählisch, T.C. Schmidt, C. Keil, J. Schönfelder, A survey on honeypot software and data analysis, arXiv:1608.06249, ACM, 2016.

[33] N. Provos, A virtual honeypot framework, in: Usenix Security Symposium, San Diego, 2004.

[34] B. Bencsáth, G. Pék, L. Buttyán, M. Félegyházi, Duqu: A Stuxnet-Like Malware Found in the Wild, Laboratory of Cryptography and System Security (CrySyS), 2011. 14 October.

[35] ENISA, The Threat From Flamer, European Network and Information Security Agency, Brussels, 2012.

[36] R. Cohen, New massive cyber-attack an 'industrial vacuum cleaner for sensitive information', Forbes (2012). 28 May. Available from: https://www.forbes.com/sites/reuvencohen/2012/05/28/new-massive-cyber-attack-an-industrial-vacuum-cleaner-for-sensitive-information/. (Accessed 24 February 2022).

[37] D. Lee, Flame: massive cyber-attack discovered, researchers say, BBC News (2012). 28 May. Available from: https://www.bbc.com/news/technology-18238326. (Accessed 24 February 2022).

[38] Microsoft, Destructive Malware Targeting Ukrainian Organizations, Microsoft Security, 2022. 15 January. Available from: https://www.microsoft.com/security/blog/2022/01/15/destructive-malware-targeting-ukrainian-organizations/. (Accessed 24 February 2022).

[39] INSIKT Group, WhisperGate malware corrupts computers in Ukraine, 2022, 28 January. Available from: https://www.recordedfuture.com/whispergate-malware-corrupts-computers-ukraine/. (Accessed 24 February 2022).

[40] CrowdStrike, Technical Analysis of the WhisperGate Malicious Bootloader, CrowdStrike Intelligence Team, 2022. 19 January. Available from: https://www.crowdstrike.com/blog/technical-analysis-of-whispergate-malware/. (Accessed 24 February 2022).

[41] F. Panagiotis, K. Taxiarxchis, K. Georgios, L. Maglaras, M. Ferrag, Intrusion detection in critical infrastructures: a literature review, Smart Cities 4 (2021) 1146–1157.

[42] University of North Dakota, 7 Types of Cyber Security Threats, University of North Dakota, 2022. Available from: https://onlinedegrees.und.edu/blog/types-of-cyber-security-threats/. (Accessed 8 March 2022).

[43] Guru Schools, 7 Types of Cyber Security Threats, Guru Schools, 2019. 5 November. Available from: https://guruschools.com/7-types-of-cyber-security-threats/. (Accessed 8 March 2022).

[44] C. Evans, 7 Types of Cyber Security Attacks With Real-Life Examples, E-Tech, 2021. 16 September. Available from: https://www.etechcomputing.com/7-types-of-cyber-security-attacks-with-real-life-examples/. (Accessed 8 March 2022).

[45] Jaro Education, 7 Types of Cyber Security Threats, Jaro Education, 2022. Available from: https://www.jaroeducation.com/blog/7-types-of-cyber-security-threats/. (Accessed 8 March 2022).

[46] G. Hastings, IBM security report: energy sector becomes UK's top target for cyberattacks as adversaries take aim at nation's critical industries, 2022, 23 February. Available from: https://uk.newsroom.ibm.com/2022-02-23-IBM-Security-Report-Energy-Sector-Becomes-UKs-Top-Target-for-Cyberattacks-as-Adversaries-Take-Aim-at-Nations-Critical-Industries?utm_medium=OSocial&utm_source=Twitter&utm_content=CAAWW&utm_id=Twitter-XforceUKI-2022-0. (Accessed 24 February 2022).

[47] IBM Security X-Force, Manufacturing Becomes the World's Most Attacked Industry, 2022, Available from: https://www.ibm.com/security/data-breach/threat-intelligence/. (Accessed 8 March 2022).

[48] ICS CERT, Threat landscape for industrial automation systems. Statistics for H1 2021, Kaspersky (2021). 9 September. Available from: https://ics-cert.kaspersky.com/publications/reports/2021/09/09/threat-landscape-for-industrial-automation-systems-statistics-for-h1-2021/. (Accessed 14 February 2022).

[49] Nozomi Networks, The Cost of OT Cybersecurity Incidents and How to Reduce Risk, Nzomi Networks, 2020.

[50] C.H. Chen, M.Y. Lin, C.C. Liu, Edge computing gateway of the industrial internet of things using multiple collaborative microcontrollers, IEEE Netw. 32 (1) (2018) 24–32.

[51] T. Gong, S. Zheng, M. Nixon, E. Rotvold, S. Han, Demo abstract: industrial IoT field gateway design for heterogeneous process monitoring and control, in: IEEE Real-Time and Embedded Technology, Porto, 2018.

[52] Shodan ICS Radar, Shodan, Available from: https://ics-radar.shodan.io/, 2022 (Accessed 11 February 2022).

[53] D. Masson, Darktrace, 2020, 6 August. Available from: https://www.darktrace.com/en/blog/darktrace-ot-threat-finds-defending-the-widening-attack-surface/. (Accessed 11 February 2022).

[54] M.J. Assante, R.M. Lee, The industrial control system cyber kill chain, Sans Institute White Paper, 2015, pp. 1–23.

[55] M. Bristow, A SANS 2021 survey: OT/ICS cybersecurity, in: Nozomi Networks, SANS Institute, 2021, pp. 1–23.

[56] M. Conti, D. Donadel, F. Turrin, A survey on industrial control system testbeds and datasets for security research, IEEE Commun. Surv. Tutor. 23 (4) (2021) 2248–2294.

[57] G.P.H. Sandaruwan, P.S. Ranaweera, V.A. Oleshchuk, PLC security and critical infrastructure protection, in: IEEE 8th International Conference on Industrial and Information Systems, Peradeniya, 2013.

[58] G. Barbieri, M. Conti, N.O. Tippenhauer, F. Turrin, Assessing the Use of Insecure ICS Protocols via IXP Network Traffic Analysis, arXiv:2007.01114, 2020.

[59] Modbus.org, MODBUS Over Serial Line: Specification & Implementation Guide V1.0, 2002. 2-12-2002 (Accessed March 2022).

[60] I. Open DeviceNet Vendor Association, Ethernet/IP quick start for vendors hand, ODVA, Ann Arbor, MI, 2008. Tech. Rep.

[61] B. Batke, J. Wilberg, D. Dube, CIP security phase 1 secure transport for EtherNet/IP, in: Industry Conference and Annual Meeting, Frisco, Texas, 2015.

[62] R. Grandgenett, R. Gandhi, W. Mahone, Exploitation of Allen Bradley's implementation of Ether-Net/IP for denial of service against industrial control systems, in: 9th International Conference on Cyber Warfare and Security, West Lafayette, Indiana, 2014.

[63] A. Ghaleb, S. Zhioua, A. Almulhem, On PLC network security, Int. J. Crit. Infrastruct. Prot. 22 (2018) 62–69.

[64] H. Hui, K. McLaughlin, S. Sezer, Vulnerability analysis of S7 PLCs: manipulating the security mechanism, Int. J. Crit. Infrastruct. Prot. 35 (2021) 100470.

[65] L. Martín-Liras, M.A. Prada, J.J. Fuertes, A. Morán, S. Alonso, M. Domínguez, Comparative analysis of the security of configuration protocols for industrial control devices, Int. J. Crit. Infrastruct. Prot. 19 (2017) 4–15.

[66] C. Lei, L. Donghong, M. Liang, The Spear to Break the Security Wall of S7CommPlus, Blackhat, 2016. Available from: https://www.blackhat.com/docs/eu-17/materials/eu-17-Lei-The-Spear-To-Break%20-The-Security-Wall-Of-S7CommPlus-wp.pdf. (Accessed 14 February 2022).

[67] S.P. McGurk, Industrial Control Systems Security, Homeland Security, 2008. December. Available from: https://csrc.nist.gov/CSRC/media/Events/ISPAB-DECEMBER-2008-MEETING/documents/ICSsecurity_ISPAB-dec2008_SPMcGurk.pdf. (Accessed 14 February 2022).

[68] CyberX, 2019 Global ICS and IIoT Risk Report, CyberX-Labs, 2019 (Accessed 11 February 2022).

[69] S. Abe, M. Fujimoto, S. Horata, Y. Uchida, T. Mitsunaga, Security threats of internet-reachable ICS, in: Annual Conference of the Society of Instrument and Control Engineers of Japan, Tsukuba, 2016.

[70] Mandiant, What About the Plant Floor? 2022, Available from: https://www.mandiant.com/resources/six-subversive-security-concerns-for-industrial-environments. (Accessed 11 February 2022).

[71] Claroty, Claroty Biannual ICS Risk & Vulnerability Report: 2H 2021, Team82, 2021. Available from: https://claroty.com/2h21-biannual-report/. (Accessed 14 February 2022).

[72] IBM Corporation, Cost of a Data Breach Report, IBM Security, Armonk, NY, 2021.

[73] J.-H. Li, Cyber security meets artificial intelligence: a survey, Front. Inform. Technol. Electron. Eng. 19 (2018) 1462–1474.

[74] H.M. Farooq, N.M. Otaibi, Optimal machine learning algorithms for cyber threat detection, in: UKSim-AMSS 20th International Conference on Computer Modelling and Simulation (UKSim), Cambridge, UK, 2018.

[75] Y. Fan, K. Tian, X. Wang, Z. Lv, J. Wang, Detecting intrusions in railway signal safety data Networks with DBSCAN-ARIMA, Front. Cyber Secur. 1286 (2020) 254–270.

[76] M. Banton, N. Shone, W. Hurst, Q. Shi, Intrusion detection using extremely limited data based on SDN, in: IEEE 10th International Conference on Intelligent Systems (IS), Varna, Bulgaria, 2020.

[77] Y. Xin, L. Kong, Z. Liu, Y. Chen, Y. Li, Machine learning and deep learning methods for cybersecurity, IEEE Access 6 (2018) 35365–35381.

[78] S.A. Salloum, M. Alshurideh, A. Elnagar, K. Shaalan, Machine learning and deep learning techniques for cybersecurity: a review, in: Proceedings of the International Conference on Artificial Intelligence and Computer Vision (AICV2020), Cairo, Egypt, 2020.

CHAPTER 13

Energy systems as a critical infrastructure: Threats, solutions, and future outlook

Tarek Alskaif[a], Miguel Ángel Pardo Picazo[b], and Bedir Tekinerdogan[a]
[a]Information Technology Group, Wageningen University & Research, Wageningen, The Netherlands
[b]Department of Civil Engineering, University of Alicante, Alicante, Spain

1 Introduction

The energy system is a complex network of energy production, transmission, distribution, and consumption components that provide electricity and heat to millions of buildings, industries, and other facilities. These elements have been designed to supply energy services to end users in a reliable and affordable manner [1, 2]. With a growing society and increased energy consumption, the size, scale, and complexity of energy systems also increase and the energy system becomes more critical. This poses national and local challenges that need to be addressed to provide access to secure, affordable, and clean energy for all consumers, both households and businesses, which are particularly addressed in the United Nations (UN) Sustainable Development Goal (SDG) 7 [3]. In addition, the energy system has been undergoing a rapid and fundamental transformation, which can be attributed to two key reasons, the global need to mitigate climate change, and the advancements in information and communication technologies (ICT).

The first has contributed to (i) a rapid development, integration, and large acceptance of relatively small-scale renewable energy generation technologies, such as wind turbines and solar photovoltaic (PV); and (ii) the electrification of numerous sectors that traditionally depend on fossil fuels, such as heating and mobility. These renewable technologies are likely to be increasingly situated in the fringes of the end user, such as rooftop PV arrays or electric vehicles (EV), where grid infrastructure might not be properly equipped to accommodate that. The development in this direction has been accelerated by international and national climate agreements, such as the Paris Agreement in 2015 [4], the Green Deal in 2020 by the European Union (EU) [5], and other national climate agreements. However, the current energy system is not fully prepared for such rapid developments and cannot easily handle intermittent renewable energy supply or increasing electric demand caused by electrifying the mobility and heating sectors.

Management and Engineering of Critical Infrastructures
https://doi.org/10.1016/B978-0-323-99330-2.00013-1
287

The second has accelerated the evolution and transformation of the electricity grid to a smarter interconnected grid that enables a bidirectional flow of power and information between utilities to consumers to and, most importantly, recognizes the user as a central stakeholder of the system [6, 7]. For instance, the past decade has seen a successful large-scale installation of smart metering infrastructure in most countries in the EU for real-time monitoring of electricity consumption in buildings [8]. Besides, the data captured by smart meters provide unique opportunities for smarter management and control of the energy system, optimization of energy flows, and enabling of new services and business opportunities. For instance, smart management can include smart charging of EVs considering different end users driving patterns and network capacity constraints [9]. Big data analytics solutions can smooth the large-scale integration of random and intermittent renewable energy sources, such as solar PV [10, 11]. Besides, the widespread use of ICT in the energy system and the availability of big data have enabled the emergence of digital twins (DT) applications in energy systems for effective and real-time monitoring, control, and autonomous decision-making [12].

However, the rapid transformation in energy systems has increased their vulnerability to not only natural disasters caused by the changing climate (e.g., heat waves, wildfires, droughts, and floods) but also to human-caused disruptions (e.g., cyber-physical attacks, human errors, and technical defects). In electricity grids, this can compromise the continuous and reliable provision of electricity supply as an essential service for end users and the economy. For instance, as the grid becomes more digitalized and interconnected, security becomes a major concern since both small and big maleficent actors may try to manipulate the grid for their benefit. As an example, Ukraine's electricity grid suffered from cyberattacks in 2015, which culminated in power outages for approximately 230,000 consumers in Ukraine for 1–6 h [13]. The vulnerability is expected to only get worse because of climate change-driven natural disasters, such as severe floods as happened in Sudan, Europe, and Asia in the past 2 years, or catastrophic droughts, heat waves, and wildfires as happened in California, the Middle East, and other parts of the world in the past decade. Despite their relatively low probability of occurrence, the impacts of such natural disasters and disruptions can be huge and not only limited to power outages for normal end users but also for critical load in the system [14, 15]. As a result, enhancing the resilience of the energy system as a critical infrastructure has become a priority and received increasing attention from the research community, industry, governments, and policymakers. Technologies, tools, and mechanisms have been proposed to address the resilience problem of critical electricity infrastructure against such high-impact low-probability events. This goes beyond the traditional energy system planning methodologies, such as network redundancy. On top of that, recent natural disasters require alternative solutions to be studied and set up for the comprehensive provision of robustness, preparedness, and recovery [16].

This chapter explores the energy system as a critical infrastructure and provides insights into the compelling urge for its resilience against cyber-physical attacks and natural disasters. The focus will be on electricity grids as one of the most vital critical infrastructures. In Section 2, an overview of criticality threats to energy systems is provided. Section 3 describes the potential technologies and solutions to enhance the resilience of electricity grids. In Section 4, the challenges are discussed and an outlook for the pathway toward more resilient electricity grids is provided. The book chapter is concluded in Section 5.

2 Sources of criticality

2.1 Natural disasters

Natural disasters, such as wildfires, hurricanes, and floods, can have a severe impact on the energy infrastructure [17]. For instance, hurricane Sandy was the most devastating storm to affect Connecticut's electricity distribution network, with more than 15,000 outages affecting more than 500,000 consumers [18]. Another example is wildfires, which burn and destroy high-voltage transmission lines. In the last few years, large wildfires have spread widely in different countries around the world, claiming many lives, destroying structures, and forcing hundreds of thousands of people to evacuate their homes. Indeed, the impact goes beyond disrupting the transmission and distribution network. Any disruption in the power infrastructure would have a huge impact on other infrastructures causing damage to national security, the economy, public health, and safety.

The National Association of Regulatory Utility Commissioners stresses that the resilience of the energy sector must be increased. Since it is complicated to quantify the costs of power outages (considering direct, indirect, and social costs), many approaches focused on strategies to minimize power outage risks [19, 20]. The characteristics of fault propagation and the severity of damage depend on the type of network. Hence, the vulnerability of a power network to natural disasters must be measured first in order to indicate which areas of the network should be protected and in what order protection measures should be undertaken [21]. The work in Ref. [14] identified the disaster characteristics, and these results are shown in Table 1.

Extreme weather events undoubtedly lead to increasingly severe and frequent power outages [22]. However, breaking down the effects of a weather event is a complicated matter. For instance, a severe tropical storm could cause damage to power lines due to associated wind and heavy rainfall but could also cause problems if floods occur. Although floods are the most common natural disasters, they are no less damaging and the power distribution network must ensure power supply during a flood.

The EU compares the natural hazards of earthquakes, floods, and climate [23]. Each weather phenomenon has its own particular characteristic that leads to the destruction of different power infrastructure equipment. Earthquakes cause ground failure and exert

Table 1 Disaster characteristics [14].

Type	Impact region	Predictability	Span (miles)	Affecting time
Hurricane	Coastal regions	24–72 h	1000	Hours to days
Tornado	Inland plains	0–2 h	5	Minutes to hours
Ice storm	High latitudes	24–72 h	1000	Hours to days
Earthquakes	Regions fault lines	Seconds or minutes	Small to large	Minutes to days
Tsunami	Coastal regions	Minutes	Small to large	Minutes to hours
Drought, wildfire	Inland regions	Days	Medium to large	Days to months

inertial shear loads on buildings. Inclement weather brings geomagnetically induced current (GIC) that damages alternating current (AC) equipment not designed to withstand direct current (DC). Finally, floods cause damage when water seeps into power components and floors. Since each hazard provokes a different voltage, it causes particular damage and affects a different set of components in the power network (see Table 2).

The time frame for infrastructure recovery can fluctuate between hazards, although it is always progressive. The rehabilitation agenda is consistent with engineering resilience [24], and the time scale and recovery time are different after each natural disaster.

Table 2 Damage types and natural hazard impacts on the electricity grid [23].

	Earthquake	Space weather	Flood
Damage types	Structural damage, inertial loading, ground failure	Damage to transmission and generation equipment from GICs for system-wide impact	Erosion and/or landslides
Contributing factors	Soil liquefaction, no warning time	Early warning possible	Early warning possible
Most vulnerable equipment	Heavy equipment (LPTs) ceramic parts (bus bars) or transformers	Equipment vulnerable to DC (transformers), equipment protected from DC excitation (tripping)	Transmission towers, substation equipment
Recovery time is driven by	Items in need of repair, access to conduct repairs	System-wide impact, delayed effects	Floodwaters recession, items in need of repair
Recovery time range	A few hours to months; most commonly, 1–4 days	Equipment tripped offline, 24 h; on damaged equipment, several months	Less than 24 h to 3 weeks, with hurricane and/or storm damage up to 5 weeks

2.2 Human-caused disruptions

Cyber-physical attacks are designed to disrupt or even damage systems, particularly critical infrastructure. In many industries, operation data are mainly sought by attackers, but in the energy industry, the goal is typically to control critical infrastructure in order to cause physical damage. Cyber-physical attacks on electricity distribution networks mainly target supervisory control and data acquisition (SCADA) systems. For example, in Iran, the vulnerability of power grids to hackers was proved when a computer worm "Stuxnet" destroyed nuclear centrifuges in 2010. It destroyed 10%–20% of its devices. In 2018, the Department of Homeland Security issued a warning that the Russian Government had engineered a series of cyberattacks on major infrastructure (nuclear power plants, water, and electricity distribution systems) [25]. In 2016, the EU adopted the Directive on the security of Network and Information Systems (NIS) and the General Data Protection Regulation (GDPR). This has set the first guidelines to protect against cyberattacks. For the energy industry, this raises many questions. For instance, is the energy infrastructure different from other sectors in terms of cybersecurity? What are the recommended cybersecurity measures for the energy infrastructure? Priorities and recommended actions have been proposed in Ref. [26], which are summarized in Table 3.

Table 3 Strategic priorities, areas, and recommended actions [26].

Strategic priorities	Strategic areas	Areas of actions
(I) Threat and risk management system	European threat and risk landscape and treatment	(1) Identification of essential services
	Identification of operators of essential services	(2) Risk analysis and treatment
	Best practice and information exchange	(3) Framework of rules for a regional cooperation
	Foster international collaboration	(4) Vulnerabilities disclosure for the energy sector
(II) Cyber response framework	Cyber response framework	(5) Define cyber response framework
	Crisis management	(6) Implement regional cooperation
(III) Improve cyber resilience	European cybersecurity maturity framework	(7) Establish a European cybersecurity maturity framework for energy
	Integrity framework for components	(8) Establish a contractual Public Private Partnership (cPPP) for supply chain integrity
	Best practice	(9) Foster European and International Collaboration
	Awareness campaign	
(IV) Capacity and competences	Capacity and competence	(10) Capacity and competence buildup

Cyber-physical attacks on the energy infrastructure are occurring more frequently and are not only focused on preventing electricity supply but also on all kinds of disruptions: from broken transmission lines to short circuits. Two elements increase the risk of attacks: (i) the increasing use of digital devices and advanced ICT, and (ii) the interconnectivity between different systems in smart grids. For example, as electricity substations are being modernized, new digital equipment is replacing their analog counterparts. Advanced monitoring devices (e.g., smart meters and PMUs), smart inverters, and charging stations are also being widely deployed rapidly in the distribution network, generating a high volume of data in high time resolution (i.e., high velocity) and hence making the electricity network more susceptible to cyberattack.

Electricity grid operators are responsible for anticipating possible cyber-physical attacks. Adequate knowledge of smart grids vulnerabilities can mitigate the impact of cyber-physical attacks. Several potential vulnerabilities of smart grids have been identified in prior research [27, 28]. Those can be summarized as follows:

1. *User privacy.* The end-customer information is a very sensitive parameter and must be always secured.
2. *High number of access points.* Network energy supply and demand control devices are very numerous and they all must be protected.
3. *Physical security.* Since smart grids serve as an interconnection between remote devices and systems, local control systems are required to guarantee the security of those devices.
4. *Frequent upgrading of network components.* System upgrades are a weak point.
5. *Preventing cascading failures.* It must be ensured that the impact of cascading failures in case of cyber-physical attacks or natural disasters is prevented or mitigated in the energy network. If critical devices whose output is an input to other devices are failed, data loss or cascading physical damage can cause the whole network to collapse.
6. *Dissimilarity between devices.* Communication gaps and incompatibilities lead to system security holes.
7. *Use of internet protocol (IP) and commercial hardware and software.*
8. *Increasing number of stakeholders in the energy network.*

3 Potential solutions and technologies for energy system resilience

The electricity industry and practitioners cannot effectively address the consequences of natural disasters on the electricity grid without having digital tools and system design methodologies. ICT and digital tools could help anticipate potential disasters, improving the flexibility of supply, and mitigating the risk of power failure. In this section, two main promising and complementary solutions for mitigating the impact of natural disasters and cyber-physical attacks on electricity grids are discussed, namely: (i) microgrids and (ii) DT.

3.1 Microgrids

The microgrid concept refers to a set of loads (e.g., households and EVs), distributed energy generation (e.g., solar PV panels or wind turbines), and energy storage systems (e.g., batteries and EVs), operating as a separate controllable system supplying power to its local area. Microgrids can be seen as intelligent small-scale energy systems with two particular modes of operation: stand-alone and grid-connected mode. They integrate power plants capable of satisfying local power demand of mainly critical load and possibly reinjecting surplus energy back to the main grid [29, 30]. The microgrid covers smart metering, sensors, and smart inverters capable of measuring and monitoring parameters such as power, voltage, and current. In addition, the communication infrastructure is needed to enable the components to exchange information and commands quickly and reliably. It also incorporates smart terminations and appliances capable of communicating their status and accepting commands to adjust and control their performance and service level based on user and/or utility requirements [30]. Microgrids can guarantee the continuity of energy supply during a natural disaster, especially for critical load, hence increasing their resilience to cyber-physical attacks. An example of microgrid system components is illustrated in Fig. 1.

Further, the electrification of residential, transport, heating, and agriculture, among other sectors, will present prospects of capitalizing on the flexibility of microgrids. Digitalization provides unique opportunities for smarter management of microgrids in extreme events. Such smart management can include switching off nonessential load when the network is stressed while supplying essential demand. The supply of essential loads during emergencies will also be enabled by distributed energy resources (DERs)

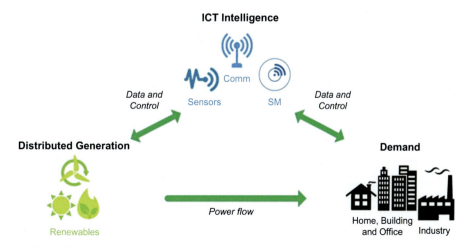

Fig. 1 Microgrid system layout.

and virtual storage capabilities from demand (e.g., battery units from EVs). This would increase the resilience of supply, as energy consumers will have their essential load supplied during high-impact events, including wildfires and other natural disasters.

Electricity grids are undergoing numerous fundamental changes and transformations [31]. The number of DERs in the system, including renewable energy generation units, EVs, charging stations, and energy storage assets, is expected to increase even further in the coming years. In line with this, the complexity of coordination and control will also increase, as will the number and variety of participants in electricity systems [32]. In parallel, electricity grids and their surrounding systems will be increasingly decentralized. Instead of today's monolithic, hierarchical, and controlled network, a modern network would be intelligent, agile, and composed of different segments that could be isolated and operate independently [31]. Local DERs and grid infrastructure are managed by smart energy management systems (EMSs) and many actors can trade energy through local electricity markets (LEMs) [33]. Also, larger grid assets such as solar farms, wind turbines, and industrial electric assets must be integrated in a similar fashion. To enable such novel energy systems, advanced ICT architecture and data management are required to house control algorithms, and to connect the different actors and owners of energy assets.

Microgrids could enable faster system restoration. The work in Ref. [34] explores the use and consequences of microgrids during natural disasters, focusing on the improvement of electricity availability by distributed generators and local energy storage. Microgrids make it possible to increase electricity availability compared to the current distribution grid. Many strategies are presented to ensure the flexible performance of the microgrid. As an example, Fig. 2 highlights the different strategies according to the time interval, predisaster, during disaster, and postdisaster.

3.2 Digital twins

Real-time monitoring and control of electricity grid components and managing the platforms and data for keeping actual energy systems operating effectively are becoming more fundamental, as does mitigating associated cyber-physical threats and data privacy risks, especially with the increasing share of DERs [36]. Several cases arise in which the possibilities and advantages of digitalization in smart grids started to become more visible. Big data analytic [10, 37] contributes to asset management, fault detection, load or renewable energy forecasting, and energy optimization, and efficiency. The work in Ref. [38] used high-resolution and large-scale data on failure and recovery to study failure impact and recovery patterns of electricity grids across multiple US service regions. Blockchain is another emerging ICT that can assist in securing transactions, data exchanges, and decentralized decision-making in the energy system, as well as optimizing power flow and energy trading [39–41]. Blockchain eliminates the need for a central controlling party on which all other actors must depend [32], which in turn reduces vulnerability against

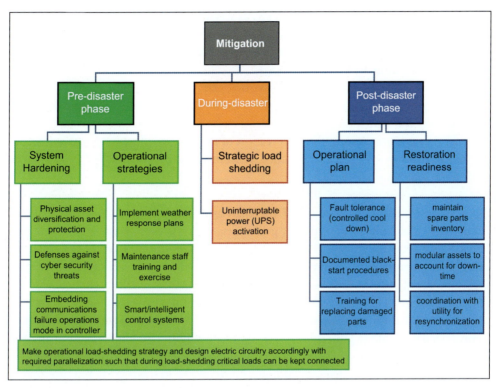

Fig. 2 Infrastructure and operational mitigation strategies for a resilient microgrid [35].

failures and cyber-physical attacks, increases transparency, and prevents potentially problematic misuse of market power by centralized actors. Internet of things (IoT) has been widely deployed in power systems to enhance flexibility, price-based demand response, and adaptive power based on consumer needs [42]. Further, the possibility to use DT as a means for creating a virtual representation of the grid, simulating a variety of scenarios, controlling decisions based on specific goals, and/or forecasting key operation variables, such as power demand and generation. A DT that resembles the physical grid makes it possible to perform experimentation and assess different scenarios that would not be possible to implement on the physical grid. DT could be empowered by artificial intelligence (AI) to detect cyberattacks, develop defense strategies, and support the human operators working in the monitoring and control rooms.

3.2.1 Digital twin for smart grids resilience
DT represents a promising technology to achieve the observability, optimal operation, sustainability, and resilience requirements of smart grids. DT was originally defined as

digital informational constructs about a physical system that could be created as an entity on its own for solving complex systems problems [43]. The definition has been extended to include more aspects. For instance, the work in Ref. [44] provides five main aspects of DT, which are summarized as follows:

- *A physical part*, that is the basis of the virtual part;
- *A virtual part*, that supports the simulation, monitoring, decision-making, and control of the physical part;
- *Data*, that is a prerequisite for the creation of new knowledge;
- *Services*, that can enhance convenience, reliability, and productivity of an engineered system; and
- *Connections*, that bridge all of the above.

There exist several applications of DT for smart grids such as the optimal design of electricity networks, simulation of power grid faults, virtual power plants (VPP), and intelligent equipment monitoring [12]. For instance, for monitoring parts of the electricity grid, the evaluation of the state of a power transformer using a DT is proposed in Ref. [45]. A DT for real-time power flow monitoring based on artificial neural networks (ANN) is proposed in Ref. [46], and a VPP prototype is developed in Ref. [47]. DT can also help advance asset management, field operations and decision support, and collaborative decision-making [48]. As examples of the possibilities for better decision-making, the opportunity to adapt a DT for power system control centers is discussed in Ref. [49]. These examples cover a wide range of applications for DT in smart grids, but more can be found in the literature, such as the review in Ref. [50]. Taking into account the advantages, DT brings to enhance the resilience of electricity grids, designing a DT for smart grids could be useful for advancing the understanding of this technology in the context of a sustainable energy transition in energy systems.

3.2.2 Digital twin design

Several and different DT architectures have been designed by industry and academia to demonstrate, develop, and introduce platforms supporting the DT concept [51]. Defining static system properties and architecture are parts of building a DT scheme as proposed in Ref. [52]. In the initial stage, there is the need to model the static properties of the system, which implies defining objectives and functional requirements as well as process planning and system requirements. The process planning is designed to clarify the process requirements, the model selection, and how data will be exchanged. As part of the architecture, there is no consensus on the number of properties and components that the design should have.

Based on the proposed DT framework in Ref. [53], it is possible to build either a control model for a DT based on a general system design approach or a technical model for their implementation. An example of the control model based on a general system design approach can be seen in Fig. 3.

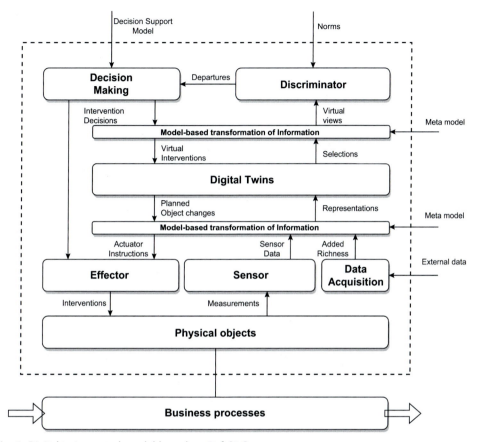

Fig. 3 Digital twin control model based on Ref. [54].

The control cycle starts with measuring the object system's state by the sensor function and with acquiring relevant external data [54]. The data are then translated into a virtual representation of the controlled object system (model-based transformation) on the basis of a meta model. The DT includes all information relevant to the supported purposes of usage (i.e., control objectives) as specified in a meta model. Depending on a specific purpose of usage, a virtual view may then filter irrelevant information and present it in such a way that it can be processed optimally by specific users (model-based transformations on the basis of a meta model). The next control function is the decision–making function, which compares a virtual view of the object system with a specific control norm. Afterward, the decision-making function selects appropriate interventions for deviations based on its decision support model, similar to conventional control systems. Lastly, the selected intervention is communicated with the effect or function, either directly or via the DT using remote actuator systems.

3.3 Solutions for cyberattacks

The strategy to defend against cyberattacks in electricity grids must contemplate security measures for (i) devices, (ii) the grid, (iii) workers and human operators, and (iv) societal awareness and policies [55].

Given that in cybersecurity, the goals are confidentiality, integrity, and availability, in the energy sector, the priority among these three depends on the specific applications. In power generation, availability and integrity are the most important. Alteration or delay of data could lead to misconfiguration of a device, which could ultimately affect system reliability. When analyzing measurement systems, confidentiality becomes the most important goal (i.e., handling sensitive data).

Several researchers developed initiatives for quickly identifying cyberattacks on electricity grids and designing actuation protocols to reduce the impact. Some studies focused on the prevention of cyberattacks by injecting data into the communication network [56] to identify potential damage that can be generated (e.g., tripping of protection relays, line disconnections, partial blackout, among others). Others implement IoT-based DT for the resilience of microgrid against cyberattacks [57].

4 Challenges and future outlook

The main challenges in implementing the above-mentioned solutions to enhance the resilience of electricity grids are listed and discussed here. This could be summarized as technology-related challenges, management, data, and governance.

4.1 Electrification and distributed generation

Grid operators, influenced by risk-averse end consumers, prefer a more stable and predictable grid operation, even if that would entail higher costs. Replacing fossil fuels with renewable energy sources (wind and solar) and the increased electrification in various sectors (i.e., industry, transport, heat, generation, and the built environment) implies an increase in electricity demand in the future. In addition, renewable energy sources are intermittent and nondispatchable by nature. This requires more flexibility from the demand side in contrast to traditional controllable and flexible fossil-fuel-based power generation. This also means that dealing with uncertainty can be a major issue in leveraging DERs for power systems resilience enhancement.

The energy transition and digitalization also mean a change in the structure of the electricity grid: from a centralized grid supplied by large power plants to a decentralized grid with small-scale power generation and a huge number of DERs. This transition poses new challenges for grid management and requires system design methodologies to accommodate the rapid changes in the energy system.

4.2 Incorporating energy storage solutions

Energy storage technologies are essential in improving the dispatchability of stochastic and intermittent renewable energy sources. For instance, in microgrids, DERs and localized energy storage technologies are controlled by an EMS that can provide different applications, such as flexibility and self-consumption from renewables [58, 59]. Hence, energy storage solutions can further enhance the resilience of microgrids and electricity grids in general.

The challenges that battery storage systems present today are the battery pack and cell prices (i.e., especially for lithium-ion), efficiency (i.e., which depends on the battery technology used, see Ref. [58]), lifetime (i.e., which depends on the battery technology used and number of charging cycles, see Ref. [58]), and criticality of raw materials used (lithium, nickel, and cobalt [60]). For instance, the production of lithium is associated with additional supply risks as its production is concentrated in a few countries. In fact, these challenges are interrelated, as the price of materials is rising, with demand far outstripping supply and while having very limited supply sources. On the other hand, the efficiency of batteries is related to the metal used, and the efficiency itself again affects the price.

The use of battery storage technologies in power systems and microgrids might be associated with economic challenges and material supply risks, especially when transitioning toward efficient lithium-based battery electricity storage. Policymakers and energy system operators and planners must account for these challenges and consider alternative electricity storage technologies in their portfolio, thus, not only focusing on battery technologies with high supply risks. On the other hand, the use of lithium-based batteries could be prioritized to parts of the electricity grid that present the highest criticality.

4.3 Coordination and management

A major challenge for grid management is to adapt to decentralized and coordinated supply and demand. The proportion of electricity generated by renewable energy sources is predicted to rise from 25% to more than 50% by 2030. Simultaneously, electricity must also be produced and delivered in sufficient quantities when there is no wind or sun. Electricity market design needs to be improved to accommodate the challenges imposed by renewable energy generation, electrification, and increasing prices of natural gas. It has to attract investment in DERs, such as energy storage, that can compensate for variable energy production. The market must likewise provide the right incentives for consumers to become more proactive and contribute to maintaining a stable and resilient electricity system.

4.4 Data

A vast amount of data is generated at different locations along the electricity grid. This data can be fully exploited using algorithms to monitor consumption, forecast generation and consumption, and control the grid, among others. The information obtained after processing this data helps reduce grid maintenance cost and anticipating potential threads and failures [61]. Developing preventive measures against cyber-physical attacks requires having a complete picture and high-resolution data of all parts of the electricity grid. Successful DT implementation requires having adequate data for robust and realistic scenario generation. This is lacking and although the data might exist, it is usually scattered over different places and owned by different stakeholders. To address this challenge, collaborative efforts from utilities, grid operators, end users, and researchers are required. Besides, the volume and velocity of big data in electricity grids challenge the ability to process it in a short period of time, whereas the variety of big data presents a further challenge to data integration. Having high-quality and trustworthy data (i.e., data veracity) can largely affect performing accurate analysis and decision-making. This makes data preprocessing a necessity for big data analytics, as missing data or incorrect data cannot be ignored. Data privacy and data protection issues must be addressed to realize the promise of big data in smart grid applications.

4.5 Layered architecture

We propose that new paradigms of the electricity grid can be conveniently described as having different dimensions or layers, which are shown in Fig. 4. The dimensions can help in better understanding the electricity grid and determining the actors and responsibilities for each part of the grid, hence improving the efficiency of actions taken under threats. This layered architecture could also be considered in the design of a DT of the electricity grid.

First, there is the *physical infrastructure*, which includes grid infrastructure and tangible physical energy assets, such as generation units and storage systems. To guide and control energy flows and respect grid constraints and limitations within the physical infrastructure, there is a need for a *digital infrastructure*, where measurement, communication, and control systems reside, big data are collected, processed, and analyzed, and information is exchanged. The digital infrastructure resides above the layer of physical infrastructure. It serves to coordinate power flows in the physical layer and enable data-driven optimal decision support. Above the digital infrastructure, there is the *economic infrastructure*, where the different actors enter into agreements and cooperation on the conditions for the exchange of electricity or electricity services. These economic aspects and purposes guide the underlying cyber-physical systems, and include market mechanisms, trading strategies, and exchange of services among different actors. Finally, we consider the fourth layer of *governance*. In this layer, agreements of governance between the different actors establish which actors operate, which part of the infrastructure, and what their mutual responsibilities are. It includes regulations, legal agreements, and the different actors in the system and the relationships between them.

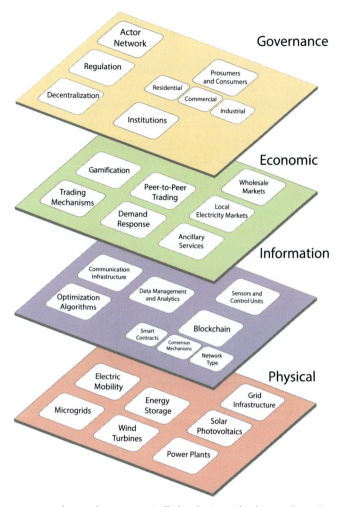

Fig. 4 The different system layers that can typically be distinguished in modern electricity grids. Each layer shows a selection of concepts that are commonly encountered in this domain.

5 Conclusions

This chapter started by describing the key sources of criticality for the energy infrastructure: natural disasters and human-caused disruptions. The former has a severe impact on infrastructure and the damage caused is shown to be dependent on the natural disaster. The severity of the latter is difficult to quantify, as they can cause small or large damage depending on the attacker's goals. In any case, it should be stressed that electricity grids are critical because they are essential for other infrastructures. Utility managers and decision-makers can find in this chapter the main challenges that the electricity grid is facing nowadays to mitigate climate change, namely: (i) the increase in intermittent and stochastic renewable energy production (e.g., wind and solar), (ii) the increasing cost of energy, and (iii) the integration of clean and distributed energy technologies and ICT.

From the studies reviewed, two main solutions were found to be most effective in increasing the resilience of electricity grids: microgrids and DT. The concept of microgrids encompasses all renewable electricity generation points and energy storage systems that can guarantee electricity supply and increase the resilience of the electricity grid during natural disasters or cyber-physical attacks. DT, on the other hand, is a promising solution to generate a well-calibrated model on which real-time monitoring and control of all grid components can be performed. This tool also makes it possible, based on historical and/or synthetic data, to quantify the potential effects of any critical phenomenon. Finally, challenges may be the largest in the electrification of microgrids, coordinating renewable stochastic energy production sources, obtaining, and managing all the data generated along the grid and incorporating energy storage solutions. This is a question of governance and politics that should be considered in future studies.

References

[1] C.R. Hudson, A.B. Badiru, Energy systems, in: Operations Research Applications, 2008, https://doi.org/10.4324/9781003156994-5. 5-1–5-30.

[2] E.E. Michaelides, Alternative Energy Sources, Springer, Berlin, Heidelberg, 2012, https://doi.org/10.1007/978-3-642-20951-2.

[3] United Nations, Sustainable development goals—energy access, 2021. https://sdgs.un.org/goals/goal7.

[4] R.L. Arantegui, A. Jäger-Waldau, Photovoltaics and wind status in the European Union after the Paris Agreement, Renew. Sustain. Energy Rev. 81 (2018) 2460–2471.

[5] C. Fetting, The European green deal, ESDN Report, December (2020).

[6] G.B. Giannakis, V. Kekatos, N. Gatsis, S.-J. Kim, H. Zhu, B.F. Wollenberg, Monitoring and optimization for power grids: a signal processing perspective, IEEE Signal Process. Mag. 30 (5) (2013) 107–128.

[7] X. Fang, S. Misra, G. Xue, D. Yang, Smart grid—the new and improved power grid: a survey, IEEE Commun. Surv. Tutor. 14 (4) (2011) 944–980.

[8] T. Alskaif, W.G.J.H.M. Van Sark, D1. 4-Smart grid roll-out and access to metering data: state-of-the-art, PARENT (PARticipatory platform for sustainable ENergy managemenT) Initiative (2016).

[9] N. Brinkel, T. AlSkaif, W. van Sark, Grid congestion mitigation in the era of shared electric vehicles, J. Energ. Storage 48 (2022) 103806.

[10] M. Kezunovic, P. Pinson, Z. Obradovic, S. Grijalva, T. Hong, R. Bessa, Big data analytics for future electricity grids, Electr. Power Syst. Res. 189 (2020) 106788.

[11] L. Visser, T. AlSkaif, W. van Sark, Operational day-ahead solar power forecasting for aggregated PV systems with a varying spatial distribution, Renew. Energy 183 (2022) 267–282.

[12] H. Pan, Z. Dou, Y. Cai, W. Li, X. Lei, D. Han, Digital twin and its application in power system, in: 2020 5th International Conference on Power and Renewable Energy (ICPRE), 2020, pp. 21–26, https://doi.org/10.1109/ICPRE51194.2020.9233278.

[13] G. Liang, S.R. Weller, J. Zhao, F. Luo, Z.Y. Dong, The 2015 Ukraine Blackout: implications for false data injection attacks, IEEE Trans. Power Syst. 32 (4) (2016) 3317–3318.

[14] Y. Wang, C. Chen, J. Wang, R. Baldick, Research on resilience of power systems under natural disasters—a review, IEEE Trans. Power Syst. 31 (2) (2015) 1604–1613.

[15] R. Moreno, M. Panteli, P. Mancarella, H. Rudnick, T. Lagos, A. Navarro, F. Ordonez, J.C. Araneda, From reliability to resilience: planning the grid against the extremes, IEEE Power Energ. Mag. 18 (4) (2020) 41–53.

[16] L. Xu, Q. Guo, Y. Sheng, S.M. Muyeen, H. Sun, On the resilience of modern power systems: a comprehensive review from the cyber-physical perspective, Renew. Sustain. Energy Rev. 152 (2021) 111642.

[17] G. Castañeda-Garza, G. Valerio-Ureña, T. Izumi, Visual narrative of the loss of energy after natural disasters, Climate 7 (10) (2019) 118.

[18] D.W. Wanik, E.N. Anagnostou, M. Astitha, B.M. Hartman, G.M. Lackmann, J. Yang, D. Cerrai, J. He, M.E.B. Frediani, A case study on power outage impacts from future hurricane sandy scenarios, J. Appl. Meteorol. Climatol. 57 (1) (2018) 51–79.

[19] A. Bhattacharyya, S. Yoon, M. Hastak, Optimal strategy selection framework for minimizing the economic impacts of severe weather induced power outages, Int. J. Disaster Risk Reduct. 60 (2021) 102265.

[20] S. Mukherjee, R. Nateghi, M. Hastak, A multi-hazard approach to assess severe weather-induced major power outage risks in the US, Reliab. Eng. Syst. Saf. 175 (2018) 283–305.

[21] D. Deka, S. Vishwanath, R. Baldick, Topological vulnerability of power grids to disasters: bounds, adversarial attacks and reinforcement, PLoS One 13 (10) (2018) e0204815.

[22] Z. Li, M. Shahidehpour, F. Aminifar, A. Alabdulwahab, Y. Al-Turki, Networked microgrids for enhancing the power system resilience, Proc. IEEE 105 (7) (2017) 1289–1310, https://doi.org/10.1109/JPROC.2017.2685558.

[23] European Commission, Joint Research Centre, E. Krausmann, G. Karagiannis, Z. Turksezer, S. Chondrogiannis, Power Grid Recovery After Natural Hazard Impact, Publications Office, 2017, https://doi.org/10.2760/87402.

[24] A.A. Ganin, E. Massaro, A. Gutfraind, N. Steen, J.M. Keisler, A. Kott, R. Mangoubi, I. Linkov, Operational resilience: concepts, design and analysis, Sci. Rep. 6 (1) (2016) 1–12.

[25] M. Weiss, M. Weiss, An assessment of threats to the American power grid, Energy Sustain. Soc. 9 (1) (2019) 1–9.

[26] Energy Expert Cyber Security Platform, Cyber security in the energy sector, EECSP Rep. (February) (2017) 74.

[27] S.A. Yadav, S.R. Kumar, S. Sharma, A. Singh, A review of possibilities and solutions of cyber attacks in smart grids, in: 2016 International Conference on Innovation and Challenges in Cyber Security (ICICCS-INBUSH), IEEE, 2016, pp. 60–63.

[28] I.L.G. Pearson, Smart grid cyber security for Europe, Energy Policy 39 (9) (2011) 5211–5218.

[29] N. Hatziargyriou, H. Asano, R. Iravani, C. Marnay, Microgrids, IEEE Power Energ. Mag. 5 (4) (2007) 78–94.

[30] H. Farhangi, The path of the smart grid, IEEE Power Energ. Mag. 8 (1) (2010) 18–28.

[31] P. Fox-Penner, Power After Carbon: Building a Clean, Resilient Grid, vol. 5, Harvard University Press, 2020, pp. 1–8.

[32] A. Ahl, M. Yarime, K. Tanaka, D. Sagawa, Review of blockchain-based distributed energy: implications for institutional development, Renew. Sustain. Energy Rev. 107 (February) (2019) 200–211, https://doi.org/10.1016/j.rser.2019.03.002.

[33] J.L. Crespo-Vazquez, T. AlSkaif, Á.M. González-Rueda, M. Gibescu, A community-based energy market design using decentralized decision-making under uncertainty, IEEE Trans. Smart Grid 12 (2) (2021).

[34] A. Kwasinski, V. Krishnamurthy, J. Song, R. Sharma, Availability evaluation of micro-grids for resistant power supply during natural disasters, IEEE Trans. Smart grid 3 (4) (2012) 2007–2018.

[35] S. Mishra, K. Anderson, B. Miller, K. Boyer, A. Warren, Microgrid resilience: a holistic approach for assessing threats, identifying vulnerabilities, and designing corresponding mitigation strategies, Appl. Energy 264 (2020) 114726.

[36] International Energy Agency, World Energy Outlook 2021, 2021. https://www.iea.org/reports/world-energy-outlook-2021.

[37] Y. Zhang, T. Huang, E.F. Bompard, Big data analytics in smart grids: a review, Energy Inform. 1 (2018) 8, https://doi.org/10.1186/s42162-018-0007-5.

[38] C. Ji, Y. Wei, H. Mei, J. Calzada, M. Carey, S. Church, T. Hayes, B. Nugent, G. Stella, M. Wallace, Large-scale data analysis of power grid resilience across multiple US service regions, Nat. Energy 1 (5) (2016) 1–8.

[39] M. Mylrea, S.N.G. Gourisetti, Blockchain for smart grid resilience: exchanging distributed energy at speed, scale and security, in: 2017 Resilience Week (RWS), September, IEEE, 2017, pp. 18–23, https://doi.org/10.1109/RWEEK.2017.8088642.

[40] T. Alskaif, J.L. Crespo-Vazquez, M. Sekuloski, G. van Leeuwen, J.P.S. Catalao, Blockchain-based fully peer-to-peer energy trading strategies for residential energy systems, IEEE Trans. Indus. Inform. 18 (2022) 231–241, https://doi.org/10.1109/TII.2021.3077008.

[41] G. van Leeuwen, T. AlSkaif, M. Gibescu, W. van Sark, An integrated blockchain-based energy management platform with bilateral trading for microgrid communities, Appl. Energy 263 (2020) 114613, https://doi.org/10.1016/j.apenergy.2020.114613.

[42] C. Bordin, A. Håkansson, S. Mishra, Smart energy and power systems modelling: an IoT and cyber-physical systems perspective, in the context of energy informatics, Procedia Comput. Sci. 176 (2020) 2254–2263, https://doi.org/10.1016/j.procs.2020.09.275.

[43] M. Zhou, J. Yan, D. Feng, Digital twin and its application to power grid online analysis, CSEE J. Power Energy Syst. 5 (2019), https://doi.org/10.17775/CSEEJPES.2018.01460.

[44] F. Tao, H. Zhang, A. Liu, A.Y.C. Nee, Digital twin in industry: state-of-the-art, IEEE Trans. Indus. Inform. 15 (2019) 2405–2415, https://doi.org/10.1109/TII.2018.2873186.

[45] Y. Yang, Z. Chen, J. Yan, Z. Xiong, J. Zhang, H. Yuan, Y. Tu, T. Zhang, State evaluation of power transformer based on digital twin, in: 2019 IEEE International Conference on Service Operations and Logistics, and Informatics (SOLI), November, IEEE, 2019, pp. 230–235, https://doi.org/10.1109/SOLI48380.2019.8955043.

[46] X. He, Q. Ai, R.C. Qiu, D. Zhang, Preliminary exploration on digital twin for power systems: challenges, framework, and applications, 2019. http://arxiv.org/abs/1909.06977.

[47] S. Bianchi, A.D. Filippo, S. Magnani, G. Mosaico, F. Silvestro, VIRTUS project: a scalable aggregation platform for the intelligent virtual management of distributed energy resources, Energies 14 (2021) 3663, https://doi.org/10.3390/en14123663.

[48] P. Palensky, M. Cvetkovic, D. Gusain, A. Joseph, Digital twins and their use in future power systems, Digit. Twin (2021), https://doi.org/10.12688/digitaltwin.17435.1.

[49] C. Brosinsky, D. Westermann, R. Krebs, Recent and prospective developments in power system control centers: adapting the digital twin technology for application in power system control centers, in: 2018 IEEE International Energy Conference (ENERGYCON), June, IEEE, 2018, pp. 1–6, https://doi.org/10.1109/ENERGYCON.2018.8398846.

[50] N. Bazmohammadi, A. Madary, J.C. Vasquez, H.B. Mohammadi, B. Khan, Y. Wu, J.M. Guerrero, Microgrid digital twins: concepts, applications, and future trends, IEEE Access 10 (2022) 2284–2302, https://doi.org/10.1109/ACCESS.2021.3138990.

[51] R. Minerva, G.M. Lee, N. Crespi, Digital twin in the IoT context: a survey on technical features, scenarios, and architectural models, Proc. IEEE 108 (2020) 1785–1824, https://doi.org/10.1109/JPROC.2020.2998530.

[52] M. Segovia, J. Garcia-Alfaro, Design, modeling and implementation of digital twins, Sensors 22 (2022), https://doi.org/10.3390/s22145396.

[53] C. Verdouw, B. Tekinerdogan, A. Beulens, S. Wolfert, Digital twins in smart farming, Agr. Syst. 189 (2021) 103046, https://doi.org/10.1016/j.agsy.2020.103046.

[54] C.N. Verdouw, A.J.M. Beulens, H.A. Reijers, J.G.A.J. van der Vorst, A control model for object virtualization in supply chain management, Comput. Ind. 68 (2015) 116–131, https://doi.org/10.1016/j.compind.2014.12.011.

[55] T. Krause, R. Ernst, B. Klaer, I. Hacker, M. Henze, Cybersecurity in power grids: challenges and opportunities, Sensors 21 (18) (2021) 6225.

[56] V.S. Rajkumar, M. Tealane, A. Ştefanov, A. Presekal, P. Palensky, Cyber attacks on power system automation and protection and impact analysis, in: 2020 IEEE PES Innovative Smart Grid Technologies Europe (ISGT-Europe), IEEE, 2020, pp. 247–254.

[57] A. Saad, S. Faddel, T. Youssef, O.A. Mohammed, On the implementation of IoT-based digital twin for networked microgrids resiliency against cyber attacks, IEEE Trans. Smart Grid 11 (6) (2020) 5138–5150.

[58] T. Terlouw, T. AlSkaif, C. Bauer, W. van Sark, Multi-objective optimization of energy arbitrage in community energy storage systems using different battery technologies, Appl. Energy 239 (2019) 356–372.

[59] T. AlSkaif, A.C. Luna, M.G. Zapata, J.M. Guerrero, B. Bellalta, Reputation-based joint scheduling of households appliances and storage in a microgrid with a shared battery, Energy Build. 138 (2017) 228–239.

[60] T. Terlouw, X. Zhang, C. Bauer, T. Alskaif, Towards the determination of metal criticality in home-based battery systems using a life cycle assessment approach, J. Clean. Prod. 221 (2019) 667–677.

[61] J.A.P. Lopes, A.G. Madureira, M. Matos, R.J. Bessa, V. Monteiro, J.L. Afonso, S.F. Santos, J.P.S. Catalão, C.H. Antunes, P. Magalhães, The future of power systems: challenges, trends, and upcoming paradigms, Wiley Interdiscip. Rev. Energy Environ. 9 (3) (2020) e368.

CHAPTER 14

Deep learning for agricultural risk management: Achievements and challenges

Saman Ghaffarian[a,b,c], Yann de Mey[a], João Valente[b], Mariska van der Voort[a], and Bedir Tekinerdogan[b]
[a]Business Economics Group, Wageningen University & Research, Wageningen, The Netherlands
[b]Information Technology Group, Wageningen University & Research, Wageningen, The Netherlands
[c]Institute for Risk and Disaster Reduction, University College London, London, United Kingdom

1 Introduction

Precision agriculture and digital twins are examples of new industrial paradigms that increasingly gain more attention in the agriculture sector. These new technological concepts are designed to support effective and timely farm management [1,2]. Data collection and processing are one of the pillars of such systems. With rapid developments of these concepts and technologies and providing real-time site data/big data, the need for automatic and effective data processing methods is increased [3,4]. Nowadays, machine learning, and in particular deep learning, methods are the state-of-the-art data processing tools in different industrial and scientific fields as well as farm management applications [5–9].

Farming systems and communities facing different challenges including economic and societal problems that may cause a risk [10]. Furthermore, the frequency and impact of those challenges have been escalated, e.g., due to rising pressure of climate change or trade wars and political boycotts [11]. Accordingly, farmers have to make decisions in different situations to maintain the functions and mitigate the impacts in the case of a risk. Adverse events, e.g., natural hazards, or any uncertainties, e.g., change in prices and costs, in an agricultural system may cause a risk. Evaluation of those risks is crucial to support the farmers as the managers to deal with them [12]. Depending on the aim and the type of the risk, various data and information can be collected and processed to support agricultural risk management (ARM). For instance, predicting the damages to crops due to an adverse event such as frost can help the farmer to be prepared for such events and reduce the possible damages and impacts [13]. Assessing the impact of disease and monitoring the status of the farm yield in such a situation can help the farmer to make decisions and reduce further impacts [14,15]. Hence, assessing and measuring risk in general or focusing on any type of risk at any stage of the agricultural cycle is an essential task in ARM [16].

Management and Engineering of Critical Infrastructures
https://doi.org/10.1016/B978-0-323-99330-2.00001-5

Machine learning (ML) has been used as an efficient data processing method in agricultural studies and different applications [7,17–20]. Recently, deep learning (DL) methods are proved to be state-of-the-art ML methods by producing accurate results for automated processing of the different data types in several science domains and agricultural studies [21–23]. DL methods also overcome other limitations of ML which is the need for feature engineering. DL has been successfully applied for risk analysis in different research domains such as computer/cybersecurity [24], disaster management [25,26], and agriculture [27]. Frequently used DL methods are convolutional neural networks (CNNs), which is the main method for image processing applications, and recurrent neural networks (RNN) for time series analysis. Several DL methods were used and developed for different applications of ARM. Here, the studies that used DL-based methods for ARM are called deep learning-based agricultural risk management (DL-ARM) studies. For example, self-attention CNN [28,29], transfer learning-based CNN [30–32], a CNN with residual connections [33,34], weakly supervised CNN [35,36], generative adversarial networks (GANs) [37], multimodels CNN [38], 3D explainable deep learning [39] were developed for plant/crop disease, pests detection, and disease severity estimations. In addition, RNN methods such as long-short term memories (LSTM) were employed for mounting behavior recognition [40], fouling and diarrhea of pigs [41], agricultural insurance risk assessment in the face of weather extremes [42], spatial–temporal prediction of drop disease severity [43], and energy, water, and food nexus modeling to support resilience assessments [44].

Advances in ARM were the focus of literature review studies in recent years that can be grouped as follows; (i) studies focused on one type of risk, e.g., plant disease detection [45]; (ii) investigating a specific application domain, e.g., economics and risk perception [46]; (iii) reviewing the methods employed to process a specific type of data, e.g., remote sensing [47]; (iv) and review the agricultural risk generally without investigating the used methods [48]. Accordingly, the use of DL methods for ARM has not yet been reviewed. Hence, the objective of this study is to elicit achievements and challenges in DL-ARM from different risk types, risk components, and used/developed DL techniques perspectives, and accordingly, provide insights and recommendations for future studies. To the best of our knowledge, this is the first systematic literature review study on the use of deep learning techniques for ARM. In addition, reviewing the methods used for all types of risks will provide insights from other disciplines to be used interchangeably. Traditional review papers summarize the published papers while a systematic literature review group and classify the papers to answer the predefined research questions. Hence, this review was conducted doing a systematic search of the publications in online databases and identify the relevant ones, then extract and synthesize the required data to respond to defined research questions.

The review is structured as follows. In Section 2, fundamental background information is provided, and basic ARM concepts and terms are defined. Section 3 explains the used systematic literature review structure in a step-by-step manner. Section 4 presents the results and discusses them in detail, and reveals the achievements and challenges. Further, the conclusions of the study are provided in Section 5.

2 Background

We define risk as the probability of occurrence of adverse events and/or any uncertainties that affect outcomes that matter. In the context of agriculture, adverse outcomes include lower incomes and yields, uncertain input/output prices, change in farm-related regulations and policies, farmer health problems, or change in the availability of financing sources [49]. Agricultural risks can be assessed according to the addressed risk types and the targeted risk phases/components. Agricultural risks can be classified into five groups based on their types [48]; (i) uncertainties in the natural growth process of agricultural products such as crops and livestock (production risks); (ii) changes in farm's financing sources and the operating cash flow (financial risks); (iii) unpredictable policy and regulation developments or changes generated by formal or informal institutions (institutional risks); (iv) problems related to individual farmers and farm households such as personal relationships and health problems (personal risks); and (v) unpredictable changes in prices, costs, and market access (market risks).

ARM can be considered as a cycle independent of risk types similar to what is defined in disaster risk management field (Fig. 1). The ARM cycle is usually comprised of four main components/phases: (i) response phase; actions are taken to reduce the further impact of a disturbance or an adverse event; (ii) recovery phase; actions are taken to return the farm to a normal condition; (iii) prevention-mitigation; actions are taken to prevent a disturbance and reducing its happening chance or limiting the diverse impacts of an

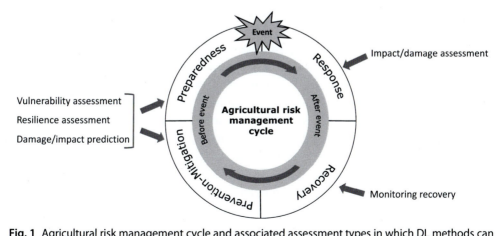

Fig. 1 Agricultural risk management cycle and associated assessment types in which DL methods can be used.

unavoidable event; (iv) preparedness; governments, individual farm holders, and other organizations develop plans to effectively anticipate and enhance the response to the impacts of likely, imminent or current events/disturbances. Two of the ARM cycle phases are related to post-event time, i.e., response and recovery, and two are related to pre-event time, i.e., prevention-mitigation and preparedness.

After an adverse event, the emerging mission is to assess the impact/damage of such an event to the farm in order to make effective decisions in the response phase and reduce further impacts. According to the initial impact assessment, the recovery process starts that may take a long time depending on the extent of the damage, farm, and risk types. Hence, this process needs to be monitored to assure the successful implementation of the plans. Assessing the current vulnerability and resilience level of the farm can provide crucial information to make plans toward mitigating risks and effective preparedness. Furthermore, modeling the risks and predicting the associated damages can provide information to enhance the preparedness of the farm. Accordingly, impact, vulnerability, and resilience assessments and predictive models are indispensable for ARM. Deep learning methods can contribute to any phase of the ARM cycle by processing the data in an automated manner for impact/damage assessment, e.g., by detecting the crop damages in the case of disease outbreaks [50], resilience and vulnerability assessments by extracting the associated indicators from data, e.g., crop yield and diversity [51], and predicting damages associated with any agricultural risk types, e.g., crop and plant disease prediction using different data sources such as climate data [52].

3 Methodology

Fig. 2 presents an overview of the review steps undertaken for the current study, adapted from Kitchenham et al. [53]. First, a set of research questions were defined according to the objective of this study (i.e., reviewing and investigating the DL-ARM). Then, the search scope and strategy were designed to define the time period and formulating the search string to find the relevant papers. The final search string was created after several trials and errors to find the relevant and remove the misleading keywords. A good search strategy brings the appropriate search results that will come to a successful conclusion in terms of sensitivity and precision rates. The final set of papers for systematic review was selected based on the exclusion criteria (EC). The ECs are defined to harness the most appropriate studies based on their contribution to the objective of this study and their quality. Rapid manual screening of the papers was performed to apply the ECs. Then, a data extraction strategy was developed to elicit the required information from each of the selected papers. In the final step, the extracted data was synthesized in data extraction form, and the results are presented and discussed. The following sections explain the execution of the steps in detail.

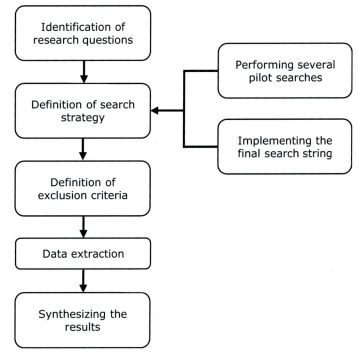

Fig. 2 The review protocol.

3.1 Research questions

Five main research questions are defined to address the objective of this study. The RQs are particularly selected to extract the state of the art and the interesting achievements in the publications as well as the remaining open questions and the challenges in this research domain. The further structured analysis in this study was built on the following RQs:

RQ1. Which types of agricultural risks were addressed with deep learning?

RQ2. What are the specific objectives in agricultural risk management that were addressed with deep learning?

RQ3. What are the adopted deep learning-based approaches for agricultural risk management?

  RQ3.1. What kind of deep learning tasks were addressed?

  RQ3.2. What kind of deep learning algorithms were used?

RQ4. What are the used data sets/types to address deep learning-based agricultural risk management?

RQ5. Which lessons can be derived from the adoption of deep learning for agricultural risk management?

RQ6. What are the existing research directions, achievements, and challenges in deep learning-based agricultural risk management?

3.2 Search strategy

Most of the systematic literature review studies have two attributes to define their search scopes; publication date and publication platforms. In this study, no limit was defined for the publication date and execute the search on the ISI Web of Knowledge online database, which indexes all the ISI journals and corresponding published papers, and thus, it includes the high-quality publications. The search query was applied to search in Title, Abstract, and Keywords of the papers. An automatic search was performed by executing the following search query on the search engine of the platform.

(("agri*" OR "farm*" OR "crop" OR "livestock" OR "dairy") AND ("risk" OR "resilien*" OR "damage" OR "impact" OR "recovery" OR "expos*" OR "vulnerability" OR "coping capacity*" OR "disease" OR "adaptive capacity") AND ("deep learning" OR "CNN" OR "convolutional neural networks"))

The search string was formulated to retrieve as much as the related publications to the defined objective and search scope. The search string comprises three primary parts that are separated by the term "AND". In the first part, the aim is to find the relevant publications in terms of their subjects including keywords for any agriculture type (e.g., dairy). In the second part, the aim is to find the relevant tasks including keywords for any type of risk (e.g., disease); in the third part, the aim is to find the papers that used relevant methods such as DL.

3.3 Study selection criteria

Initially, the selected papers were manually filtered, which were the result of executing the search query, to select the final list of the most suitable papers. General exclusion criteria and particularly defined ones according to the objective of this study were employed (Table 1). This was done based on screening the papers mostly reading their abstract and introductions sections.

Table 1 Exclusion criteria.

ID	Criterion
EC1.	Papers in which the full text is unavailable
EC2.	Papers are not written in English
EC3.	Papers are not aiming to directly contribute to agriculture risk management (ARM)
EC4.	Papers do not explicitly discuss the ARM or connect the study to ARM
EC5.	Papers do not directly use DL methods
EC6.	Papers do not validate the proposed study
EC7.	Papers that provide a general summary without a clear contribution
EC8.	Review and editorial papers

3.4 Data extraction

The data extraction step is to find out how the research questions can be answered for the selected papers. A preliminary data extraction form was created based on the initial screening of the papers in order to retrieve all the information required to answer the research questions. The preliminary data extraction form comprises a set of attributes such as identification of the study, publisher, publication year. The final data extraction form contains more details such as the specific objective addressed in the papers and the used DL algorithms and was created and filled after carefully reviewing the selected papers.

3.5 Data synthesis

Synthesizing the extracted data from the selected papers is the most important step of the systematic literature review study. In this step, the research questions are answered and the extracted data and the results are presented. Hence, the results are summarized and visualized, and the papers are grouped into defined classes for each research question. In addition, the results are discussed in detail and the important points are figured out from the selected studies. Finally, the current research directions, achievements, challenges, and recommendations for future studies are provided.

4 Results and discussions

In total, 203 papers were extracted using the search strategy, and then according to the exclusions criteria, 100 papers were selected for the detailed review. In this section, an overview of the main statistics about the selected papers is provided. Then, the detailed results associated with each research question are presented and discussed in the subsequent sections.

4.1 Overview of the reviewed studies

The publisher and journal names and the publication years of the selected 100 studies were extracted. Researchers started using deep learning for ARM in 2016, and accordingly, the selected papers were published from 2016 till 2020. Fig. 3 illustrates the year-wise distribution of the selected papers. There is a significant increase in the number of DL-ARM papers in 2019, which continued in 2020 considering that the search from the online database was conducted in August 2020. Elsevier, MDPI, IEEE, and Springer were published most of the papers in DL-ARM (Fig. 4). In total, 51 different journals published the papers. The most popular journal is "Computers and Electronics in Agriculture" journal with 24 publications while the second popular one is "IEEE Access" with 8 publications (Fig. 5). In addition, 37 journals have only published one of the selected papers. These statistics show the diversity of the perspectives and research

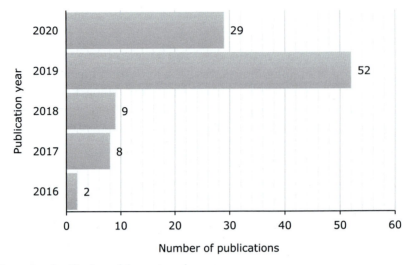

Fig. 3 Year-wise classification of the reviewed papers.

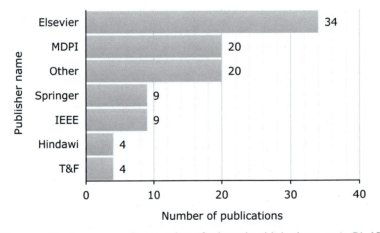

Fig. 4 Publishers and their corresponding number of selected published papers in DL-ARM.

domains in tackling and addressing the ARM using DL methods. This also shows that DL-ARM is a multidisciplinary and interdisciplinary field of study.

4.2 RQ1. Which types of agricultural risks were addressed with deep learning?

The selected papers are grouped based on their risk types, according to Komarek et al. [48] as explained in Section 2. Mainly, the production risks are addressed using DL

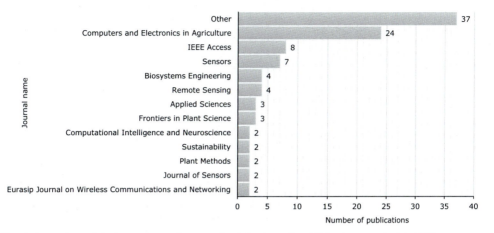

Fig. 5 Journals and their corresponding number of selected published papers in DL-ARM.

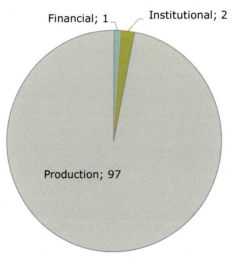

Fig. 6 The addressed agricultural risk types in the selected papers.

methods, and only two and one paper from the selected publications addressed financial and institutional risks (Fig. 6).

4.3 RQ2. What are the specific objectives in agricultural risk management that addressed with deep learning?

Initially, the papers were reviewed to extract the general objectives of the papers, and the following distinct objectives were detected. The objectives are mostly related to response phase of the ARM cycle presented in Section 2 by detecting damages. In addition, DL methods have been used to predict impacts and damages contributing to risk prevention–

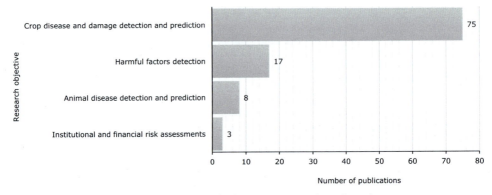

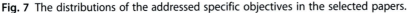

Fig. 7 The distributions of the addressed specific objectives in the selected papers.

mitigation and preparedness phases of the ARM cycle, while only a few papers addressed the ARM by conducting a general risk assessment. Accordingly, the papers were grouped to extract their frequency and distributions in the defined objectives.

A. Crop disease and damage detection and prediction

B. Harmful factors detection

C. Animal disease detection and prediction

D. Institutional and financial risk assessments

Fig. 7 shows the number of papers associated with each of the defined groups. Crop disease and damage detection and prediction with DL are predominantly studied in the publications. Harmful factors, e.g., pests, detections topic studied in 17 publications of the selected papers. Animal-based agriculture and institutional and financial risk assessments are only covered in 8 and 3 publications, respectively.

4.4 RQ3. What are the adopted deep learning-based approaches for agricultural risk management?

DL approaches can be grouped in two manners; firstly on the addressed DL task, secondly on the used DL algorithm. Accordingly, in this subsection, the main question is answered by investigating the addressed DL types, and employed DL algorithms, by answering the following subresearch questions.

4.4.1 RQ3.1. What kind of deep learning types are addressed?

Classification is the main applied task in DL-ARM studies, being addressed in 91% of the papers, whereas there are 8% of the selected papers conducted regression methods for processing the data. Only one paper used reinforcement learning for institutional and financial risk assessment. Classification methods are mostly (69 papers) employed to perform crop disease and damage detection and predictions (Fig. 8).

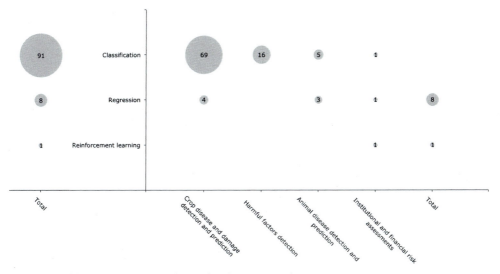

Fig. 8 The addressed DL types and specific objectives in the papers.

4.4.2 RQ3.2. What kind of deep learning algorithms are used?

Fig. 9 shows the number of papers employed in each DL algorithm. Accordingly, convolutional neural network (CNN) algorithm is the predominant DL method employed for ARM, which applied in 89 out of 100 reviewed papers [54–58]. The second most frequently used DL algorithm is recurrent neural networks (RNN) [43] including long-short term memories (LSTM) algorithm with six papers [41]. Generative adversarial networks (GANs) [37], DL-based regression [59], capsule networks (CapsNet) [60],

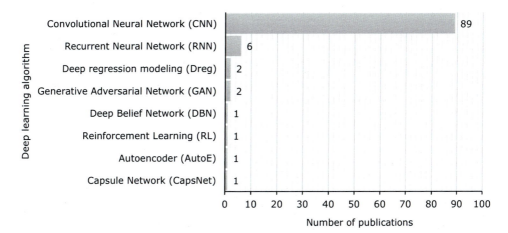

Fig. 9 The employed DL algorithms in the selected papers.

autoencoders (AutoE) [61], reinforcement learning (RL) [44], and deep belief networks (DBNs) [42] are the other used DL algorithms in the publications for ARM.

4.5 RQ4. What are the used data sets/types to address deep learning-based agricultural risk management?

Imaging sensors data including images, e.g., UAV [62], mobile phone images [63], and videos [40], are by far the most popular data sets utilized (90 papers) (Fig. 10). Nonsensor field data including fieldwork and survey data are the second main dataset with 9 papers used in their studies [14,64]. Climate data (e.g., climate and weather data) [42], nonimaging sensors data (e.g., sensor data for a dairy farm, GPS data) [65], and Earth data (e.g., hydrological and soil data) [44] are the other datasets used in the selected papers.

4.6 RQ5. Which lessons can be derived from the adoption of deep learning for agricultural risk management?

To have a structured review and associate the findings and limitations of the studies to the related specific objectives, the papers were reviewed within the defined groups in the previous research question (RQ2).

4.6.1 Crop disease and damage detection and prediction

As explained in Section 4.3, most of the selected papers fall in group A. They mainly focused on the development of DL methods to detect diseases in different crop types, e.g., different fruit types or general banana or general leaf disease detection for several crop types (Table 2).

A few studies conducted damage detection in the agricultural area due to cold [96], frost [13], and the wild boar [97] using a DL approach. Wang et al. [98] developed a DL

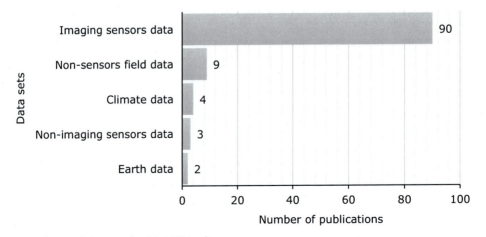

Fig. 10 The used datasets for DL-ARM in the papers.

Table 2 Addressed crop types with DL-based disease detection methods.

Crop type	References
Apple	[37]
Banana	[66]
Citrus	[67,68]
Coffee leaf	[65,69]
General leaf for several crop types	[29,30,54,60,70–74]
Groundnut	[75]
Maize	[58,76]
Mango	[77]
Millet	[32]
Oilseed	[78,79]
Onion	[80]
Other fruits	[81,82]
Pepper	[38]
Potato	[83–85]
Rice	[31,86,87]
Seasonal crops	[61]
Soybean	[88]
Sweet pepper	[89]
Tomato	[15,28,50,90,91]
Vine	[62,92]
Wheat	[93–95]

method to detect internal mechanical damages to blueberries. Furthermore, the DL methods are used to predict crop disease, i.e., general crop disease prediction [14], predict disease severity [43], and seasonal crop disease prediction [61]. All of the studies applied the DL methods on imaging sensor-based datasets including UAV [62,99], hyperspectral [39,100], plant leaf images [101,102], and mobile phone images [63,103].

Most of the developed DL methods in this group are based on CNN methods. Accordingly, several methods with different neural network designs are developed for crop disease detection (Table 3). Early studies in CNN for ARM employed well-known network architectural designs such as CaffeNet, AlexNet, and VGG16 and all demonstrated the superiority of CNN-based methods in such applications.

Afterward, in order to increase the accuracy and speed of the detection and prediction methods, more advanced networks are designed (Table 3) and other techniques are developed such as transfer learning [31,32,78], shallow [107], multimodal [38], multitask [69], attention embedded [28,29,108] CNNs. The latest versions of the CNN methods are mainly developed to increase the accuracy of the results and/or speed of the processing time for a specific crop type or in general for a specific image acquisition platform. Moreover, the capsule network (CapsNet) model as a promising method for image

Table 3 Used CNN architectures for crop disease and damage detection.

CNN architecture	References
3D CNN	[39]
AlexNet	[50,56,58,90]
CaffeNet	[102]
CycleGANs	[37]
Fast RCNN	[50]
GANs	[104]
GoogLeNet	[56,90]
LeNet	[62]
Mask RCNN	[105]
MaxNet	[78,79]
MobileNet	[66,103]
PDSE-S-2-Net	[106]
ResNet	[34,57,66,82]
ResNeXt	[98]
SoyNet	[88]
VGG16	[50,58]
YOLOv3	[67]

processing applications is developed to overcome limitations of the CNN methods, and in a recent study [60], it obtained a comparable performance to state-of-the-art CNN methods in image-based plant disease classification. Moreover, RNN and LSTM methods are used to predict diseases and damage to crops [43,109].

4.6.2 Harmful factors detection

Harmful factors such as pests and weeds may cause production loss in crop agriculture, and thus, detecting such factors can help the framer to make decisions to respond and further mitigate the risk. To do so, mainly CNN-based methods are developed and employed. Pests in crops are detected using different designs of the CNN methods such as GoogLeNet [110,111], AlexNet [110,112], ResNet [33,112], R-CNN [50,89], Inception [66], MobileNet [66], DenseNet [36], and hybrid methods [71,113] from field images. Weed detection methods are based on object detection (YOLOv3) [114], fully connected CNN classification (SegNet) [115], and transfer learning-aided CNN model [115].

Cotton spider mites can cause damages to cotton leaves and change their characteristics; accordingly, extracting them is important for cotton farmers. Yang et al. [116] used MobileNetV1 to detect such damages and obtain accurate results. In a different study, Carolina Geranium as a serious concern for strawberry farms is detected employing several DL methods, namely GoogLeNet, VGGNet, and DetectNet, and finally, the

DetectNet produced the highest accuracy value [117]. Furthermore, harmful spores are successfully detected using a hybrid approach consists of DeepLabv3 and constrained focal loss models [118].

4.6.3 Animal disease detection and prediction

Early disease detection and prediction are also crucial in animal agriculture. Recently, researchers started using DL methods to process data collected from dairy [119,120], pig [40,41,55], poultry [59,121], and swine [122] farms using different DL-based methods such as CNN, deep regression, RNN, and LSTM algorithms. For example, Li et al. [40] used mask R-CNN to detect mounting behavior of pigs, which is an early indicator of damage such as lameness and fractures, from an overlooked video camera data. They showed that their proposed method produced high accuracy in pig segmentation and mounting behavior detection. Pecking activity in poultry farms can be used as an indicator of early warning for cannibalism. An automatic CNN-based method is proposed to detect turkeys pecking activity from acoustic data [121]. In dairy cows, body condition scores are used for determining animal welfare, disease detection, and milk production performances. Sun et al. [120] developed a CNN model to automatically estimate the body condition score for dairy cows from images with a robust performance. In a different study, Augusta et al. [122] investigated the machine learning methods including DL to simulate diseases in the population of swine farms. This model allows the users to do a fast inference in estimating the parameters of the emerging epidemics. They showed that the DL model produces high-accuracy values even on a sparser population.

4.6.4 Institutional and financial risk assessments

Unpredictable changes in policy and regulations or uncertainties in farmers' financing sources may cause risk in farms. Hence, assessing these types of risks can help the decisions making and farm management. There are a few studies that investigated such risks in farms using deep learning methods. Espejo-Garcia et al. [64] developed an end-to-end sequence labeler using deep learning methods such as LSTM to extract different information to develop regulations, e.g., determine which pesticide can be applied to a crop. Govindan and Al-Ansari [44] developed a DL-based decision framework for enhancing the resilience of energy, water, and food nexus in risky environments. They particularly studied the water-food nexus in a weather volatility condition for agricultural operations. They showed that the proposed model can help effective decision making in such adverse events to support effective ARM. In a different study, Ghahari et al. [42] developed a DL method, in particular deep belief network (DBF), to assess the climate-related risks in agriculture and then used it to assess the agricultural insurance risk. They demonstrated that their method can produce higher prediction accuracy compared to conventional methods especially using heterogeneous data.

4.7 RQ6. What are the current research directions, achievements, and challenges in deep learning-based agricultural risk management?

The selected papers are thoroughly analyzed and classified within the defined research questions in previous sections. Based on the classification of the studies in Section 4.3, the research directions, achievements, and challenges within each objectives are described, and emerging topics and the challenges that are not addressed in current publications are discussed to provide insights and recommendations for future studies.

4.7.1 Crop disease and damage detection and prediction

This research direction focused mainly on crop disease and damages detection from imaging sensors data. Accordingly, several DL methods are developed and successfully detected disease and damages to crop agriculture. In addition, a few studies focus on predicting the damages. It is observed that there are three research directions in this topic; (1) developing new DL methods, in particular CNN, to increase the accuracy of the current detection works; (2) adopt and apply current DL methods to solve new problems such as detecting the new type of diseases in crops; and (3) using DL for predicting damages. Quite a few reliable CNN-based methods are developed using the state-of-the-art methods such as GANs and already achieved high accuracy rates. Also, more recently the explainable AI techniques employed in this topic to improve the accuracy of the current methods [29] or explain the importance of the parameters in detecting diseases [39]. Yet, developing an automated method to detect disease and damages in complicated datasets, which involves images in dense agricultural areas or different environmental conditions remain challenging. Furthermore, predicting damages and disease yet need more studies to support farmers in making decisions, so far, only a few studies exist on predicting frost and cold damages based on forecasting environmental parameters.

4.7.2 Harmful factors detection

Pests, weeds, and other harmful factors were detected using CNN models from images. This research direction aims to improve the current methods to increase the accuracy of the detection work and develop methods to detect other harmful factors such as spores. The currently developed methods for pests and weeds detection achieved a comparable performance; however, there is a need for more studies to investigate and develop methods to detect such factors in challenging conditions with high accuracy rates.

4.7.3 Animal disease detection and prediction

DL methods were adopted and used to predict and detect diseases in pig, dairy, and poultry farms. In all of the studies, it is shown that the DL methods achieve the state-of-the-art performance; however, yet only a few studies used DL methods in animal ARM. Since only recently DL entered this research topic, it will continue to be used and adapted in this domain. In addition, currently used conventional machine learning methods can be replaced with DL methods for other animal disease detection and prediction objectives. Although a few DL methods are developed to detect animal behaviors on the farm, there still a need for more studies to investigate and detect the behavioral patterns for different farms, and develop robust methods. There are different types of data, e.g., in a dairy farm, and to achieve a reliable prediction outcome one of the challenges is to integrate them into a single model.

4.7.4 Institutional and financial risk assessments

This is the least studied objective in DL–ARM. Only one of the three identified papers is directly related to this field; however, they provide some clues on how to employ DL methods for institutional and financial risk assessments. From a general point of view, using DL methods for institutional and financial analysis is currently a challenging topic. This also reflects on the use of DL methods for ARM. However, DL techniques are proved to be a robust predictive method, showing the potential of using such methods in this research field.

4.8 Discussion and insights from wider ARM and DL perspectives

In previous sections, the results are discussed under the umbrella of the existing research directions and the groupings are based on the current literature, and thus, those do not cover the entire ARM domain problems and challenges, which can be addressed using DL methods. Hence, in this subsection, the results are further discussed and insights from the wider ARM and DL domain are provided to come up with guidelines for future studies.

As stated before, most of the studies focused on production risk assessment, and only a few of them did institutional or financial risk assessments. Furthermore, there are other risk types, i.e., market and personal risks, which are not yet investigated with DL methods [48]. Therefore, there is a need for DL methods to do comprehensive risk assessments considering all types of risks. One possible research direction can be to replace the traditional machine learning methods which have been used to address other risk types, e.g., financial [123,124], institutional [125], market [126], and personal [127] risks with novel DL methods. Moreover, farmers need decision-making tools rather than only data processing ones, which can be possible by combining different DL methods.

Risk consists of four main components, response, recovery, mitigation, and preparedness as explained in Section 2. Currently, the majority of the papers in DL-ARM focused on the response phase doing damage and disease detections. However, other components are also significant for risk analysis and farm decision making, such as resilience [10] and vulnerability [128,129] assessments. Also, in this case, this can be done by improving the already implemented conventional machine learning methods or address the open challenges in the field.

DL methods are known as black-box methods and are not easily interpreted. Hence, researchers cannot fully understand the inside of the DL and thus cannot find reasons behind the produced/predicted results. Furthermore, DL methods are computationally expensive and require a big amount of data to produce accurate results.

Moreover, using new DL techniques, i.e., explainable AI, can provide new opportunities to address ARM, such as improving the prediction precision [29] or sorting the influencing parameters according to their importance in addition to prediction [39].

4.9 Limitations

Some limitations may bias every systematic literature review study such as publication bias, data extraction, and classification. The main limitations and the threats to validity of this study are sorted and discussed under conclusions, construct, internal, and external validities titles.

Conclusion validity: The current review was conducted based on the accepted structure and protocol for systematic literature review studies [53]. Accordingly, all the steps such as defining the research questions, search strategies, exclusion criteria, and synthesizing results were formulated and performed based on the widely used structure. In addition, the defined search string, and data extraction form (Appendix) in the paper, and the full list of the extracted publications (supplementary materials) are provided and attached to the paper. Hence, one can easily reproduce the results using that information.

Construct validity: In the current study, the aim was to find high-quality studies. Only the ISI Web of Knowledge was employed as the only source to find the relevant publications. Accordingly, relevant publications might not be indexed in this platform/database and thus not included in this review. Using this database as the only source of publications may lead to missing other relevant publications that are not included in this study. However, since this study aimed to provide an overview of high-quality publications, indexing in an ISI journal is an accepted way that was used to find and extract the relevant high-quality papers. There might be missing terms that may impact the final result. However, the search has been tried to be kept broad (the initial number of selected papers was 203), and the search query was refined several times to reduce the possibility of missing any relevant study. Hence, the impact of missing any relevant papers in the final results is low.

Internal validity: The research questions were particularly designed and formulated to investigate and extract all the required information and components for DL-ARM. The research questions are based on the precisely defined risk types, risk components, Dl methods, and other necessary information. Accordingly, the findings of this study are properly explained and linked to the extracted results.

External validity: This study reviewed the publications which employed DL methods for ARM in terms of different risk types and components. However, all of the existing DL methods and possible agricultural risk types and components are not addressed in the papers and therefore not included and discussed in this study. However, insights from wider points of view from computer science/deep learning, and agricultural risk management domains are provided and discussed in a separate section (Section 4.8).

5 Conclusions

The main goal of this study was to identify the achievements and challenges in the use of deep learning (DL) methods for agricultural risk management (ARM). In total, 203 publications were initially extracted (from those 100 publications were selected for a detailed review) in DL–ARM subject using the defined search string from ISI Web of Knowledge and implementing the exclusion criteria. The selected papers are classified to answer the research questions, and also, further analyses are conducted to provide research directions, achievements, and challenges. In conclusion, it is observed that there is a substantial increase in the number of studies that use DL for ARM in 2019 and 2020. This can be a result of general technological developments in monitoring (sensors) hardware and systems to collect required data, and computer facilities to perform computationally expensive DL methods. However, this increase is mainly in the field of production risk assessment by detecting damages and diseases in crop- and animal-based agriculture. In crop agriculture, considering the recent advances in the precision agriculture field, new imaging and other sensors data are collected using different platforms, e.g., UAVs [130], and thus, provide opportunities for the researchers to develop DL (predominantly CNN) methods for crop disease, damage, and harmful factors detection [20,131]. In addition, the problem statement in crop production risk assessment is easier than others, for example, extracting a new type of disease only using the current state-of-the-art methods. However, it is not that easy to collect data to evaluate the other risk types such as financial and institutional risks, and understanding the risks and the data to be processed with DL is more complicated than a computer vision-based approach. Nevertheless, there is a need to develop DL-ARM considering other risk components, e.g., resilience and vulnerability, and other risk types including financial and market risk assessments. As future work, DL can be used to extract vulnerability and resilience indicators such as crop diversity and availability of arable lands [10,51]. Despite recent efforts in explaining the DL methods, yet they are not fully responsible and interpretable methods. Hence, they cannot be used for finding the reasons behind their prediction results. In addition, DL methods do not provide accurate results when only small datasets are available and accordingly make it challenging to use such methods in some ARM applications. In this case, new techniques such as transfer learning [132] can be used in the future to improve the performance of the DL methods for ARM where extensive datasets are not available to train the DL methods.

Appendix. Data extraction form

Table A.1 Data extraction form.

#	Extraction element	Contents
General information		
1	ID	Unique ID for the study
2	Title	Full title of the article
3	Authors	The authors of the article
4	Year	The publication year
5	Publisher name	The publisher name (e.g., Elsevier)
6	Journal name	The journal name (e.g., Journal of Dairy Science)
Study description		
7	Agricultural risk type	☐Production risk ☐Market risk ☐Institutional risk ☐Personal risk ☐Financial risk
8	Main objective of the study	☐Crop disease and damage detection and prediction ☐Harmful factors detection ☐Animal disease detection and prediction ☐Institutional and financial risk assessment
9	Details about the study	E.g., any interesting findings or problems
10	Directly address ARM	☐Yes ☐ No
11	Deep learning type	☐Classification ☐Regression ☐Reinforcement learning
12	Deep learning algorithm	☐CNN ☐RNN ☐Dreg ☐GAN ☐DBN ☐RL ☐AutoE ☐CapsNet
13	Data type	☐Imaging sensors data ☐Non-imaging sensors data ☐Earth data ☐Non-sensor field data ☐Climate data ☐Socio-economic data ☐Other
14	Additional notes	E.g., the opinions of the reviewer about the study

Data availability statement

The full list of the reviewed publications is provided in a supplementary file.

Declaration of competing interest

The authors declare no conflict of interest.

References

[1] I. Cisternas, I. Velásquez, A. Caro, A. Rodríguez, Systematic literature review of implementations of precision agriculture, Comput. Electron. Agric. 176 (2020) 105626, https://doi.org/10.1016/j.compag.2020.105626.

[2] C. Pylianidis, S. Osinga, I.N. Athanasiadis, Introducing digital twins to agriculture, Comput. Electron. Agric. 184 (2021) 105942, https://doi.org/10.1016/j.compag.2020.105942.

[3] C. Verdouw, B. Tekinerdogan, A. Beulens, S. Wolfert, Digital twins in smart farming, Agric. Syst. 189 (2021) 103046, https://doi.org/10.1016/j.agsy.2020.103046.

[4] S. Wolfert, L. Ge, C. Verdouw, M.-J. Bogaardt, Big data in smart farming—a review, Agric. Syst. 153 (2017) 69–80, https://doi.org/10.1016/j.agsy.2017.01.023.

[5] J. Conesa-Muñoz, J. Valente, J. Del Cerro, A. Barrientos, A. Ribeiro, A multi-robot sense-act approach to Lead to a proper acting in environmental incidents, Sensors 16 (2016) 1269, https://doi.org/10.3390/s16081269.

[6] A. Koirala, K.B. Walsh, Z. Wang, C. McCarthy, Deep learning for real-time fruit detection and orchard fruit load estimation: benchmarking of 'MangoYOLO', Precis. Agric. 20 (2019) 1107–1135, https://doi.org/10.1007/s11119-019-09642-0.

[7] K.G. Liakos, P. Busato, D. Moshou, S. Pearson, D. Bochtis, Machine learning in agriculture: a review, Sensors 18 (2018) 2674, https://doi.org/10.3390/s18082674.

[8] J. Valente, R. Almeida, L. Kooistra, A comprehensive study of the potential application of flying ethylene-sensitive sensors for ripeness detection in apple orchards, Sensors 19 (2019) 372, https://doi.org/10.3390/s19020372.

[9] J. Valente, L. Kooistra, S. Mücher, Fast classification of large germinated fields via high-resolution UAV imagery, IEEE Robot. Autom. Lett. 4 (2019) 3216–3223, https://doi.org/10.1109/LRA.2019.2926957.

[10] M.P.M. Meuwissen, P.H. Feindt, A. Spiegel, C.J.A.M. Termeer, E. Mathijs, Y. de Mey, R. Finger, A. Balmann, E. Wauters, J. Urquhart, M. Vigani, K. Zawalińska, H. Herrera, P. Nicholas-Davies, H. Hansson, W. Paas, T. Slijper, I. Coopmans, W. Vroege, A. Ciechomska, F. Accatino, B. Kopainsky, P.-M. Poortvliet, J.J.L. Candel, D. Maye, S. Severini, S. Senni, B. Soriano, C.-J. Lagerkvist, M. Peneva, C. Gavrilescu, P. Reidsma, A framework to assess the resilience of farming systems, Agric. Syst. 176 (2019) 102656, https://doi.org/10.1016/j.agsy.2019.102656.

[11] D. Maye, H. Chiswell, M. Vigani, J. Kirwan, 'Present realities' and the need for a 'lived experience' perspective in Brexit Agri-food governance, Space Polity 22 (2018) 270–286, https://doi.org/10.1080/13562576.2018.1519390.

[12] F. van Winsen, Y. de Mey, L. Lauwers, S. Van Passel, M. Vancauteren, E. Wauters, Determinants of risk behaviour: effects of perceived risks and risk attitude on farmer's adoption of risk management strategies, J. Risk Res. 19 (2016) 56–78, https://doi.org/10.1080/13669877.2014.940597.

[13] R.M.A. Latif, S.B. Brahim, S. Saeed, L.B. Imran, M. Sadiq, M. Farhan, Integration of Google play content and frost prediction using CNN: scalable IoT framework for big data, IEEE Access 8 (2020) 6890–6900, https://doi.org/10.1109/access.2019.2963590.

[14] S. Lee, Y. Jeong, S. Son, B. Lee, A self-predictable crop yield platform (SCYP) based on crop diseases using deep learning, Sustainability 11 (2019), https://doi.org/10.3390/su11133637.

[15] S. Verma, A. Chug, A.P. Singh, Application of convolutional neural networks for evaluation of disease severity in tomato plant, J. Discret. Math. Sci. Cryptogr. 23 (2020) 273–282, https://doi.org/10.1080/09720529.2020.1721890.

[16] D.L. Hoag, Applied Risk Management in Agriculture, CRC Press, 2010, https://doi.org/10.1201/b15855.

[17] F. Emmert-Streib, Z. Yang, H. Feng, S. Tripathi, M. Dehmer, An introductory review of deep learning for prediction models with big data, Front. Artif. Intell. 3 (2020), https://doi.org/10.3389/frai.2020.00004.

[18] S. Ghaffarian, M. Turker, An improved cluster-based snake model for automatic agricultural field boundary extraction from high spatial resolution imagery, Int. J. Remote Sens. 40 (2019) 1217–1247, https://doi.org/10.1080/01431161.2018.1524178.

[19] Z. Kang, C. Catal, B. Tekinerdogan, Machine learning applications in production lines: a systematic literature review, Comput. Ind. Eng. 149 (2020) 106773, https://doi.org/10.1016/j.cie.2020.106773.

[20] J. Siebring, J. Valente, M.H. Domingues Franceschini, J. Kamp, L. Kooistra, Object-based image analysis applied to low altitude aerial imagery for potato plant trait retrieval and pathogen detection, Sensors 19 (2019) 5477, https://doi.org/10.3390/s19245477.

[21] A. Kamilaris, F.X. Prenafeta-Boldú, Deep learning in agriculture: a survey, Comput. Electron. Agric. 147 (2018) 70–90, https://doi.org/10.1016/j.compag.2018.02.016.

[22] A. Kaya, A.S. Keceli, C. Catal, H.Y. Yalic, H. Temucin, B. Tekinerdogan, Analysis of transfer learning for deep neural network based plant classification models, Comput. Electron. Agric. 158 (2019) 20–29, https://doi.org/10.1016/j.compag.2019.01.041.

[23] L. Santos, F.N. Santos, P.M. Oliveira, P. Shinde, Deep learning applications in agriculture: a short review, in: Robot 2019: Fourth Iberian Robotics Conference, Springer International Publishing, Cham, 2020, pp. 139–151.

[24] Y.-H. Choi, P. Liu, Z. Shang, H. Wang, Z. Wang, L. Zhang, J. Zhou, Q. Zou, Using deep learning to solve computer security challenges: a survey, Cybersecurity 3 (2020) 15, https://doi.org/10.1186/s42400-020-00055-5.

[25] S. Ghaffarian, N. Kerle, E. Pasolli, J. Jokar Arsanjani, Post-disaster building database updating using automated deep learning: an integration of pre-disaster OpenStreetMap and multi-temporal satellite data, Remote Sens. 11 (2019) 2427, https://doi.org/10.3390/rs11202427.

[26] F. Nex, D. Duarte, F.G. Tonolo, N. Kerle, Structural building damage detection with deep learning: assessment of a state-of-the-art CNN in operational conditions, Remote Sens. 11 (2019) 2765, https://doi.org/10.3390/rs11232765.

[27] A. Bauer, A.G. Bostrom, J. Ball, C. Applegate, T. Cheng, S. Laycock, S.M. Rojas, J. Kirwan, J. Zhou, Combining computer vision and deep learning to enable ultra-scale aerial phenotyping and precision agriculture: a case study of lettuce production, Hortic. Res. 6 (2019) 70, https://doi.org/10.1038/s41438-019-0151-5.

[28] R. Karthik, M. Hariharan, S. Anand, P. Mathikshara, A. Johnson, R. Menaka, Attention embedded residual CNN for disease detection in tomato leaves, Appl. Soft Comput. 86 (2020), https://doi.org/10.1016/j.asoc.2019.105933.

[29] W.H. Zeng, M. Li, Crop leaf disease recognition based on self-attention convolutional neural network, Comput. Electron. Agric. 172 (2020), https://doi.org/10.1016/j.compag.2020.105341.

[30] J.D. Chen, J.X. Chen, D.F. Zhang, Y.D. Sun, Y.A. Nanehkaran, Using deep transfer learning for image-based plant disease identification, Comput. Electron. Agric. 173 (2020), https://doi.org/10.1016/j.compag.2020.105393.

[31] J.D. Chen, D.F. Zhang, Y.A. Nanehkaran, D.L. Li, Detection of rice plant diseases based on deep transfer learning, J. Sci. Food Agric. 100 (2020) 3246–3256, https://doi.org/10.1002/jsfa.10365.

[32] S. Coulibaly, B. Kamsu-Foguem, D. Kamissoko, D. Traore, Deep neural networks with transfer learning in millet crop images, Comput. Ind. 108 (2019) 115–120, https://doi.org/10.1016/j.compind.2019.02.003.

[33] X. Cheng, Y.H. Zhang, Y.Q. Chen, Y.Z. Wu, Y. Yue, Pest identification via deep residual learning in complex background, Comput. Electron. Agric. 141 (2017) 351–356, https://doi.org/10.1016/j.compag.2017.08.005.

[34] H.X. Yang, L.T. Gao, N.S. Tang, P. Yang, Experimental analysis and evaluation of wide residual networks based agricultural disease identification in smart agriculture system, EURASIP J. Wirel. Commun. Netw. 2019 (2019), https://doi.org/10.1186/s13638-019-1613-z.

[35] Y. Wu, L.H. Xu, Crop organ segmentation and disease identification based on weakly supervised deep neural network, Agronomy 9 (2019), https://doi.org/10.3390/agronomy9110737.

[36] S.L. Xing, M. Lee, K.K. Lee, Citrus pests and diseases recognition model using weakly dense connected convolution network, Sensors 19 (2019), https://doi.org/10.3390/s19143195.

[37] Y.N. Tian, G.D. Yang, Z. Wang, E. Li, Z.Z. Liang, Detection of apple lesions in orchards based on deep learning methods of CycleGAN and YOLOV3-dense, J. Sens. 2019 (2019), https://doi.org/10.1155/2019/7630926.

[38] Q.F. Wu, M.M. Ji, Z. Deng, Automatic detection and severity assessment of pepper bacterial spot disease via MultiModels based on convolutional neural networks, Int. J. Agric. Environ. Inf. Syst. 11 (2020) 29–43, https://doi.org/10.4018/ijaeis.2020040103.

[39] K. Nagasubramanian, S. Jones, A.K. Singh, S. Sarkar, A. Singh, B. Ganapathysubramanian, Plant disease identification using explainable 3D deep learning on hyperspectral images, Plant Methods 15 (2019), https://doi.org/10.1186/s13007-019-0479-8.

[40] D. Li, Y.F. Chen, K.F. Zhang, Z.B. Li, Mounting behaviour recognition for pigs based on deep learning, Sensors 19 (2019), https://doi.org/10.3390/s19224924.

[41] Y. Domun, L.J. Pedersen, D. White, O. Adeyemi, T. Norton, Learning patterns from time-series data to discriminate predictions of tail-biting, fouling and diarrhoea in pigs, Comput. Electron. Agric. 163 (2019), https://doi.org/10.1016/j.compag.2019.104878.

[42] A. Ghahari, N.K. Newlands, V. Lyubchich, Y.R. Gel, Deep learning at the interface of agricultural insurance risk and spatio-temporal uncertainty in weather extremes, N. Am. Actuar. J. 23 (2019) 535–550, https://doi.org/10.1080/10920277.2019.1633928.

[43] W. Xu, Q.L. Wang, R.Y. Chen, Spatio-temporal prediction of crop disease severity for agricultural emergency management based on recurrent neural networks, GeoInformatica 22 (2018) 363–381, https://doi.org/10.1007/s10707-017-0314-1.

[44] R. Govindan, T. Al-Ansari, Computational decision framework for enhancing resilience of the energy, water and food nexus in risky environments, Renew. Sust. Energ. Rev. 112 (2019) 653–668, https://doi.org/10.1016/j.rser.2019.06.015.

[45] S. Nigam, R. Jain, Plant disease identification using deep learning: a review, Indian J. Agric. Sci. 90 (2020) 249–257.

[46] P. Iyer, M. Bozzola, S. Hirsch, M. Meraner, R. Finger, Measuring farmer risk preferences in Europe: a systematic review, J. Agric. Econ. 71 (2020) 3–26, https://doi.org/10.1111/1477-9552.12325.

[47] J.A. García-Berná, S. Ouhbi, B. Benmouna, G. García-Mateos, J.L. Fernández-Alemán, J.M. Molina-Martínez, Systematic mapping study on remote sensing in agriculture, Appl. Sci. 10 (2020) 3456, https://doi.org/10.3390/app10103456.

[48] A.M. Komarek, A. De Pinto, V.H. Smith, A review of types of risks in agriculture: what we know and what we need to know, Agric. Syst. 178 (2020) 102738, https://doi.org/10.1016/j.agsy.2019.102738.

[49] J.B. Hardaker, G. Lien, J.R. Anderson, R. Huirne, Coping With Risk in Agriculture: Applied Decision Analysis, third ed., CABI Publishing, Wallingford, UK, 2015, https://doi.org/10.1079/9781780645742.0000.

[50] A. Fuentes, S. Yoon, S.C. Kim, D.S. Park, A robust deep-learning-based detector for real-time tomato plant diseases and pests recognition, Sensors 17 (2017), https://doi.org/10.3390/s17092022.

[51] T.-S. Neset, L. Wiréhn, T. Opach, E. Glaas, B.-O. Linnér, Evaluation of indicators for agricultural vulnerability to climate change: the case of Swedish agriculture, Ecol. Indic. 105 (2019) 571–580, https://doi.org/10.1016/j.ecolind.2018.05.042.

[52] G. Fenu, F.M. Malloci, Forecasting plant and crop disease: an explorative study on current algorithms, Big Data Cogn. Comput. 5 (2021) 2.

[53] B. Kitchenham, O. Pearl Brereton, D. Budgen, M. Turner, J. Bailey, S. Linkman, Systematic literature reviews in software engineering—a systematic literature review, Inf. Softw. Technol. 51 (2009) 7–15, https://doi.org/10.1016/j.infsof.2008.09.009.

[54] T. Fang, P. Chen, J. Zhang, B. Wang, Crop leaf disease grade identification based on an improved convolutional neural network, J. Electron. Imaging 29 (2020), https://doi.org/10.1117/1.Jei.29.1.013004.

[55] M. Marsot, J.Q. Mei, X.C. Shan, L.Y. Ye, P. Feng, X.J. Yan, C.F. Li, Y.F. Zhao, An adaptive pig face recognition approach using convolutional neural networks, Comput. Electron. Agric. 173 (2020), https://doi.org/10.1016/j.compag.2020.105386.

[56] S.P. Mohanty, D.P. Hughes, M. Salathe, Using deep learning for image-based plant disease detection, Front. Plant Sci. 7 (2016), https://doi.org/10.3389/fpls.2016.01419.

[57] A. Picon, M. Seitz, A. Alvarez-Gila, P. Mohnke, A. Ortiz-Barredo, J. Echazarra, Crop conditional convolutional neural networks for massive multi-crop plant disease classification over cell phone acquired images taken on real field conditions, Comput. Electron. Agric. 167 (2019), https://doi.org/10.1016/j.compag.2019.105093.

[58] X.H. Zhang, Y. Qiao, F.F. Meng, C.G. Fan, M.M. Zhang, Identification of maize leaf diseases using improved deep convolutional neural networks, IEEE Access 6 (2018) 30370–30377, https://doi.org/10.1109/access.2018.2844405.

[59] C. Fang, J.D. Huang, K.X. Cuan, X.L. Zhuang, T.M. Zhang, Comparative study on poultry target tracking algorithms based on a deep regression network, Biosyst. Eng. 190 (2020) 176–183, https://doi.org/10.1016/j.biosystemseng.2019.12.002.

[60] S. Verma, A. Chug, A.P. Singh, Exploring capsule networks for disease classification in plants, J. Stat. Manag. Syst. 23 (2020) 307–315, https://doi.org/10.1080/09720510.2020.1724628.

[61] A. Khamparia, G. Saini, D. Gupta, A. Khanna, S. Tiwari, V.H.C. de Albuquerque, Seasonal crops disease prediction and classification using deep convolutional encoder network, Circuits Syst. Signal Process. 39 (2020) 818–836, https://doi.org/10.1007/s00034-019-01041-0.

[62] M. Kerkech, A. Hafiane, R. Canals, Deep leaning approach with colorimetric spaces and vegetation indices for vine diseases detection in UAV images, Comput. Electron. Agric. 155 (2018) 237–243, https://doi.org/10.1016/j.compag.2018.10.006.

[63] A. Picon, A. Alvarez-Gila, M. Seitz, A. Ortiz-Barredo, J. Echazarra, A. Johannes, Deep convolutional neural networks for mobile capture device-based crop disease classification in the wild, Comput. Electron. Agric. 161 (2019) 280–290, https://doi.org/10.1016/j.compag.2018.04.002.

[64] B. Espejo-Garcia, F.J. Lopez-Pellicer, J. Lacasta, R.P. Moreno, F.J. Zarazaga-Soria, End-to-end sequence labeling via deep learning for automatic extraction of agricultural regulations, Comput. Electron. Agric. 162 (2019) 106–111, https://doi.org/10.1016/j.compag.2019.03.027.

[65] D. Velasquez, A. Sanchez, S. Sarmiento, M. Toro, M. Maiza, B. Sierra, A method for detecting coffee leaf rust through wireless sensor networks, remote sensing, and deep learning: case study of the Caturra variety in Colombia, Appl. Sci. 10 (2020), https://doi.org/10.3390/app10020697.

[66] M.G. Selvaraj, A. Vergara, H. Ruiz, N. Safari, S. Elayabalan, W. Ocimati, G. Blomme, AI-powered banana diseases and pest detection, Plant Methods 15 (2019), https://doi.org/10.1186/s13007-019-0475-z.

[67] Y. Ampatzidis, V. Partel, UAV-based high throughput phenotyping in citrus utilizing multispectral imaging and artificial intelligence, Remote Sens. 11 (2019), https://doi.org/10.3390/rs11040410.

[68] V. Partel, L. Nunes, P. Stansly, Y. Arnpatzidis, Automated vision-based system for monitoring Asian citrus psyllid in orchards utilizing artificial intelligence, Comput. Electron. Agric. 162 (2019) 328–336, https://doi.org/10.1016/j.compag.2019.04.022.

[69] J.G.M. Esgario, R.A. Krohling, J.A. Ventura, Deep learning for classification and severity estimation of coffee leaf biotic stress, Comput. Electron. Agric. 169 (2020), https://doi.org/10.1016/j.compag.2019.105162.

[70] J.S.H. Al-bayati, B.B. Ustundag, Evolutionary feature optimization for plant leaf disease detection by deep neural networks, Int. J. Comput. Intell. Syst. 13 (2020) 12–23, https://doi.org/10.2991/ijcis.d.200108.001.

[71] T.K. Fan, J. Xu, Image classification of crop diseases and pests based on deep learning and fuzzy system, Int. J. Data Warehous. Min. 16 (2020) 34–47, https://doi.org/10.4018/ijdwm.2020040103.

[72] M. Francis, C. Deisy, Mathematical and visual understanding of a deep learning model towards m-agriculture for disease diagnosis, Arch. Comput. Methods Eng. (2021), https://doi.org/10.1007/s11831-020-09407-3.

[73] F.A. Khan, A.A. Ibrahim, A.M. Zeki, Environmental monitoring and disease detection of plants in smart greenhouse using internet of things, J. Phys. Commun. 4 (2020), https://doi.org/10.1088/2399-6528/ab90c1.

[74] S.H. Lee, H. Goeau, P. Bonnet, A. Joly, New perspectives on plant disease characterization based on deep learning, Comput. Electron. Agric. 170 (2020), https://doi.org/10.1016/j.compag.2020.105220.

[75] M.P. Vaishnnave, K.S. Devi, P. Ganeshkumar, Automatic method for classification of groundnut diseases using deep convolutional neural network, Soft. Comput. (2020), https://doi.org/10.1007/s00500-020-04946-0.

[76] R.A. Priyadharshini, S. Arivazhagan, M. Arun, A. Mirnalini, Maize leaf disease classification using deep convolutional neural networks, Neural Comput. Applic. 31 (2019) 8887–8895, https://doi.org/10.1007/s00521-019-04228-3.

[77] U.P. Singh, S.S. Chouhan, S. Jain, S. Jain, Multilayer convolution neural network for the classification of mango leaves infected by anthracnose disease, IEEE Access 7 (2019) 43721–43729, https://doi.org/10.1109/access.2019.2907383.

[78] A. Abdalla, H.Y. Cen, L. Wan, R. Rashid, H.Y. Weng, W.J. Zhou, Y. He, Fine-tuning convolutional neural network with transfer learning for semantic segmentation of ground-level oilseed rape images in

a field with high weed pressure, Comput. Electron. Agric. 167 (2019), https://doi.org/10.1016/j. compag.2019.105091.

[79] J.A. Xia, Y.W. Yang, H.X. Cao, W.X. Zhang, L. Xu, Q. Wan, Y.Q. Ke, W.Y. Zhang, D.K. Ge, B. Huang, Hyperspectral identification and classification of oilseed rape waterlogging stress levels using parallel computing, IEEE Access 6 (2018) 57663–57675, https://doi.org/10.1109/access.2018. 2873689.

[80] W.S. Kim, D.H. Lee, Y.J. Kim, Machine vision-based automatic disease symptom detection of onion downy mildew, Comput. Electron. Agric. 168 (2020), https://doi.org/10.1016/j.compag.2019. 105099.

[81] M.A. Khan, T. Akram, M. Sharif, M. Awais, K. Javed, H. Ali, T. Saba, CCDF: automatic system for segmentation and recognition of fruit crops diseases based on correlation coefficient and deep CNN features, Comput. Electron. Agric. 155 (2018) 220–236, https://doi.org/10.1016/j.compag.2018. 10.013.

[82] M. Rahnemoonfar, C. Sheppard, Deep count: fruit counting based on deep simulated learning, Sensors 17 (2017), https://doi.org/10.3390/s17040905.

[83] D. Bienkowski, M.J. Aitkenhead, A.K. Lees, C. Gallagher, R. Neilson, Detection and differentiation between potato (Solanum tuberosum) diseases using calibration models trained with non-imaging spectrometry data, Comput. Electron. Agric. 167 (2019), https://doi.org/10.1016/j.compag.2019.105056.

[84] S. Marino, P. Beauseroy, A. Smolarz, Weakly-supervised learning approach for potato defects segmentation, Eng. Appl. Artif. Intell. 85 (2019) 337–346, https://doi.org/10.1016/j.engappai.2019.06.024.

[85] D. Oppenheim, G. Shani, O. Erlich, L. Tsror, Using deep learning for image-based potato tuber disease detection, Phytopathology 109 (2019) 1083–1087, https://doi.org/10.1094/phyto-08-18-0288-r.

[86] C.R. Rahman, P.S. Arko, M.E. Ali, M.A.I. Khan, S.H. Apon, F. Nowrin, W. Abu, Identification and recognition of rice diseases and pests using convolutional neural networks, Biosyst. Eng. 194 (2020) 112–120, https://doi.org/10.1016/j.biosystemseng.2020.03.020.

[87] C.Q. Zhou, H.B. Ye, J. Hu, X.Y. Shi, S. Hua, J.B. Yue, Z.F. Xu, G.J. Yang, Automated counting of Rice panicle by applying deep learning model to images from unmanned aerial vehicle platform, Sensors 19 (2019), https://doi.org/10.3390/s19143106.

[88] A. Karlekar, A. Seal, SoyNet: soybean leaf diseases classification, Comput. Electron. Agric. 172 (2020), https://doi.org/10.1016/j.compag.2020.105342.

[89] T.L. Lin, H.Y. Chang, K.H. Chen, The pest and disease identification in the growth of sweet peppers using faster R-CNN and mask R-CNN, J. Internet Technol. 21 (2020) 605–614, https://doi.org/ 10.3966/160792642020032102027.

[90] M. Brahimi, K. Boukhalfa, A. Moussaoui, Deep learning for tomato diseases: classification and symptoms visualization, Appl. Artif. Intell. 31 (2017) 299–315, https://doi.org/10.1080/08839514. 2017.1315516.

[91] L. Zhang, J.D. Jia, Y. Li, W.L. Gao, M.J. Wang, Deep learning based rapid diagnosis system for identifying tomato nutrition disorders, KSII Trans. Internet Inf. Syst. 13 (2019) 2012–2027, https://doi. org/10.3837/tiis.2019.04.015.

[92] F. Rancon, L. Bombrun, B. Keresztes, C. Germain, Comparison of SIFT encoded and deep learning features for the classification and detection of esca disease in Bordeaux vineyards, Remote Sens. 11 (2019), https://doi.org/10.3390/rs11010001.

[93] Z.Q. Lin, S.M. Mu, F. Huang, K.A. Mateen, M.J. Wang, W.L. Gao, J.D. Jia, A unified matrix-based convolutional neural network for fine-grained image classification of wheat leaf diseases, IEEE Access 7 (2019) 11570–11590, https://doi.org/10.1109/access.2019.2891739.

[94] J. Lu, J. Hu, G.N. Zhao, F.H. Mei, C.S. Zhang, An in-field automatic wheat disease diagnosis system, Comput. Electron. Agric. 142 (2017) 369–379, https://doi.org/10.1016/j.compag.2017.09.012.

[95] T. Su, S. Mu, A. Shi, Z. Cao, M. Dong, A CNN-LSVM model for imbalanced images identification of wheat leaf, Neural Netw. World 29 (2019) 345–361, https://doi.org/10.14311/nnw.2019.29.021.

[96] W. Yang, C. Yang, Z.Y. Hao, C.Q. Xie, M.Z. Li, Diagnosis of plant cold damage based on hyperspectral imaging and convolutional neural network, IEEE Access 7 (2019) 118239–118248, https:// doi.org/10.1109/access.2019.2936892.

[97] W.S. Lee, S.O. Kim, Y. Kim, J.H. Kim, G.S. Jang, Maximum entropy modeling of farmland damage caused by the wild boar (sus scrofa), Appl. Ecol. Environ. Res. 16 (2018) 1101–1117, https://doi.org/ 10.15666/aeer/1602_11011117.

[98] Z.D. Wang, M.H. Hu, G.T. Zhai, Application of deep learning architectures for accurate and rapid detection of internal mechanical damage of blueberry using hyperspectral transmittance data, Sensors 18 (2018), https://doi.org/10.3390/s18041126.

[99] H.S. Huang, J.Z. Deng, Y.B. Lan, A.Q. Yang, L. Zhang, S. Wen, H.H. Zhang, Y.L. Zhang, Y.S. Deng, Detection of Helminthosporium leaf blotch disease based on UAV imagery, Appl. Sci. 9 (2019), https://doi.org/10.3390/app9030558.

[100] X. Zhang, L.X. Han, Y.Y. Dong, Y. Shi, W.J. Huang, L.H. Han, P. Gonzalez-Moreno, H.Q. Ma, H.C. Ye, T. Sobeih, A deep learning-based approach for automated yellow rust disease detection from high-resolution hyperspectral UAV images, Remote Sens. 11 (2019), https://doi.org/10.3390/rs11131554.

[101] M.H. Saleem, J. Potgieter, K.M. Arif, Plant disease detection and classification by deep learning, Plants 8 (2019), https://doi.org/10.3390/plants8110468.

[102] S. Sladojevic, M. Arsenovic, A. Anderla, D. Culibrk, D. Stefanovic, Deep neural networks based recognition of plant diseases by leaf image classification, Comput. Intell. Neurosci. (2016), https://doi.org/10.1155/2016/3289801.

[103] A. Ramcharan, P. McCloskey, K. Baranowski, N. Mbilinyi, L. Mrisho, M. Ndalahwa, J. Legg, D.P. Hughes, A mobile-based deep learning model for cassava disease diagnosis, Front. Plant Sci. 10 (2019), https://doi.org/10.3389/fpls.2019.00272.

[104] C. Douarre, C.F. Crispim-Junior, A. Gelibert, L. Tougne, D. Rousseau, Novel data augmentation strategies to boost supervised segmentation of plant disease, Comput. Electron. Agric. 165 (2019), https://doi.org/10.1016/j.compag.2019.104967.

[105] E.L. Stewart, T. Wiesner-Hanks, N. Kaczmar, C. DeChant, H. Wu, H. Lipson, R.J. Nelson, M.A. Gore, Quantitative phenotyping of northern leaf blight in UAV images using deep learning, Remote Sens. 11 (2019), https://doi.org/10.3390/rs11192209.

[106] Q.K. Liang, S. Xiang, Y.C. Hu, G. Coppola, D. Zhang, W. Sun, (PDSE)-S-2-Net: computer-assisted plant disease diagnosis and severity estimation network, Comput. Electron. Agric. 157 (2019) 518–529, https://doi.org/10.1016/j.compag.2019.01.034.

[107] G. Wang, Y. Sun, J.X. Wang, Automatic image-based plant disease severity estimation using deep learning, Comput. Intell. Neurosci. (2017), https://doi.org/10.1155/2017/2917536.

[108] Q.J. Guan, Y.P. Huang, Z. Zhong, Z.D. Zheng, L. Zheng, Y. Yang, Thorax disease classification with attention guided convolutional neural network, Pattern Recogn. Lett. 131 (2020) 38–45, https://doi.org/10.1016/j.patrec.2019.11.040.

[109] Y.N. Jeong, S. Son, S. Lee, B. Lee, A total crop-diagnosis platform based on deep learning models in a natural nutrient environment, Appl. Sci. 8 (2018), https://doi.org/10.3390/app8101992.

[110] A. Gutierrez, A. Ansuategi, L. Susperregi, C. Tubio, I. Rankic, L. Lenza, A benchmarking of learning strategies for pest detection and identification on tomato plants for autonomous scouting robots using internal databases, J. Sens. 2019 (2019), https://doi.org/10.1155/2019/5219471.

[111] Y.F. Li, H.X. Wang, L.M. Dang, A. Sadeghi-Niaraki, H. Moon, Crop pest recognition in natural scenes using convolutional neural networks, Comput. Electron. Agric. 169 (2020), https://doi.org/10.1016/j.compag.2019.105174.

[112] W.L. Lie, P. Chen, B. Wang, C.J. Xie, Automatic localization and count of agricultural crop pests based on an improved deep learning pipeline, Sci. Rep. 9 (2019), https://doi.org/10.1038/s41598-019-43171-0.

[113] R. Li, R.J. Wang, J. Zhang, C.J. Xie, L. Liu, F.Y. Wang, H.B. Chen, T.J. Chen, H.Y. Hu, X.F. Jia, M. Hu, M. Zhou, D.S. Li, W.C. Liu, An effective data augmentation strategy for CNN-based Pest localization and recognition in the field, IEEE Access 7 (2019) 160274–160283, https://doi.org/10.1109/access.2019.2949852.

[114] V. Partel, C. Kakarla, Y. Ampatzidis, Development and evaluation of a low-cost and smart technology for precision weed management utilizing artificial intelligence, Comput. Electron. Agric. 157 (2019) 339–350, https://doi.org/10.1016/j.compag.2018.12.048.

[115] I. Sa, Z.T. Chen, M. Popovic, R. Khanna, F. Liebisch, J. Nieto, R. Siegwart, weedNet: dense semantic weed classification using multispectral images and MAV for smart farming, IEEE Robot. Autom. Lett. 3 (2018) 588–595, https://doi.org/10.1109/lra.2017.2774979.

[116] L.L. Yang, J. Luo, Z.P. Wang, Y. Chen, C.C. Wu, Research on recognition for cotton spider mites' damage level based on deep learning, Int. J. Agric. Biol. Eng. 12 (2019) 129–134, https://doi.org/10.25165/j.ijabe.20191206.4816.

[117] S.M. Sharpe, A.W. Schumann, N.S. Boyd, Detection of Carolina geranium (Geranium carolinianum) growing in competition with strawberry using convolutional neural networks, Weed Sci. 67 (2019) 239–245, https://doi.org/10.1017/wsc.2018.66.

[118] Y.C. Zhao, F.S. Lin, S.G. Liu, Z.H. Hu, H. Li, Y. Bai, Constrained-focal-loss based deep learning for segmentation of spores, IEEE Access 7 (2019) 165029–165038, https://doi.org/10.1109/access.2019.2953085.

[119] M. Ebrahimi, M. Mohammadi-Dehcheshmeh, E. Ebrahimie, K.R. Petrovski, Comprehensive analysis of machine learning models for prediction of sub-clinical mastitis: deep learning and gradient-boosted trees outperform other models, Comput. Biol. Med. 114 (2019), https://doi.org/10.1016/j.compbiomed.2019.103456.

[120] Y.K. Sun, P.J. Huo, Y.J. Wang, Z.Q. Cui, Y. Li, B.S. Dai, R.Z. Li, Y.G. Zhang, Automatic monitoring system for individual dairy cows based on a deep learning framework that provides identification via body parts and estimation of body condition score, J. Dairy Sci. 102 (2019) 10140–10151, https://doi.org/10.3168/jds.2018-16164.

[121] A. Nasirahmadi, J. Gonzalez, B. Sturm, O. Hensel, U. Knierim, Pecking activity detection in group-housed turkeys using acoustic data and a deep learning technique, Biosyst. Eng. 194 (2020) 40–48, https://doi.org/10.1016/j.biosystemseng.2020.03.015.

[122] C. Augusta, R. Deardon, G. Taylor, Deep learning for supervised classification of spatial epidemics, Spat. Spatiotemporal Epidemiol. 29 (2019) 187–198, https://doi.org/10.1016/j.sste.2018.08.002.

[123] M.H.D. Ribeiro, L.D. Coelho, Ensemble approach based on bagging, boosting and stacking for short-term prediction in agribusiness time series, Appl. Soft Comput. 86 (2020), https://doi.org/10.1016/j.asoc.2019.105837.

[124] X. Zhong, S. Zhou, Risk analysis method of bank microfinance based on multiple genetic artificial neural networks, Neural Comput. Applic. 32 (2020) 5367–5377, https://doi.org/10.1007/s00521-019-04683-y.

[125] L.X. Liu, X.L. Zhan, Analysis of financing efficiency of Chinese agricultural listed companies based on machine learning, Complexity 2019 (2019), https://doi.org/10.1155/2019/9190273.

[126] M. Park, Y.H.H. Jin, D.A. Bessler, The impacts of animal disease crises on the Korean meat market, Agric. Econ. 39 (2008) 183–195, https://doi.org/10.1111/j.1574-0862.2008.00325.x.

[127] E. Elahi, W.J. Cui, H.M. Zhang, M. Nazeer, Agricultural intensification and damages to human health in relation to agrochemicals: application of artificial intelligence, Land Use Policy 83 (2019) 461–474, https://doi.org/10.1016/j.landusepol.2019.02.023.

[128] A. Chandra, K.E. McNamara, P. Dargusch, A.M. Caspe, D. Dalabajan, Gendered vulnerabilities of smallholder farmers to climate change in conflict-prone areas: a case study from Mindanao, Philippines, J. Rural. Stud. 50 (2017) 45–59, https://doi.org/10.1016/j.jrurstud.2016.12.011.

[129] K. Kantamaneni, L. Rice, K. Yenneti, L.C. Campos, Assessing the vulnerability of agriculture systems to climate change in coastal areas: a novel index, Sustainability 12 (2020), https://doi.org/10.3390/su12114771.

[130] O.E. Apolo-Apolo, M. Pérez-Ruiz, J. Martínez-Guanter, J. Valente, A cloud-based environment for generating yield estimation maps from apple orchards using UAV imagery and a deep learning technique, Front. Plant Sci. 11 (2020), https://doi.org/10.3389/fpls.2020.01086.

[131] J. Valente, B. Sari, L. Kooistra, H. Kramer, S. Mücher, Automated crop plant counting from very high-resolution aerial imagery, Precis. Agric. 21 (2020) 1366–1384, https://doi.org/10.1007/s11119-020-09725-3.

[132] C. Tan, F. Sun, T. Kong, W. Zhang, C. Yang, C. Liu, A Survey on Deep Transfer Learning, Springer International Publishing, Cham, 2018, pp. 270–279, https://doi.org/10.1007/978-3-030-01421-6.

Index

Note: Page numbers followed by *f* indicate figures and *t* indicate tables.

Printed in the United States
by Baker & Taylor Publisher Services